The IBP survey of conservation sites: an experimental study

THE INTERNATIONAL BIOLOGICAL PROGRAMME

The International Biological Programme was established by the International Council of Scientific Unions in 1964 as a counterpart of the International Geophysical Year. The subject of the IBP was defined as 'The Biological Basis of Productivity and Human Welfare', and the reason for its establishment was recognition that the rapidly increasing human population called for a better understanding of the environment as a basis for the rational management of natural resources. This could be achieved only on the basis of scientific knowledge, which in many fields of biology and in many parts of the world was felt to be inadequate. At the same time it was recognised that human activities were creating rapid and comprehensive changes in the environment. Thus, in terms of human welfare, the reason for the IBP lay in its promotion of basic knowledge relevant to the needs of man.

The IBP provided the first occasion on which biologists throughout the world were challenged to work together for a common cause. It involved an integrated and concerted examination of a wide range of problems. The Programme was co-ordinated through a series of seven sections representing the major subject areas of research. Four of these sections were concerned with the study of biological productivity on land, in freshwater, and in the seas, together with the processes of photosynthesis and nitrogen fixation. Three sections were concerned with adaptability of human populations, conservation of ecosystems and the use of biological resources.

After a decade of work, the Programme terminated in June 1974 and this series of volumes brings together, in the form of syntheses, the results of national and international activities.

INTERNATIONAL BIOLOGICAL PROGRAMME 24

The IBP survey of conservation sites: an experimental study

EDITED BY

A. R. Clapham

Professor Emeritus of Botany
Sheffield University

CAMBRIDGE UNIVERSITY PRESS
CAMBRIDGE
LONDON · NEW YORK · NEW ROCHELLE
MELBOURNE · SYDNEY

CAMBRIDGE UNIVERSITY PRESS
Cambridge, New York, Melbourne, Madrid, Cape Town,
Singapore, São Paulo, Delhi, Tokyo, Mexico City

Cambridge University Press
The Edinburgh Building, Cambridge CB2 8RU, UK

Published in the United States of America by Cambridge University Press, New York

www.cambridge.org
Information on this title: www.cambridge.org/9780521274296

First published 1980
First paperback edition 2011

A catalogue record for this publication is available from the British Library

Library of Congress Cataloguing in Publication data

Main entry under title:

The IBP survey of conservation sites.

(International Biological Programme; 24)

Bibliography: p.

Includes index.

1. Ecological surveys. 2. Nature conservation.
3. International Biological Programme.
4. Sampling (Statistics) I. Clapham, Arthur Roy.
II. Series.

QH541.15.S95I18 574.5′01′82 79-50233

ISBN 978-0-521-22697-4 Hardback
ISBN 978-0-521-27429-6 Paperback

Contents

Table des matières

Содержание

Contenido

Contributors

Chief Editor

A. R. Clapham, 'The Parrock', Arkholme, Carnforth, Lancashire, LA6 1AU, UK

Editorial Committee

R. Dasmann, IUCN, 1110 Morges, Switzerland

R. J. de Boer, Nieuwlandse Weg, Abbenbroek, Netherlands

M. Kassas, Department of Botany, University of Cairo, Giza, Egypt

Th. Monod, Muséum National d'Histoire Naturelle, 57 Rue Cuvier, Paris 5, France

D. Mueller-Dombois, Department of Botany, University of Hawaii at Manoa, 3190 Maile Way, Honolulu, Hawaii 96822, USA

E. M. Nicholson, Convener, The Nature Conservancy, 19 Belgrave Square, London SW1, UK

L. J. Webb, CSIRO, Rain Forest Ecology Section, Long Pocket Laboratories, Private Bag No. 3, Indooroopilly, Queensland, Australia

P. Vtorov, Academy of Veterinary Sciences, Moscow 378, Kusminski, USSR

Authors

R. J. de Boer, Nieuwlandse Weg, Abbenbroek, Netherlands

A. R. Clapham, 'The Parrock', Arkholme, Carnforth, Lancashire, LA6 1AU, UK

H. Ellenberg, Systematisch-Geobotanisches Institut, The University, Göttingen, Federal Republic of Germany

R. F. Isbell, CSIRO, Division of Soils, Davies Laboratory, Townsville, Queensland, Australia

D. Mueller-Dombois, Department of Botany, University of Hawaii at Manoa, 3190 Maile Way, Honolulu, Hawaii 96822, USA

E. M. Nicholson, 13 Upper Cheyne Row, London SW3, UK

G. L. Radford, Institute of Terrestrial Ecology, Bangor Research Station, Penrhos Road, Bangor, Gwynedd LL57 2LQ, UK

L. J. Webb, CSIRO, Rain Forest Ecology Section, Long Pocket Laboratories, Private Bag No. 3, Indooroopilly, Queensland, Australia

Foreword

The conservation of wildlife and wild places began as an attempt to protect wild mammals and birds from over-hunting, and vegetation of importance to man from over-cropping. It was not until around the beginning of this century that groups of individuals got together to consider the preservation or conservation of wildlife for its own sake and organizations such as the Audubon Society in the United States and the Society for the Preservation of the Wild Fauna of the Empire (now the Fauna Society) in Great Britain came to the fore. International ideas on conservation had been stimulated at a conference in Paris during 1931, but it was not until just before the second world war that an International Office for the Protection of Nature was established in Brussels, subsequently to evolve into the International Union for Conservation of Nature and Natural Resources (IUCN), based in Switzerland. For some years action focussed on individual species of animals or plants, for not until after the second world war did the concept of the ecosystem become important in conservation thinking. The idea developed that plants and animals could generally look after themselves if an adequate slice of their environment could be retained in its natural state. Since then the environmental revolution, and with it the creation of National Parks, Nature Reserves and other sites of conservation, has progressed very rapidly. The scientific concepts which underlie such practical activities have evolved in parallel. Indeed the rapidity with which the science of conservation is changing explains perhaps why this particular volume in the international synthesis of the IBP has been one of the more difficult to prepare and therefore one of the later to be issued.

In the early 1960s when the seven sections of IBP were getting established and formulating their programmes the Terrestrial Conservation Section (CT) was created at the request of Jean Baer, then President of IUCN, to initiate a world-wide study of the great variety of terrestrial ecosystems, leading to selection of representative samples of all the more important ones. Not only would this place the world conservation movement on a surer scientific foundation but it would also be an essential part of the study of biological productivity which was the central objective of the IBP. By close collaboration with other sections, particularly that on terrestrial productivity (PT), a series of 'benchmarks' could be set up against which to measure changes in ecosystems, either natural or man-made, which were becoming unavoidable in most parts of the world. Such a survey would require accurate data to be collected in such a form that one site could be compared with another, so that an organization of field workers would need to be backed by a well equipped and efficient data centre. Much of the work had to be related to an appropriate classification of vegetational types, and this involved a comparison between and an attempt to harmonize the many systems which had been proposed by prominent workers in this field. The data centre was the more easy to arrange and, though difficulty was experienced in mobilizing biologists to help with field work, the results will surely stand as an important landmark when the full history of the environmental revolution comes to be written. The 'check-sheet survey' as it came to be known, provided the core of CT. In some countries whose ecologists worked hard the process has been followed through from initial survey to the selection of appropriate sites, the establishment of reserves, and even to legislation in order to ensure their continuity and supervision.

The choice of Professor Clapham as general editor of this volume was particularly fortunate. As a well-known botanist he has travelled widely and has had the opportunity to study at first hand many of the sites and situations which had to be considered. Moreover, as chairman of the United Kingdom National Committee for the IBP he had experience of all seven sections into which the programme as a whole was divided, and his tact was well known in bringing disparate viewpoints into harmony. Backed by a strong international committee and a team of contributors he has done that in this volume, and near its end he has indicated useful guidelines on how this important work should be pursued in future. George Peterken and Geoffrey Radford were closely associated with the design of the check-sheet and in setting up the data base. The account of the survey's achievements and shortcomings is also a major contribution. Many others who were involved in this work also deserve congratulations on bringing it fully into line with current thinking, bearing in mind that the ideas on which it is based had to be formulated as part of the overall plan of IBP as long ago as 1964.

Max Nicholson who, as convener of Section CT, was the initiator and driving force behind its work, writes in his introductory chapter of the disappointment of many potential helpers that the financing of IBP was so limited, especially in the international context. But the idea that there might be a large international fund to grant-aid research undertaken within the programme was discarded very early. Reliance for research funds had to be on national sources, some of which proved in the event to be substantial, whereas the strictly limited international finance was devoted solely to planning and guidance, to the exchange of information through personal contacts, meetings and publications, and to the co-ordination and assessment of results. Nicholson regrets also that the aim of a world strategy for ecologically based conservation was not achieved during the decade of IBP. But in 1978, four years after the programme was formally concluded, some of the leading CT personalities were involved in the 14th General Assembly of the IUCN held at Ashkhabad in the USSR, when a world conservation strategy was promulgated. The results of IBP are being woven, not only into that strategy, but also into the Man and Biosphere Programme (MAB) under the auspices of UNESCO. Thus we proceed to the future of the conservation movement of which the world was never in greater need.

March 1979 E. B. Worthington

Introduction

E. M. NICHOLSON

It is clear from the lucid accounts of the origin and early history of the International Biological Programme already published in the first volume of this series (Worthington, 1975) that the idea was triggered off extraneously by the success of the International Geophysical Year (IGY), and that it was fostered not in response to widespread demand but through the vision of a very few leaders. The original concept of these leaders proved to be impracticable. First to be abandoned was the nucleic acid field, followed by genetics and then by the theme of fundamental biology in relation to food production and human health. Only after three years of abortive discussions did 'Man and Ecology' emerge as a practicable and timely focal point, under the guidance of the late C. H. Waddington, then President of the International Union of Biological Societies (IUBS).

By a coincidence the Chairman of the Division of Zoology of IUBS, the late Jean Baer, being also President of the International Union for the Conservation of Nature (IUCN), invited the Planning Committee of IBP to meet at its new headquarters at Morges on Lake Geneva on a day when the late Edward Graham of Washington, DC, Heinz Ellenberg, then of Zurich, and I from London happened to be visiting there, and we were invited to join the discussion. Jean Baer had already been pressing conservation as a suitable topic since the previous spring, without either attracting much support or suffering its elimination with the other early starters. As host for the day he prevailed on the Committee to adopt terrestrial conservation as one of the topics, and on Edward Graham and myself to take on its planning.

At that time Waddington's idea that ecology should be looked on as a matter of energy throughput and processing was strange and largely unwelcome to biologists. Not for another four years did it really take root and flourish on both sides of the Atlantic. Although within the perspective and programme of the British Nature Conservancy it integrated fairly easily with ecology applied in conservation, that kinship was not readily recognized elsewhere in the worlds either of biology or of conservation. It followed that the intellectual and social isolation of the Conservation of Terrestrial Communities (CT) Section, which had been inherent from the outset, became steadily more pronounced as new laboratory-based scientists flocked in and were converted to the new gospel of biological productivity.

The idea of massively expanding ecological research to complement the ethical and social forces dedicated to conservation of nature along less scientific lines, proved even more difficult to put over than that of biological productivity. By far the greater part of the resources, institutions and manpower deployed throughout the world on conservation were indifferent, or even suspicious and mildly hostile, towards the injection of ecological principles, methods and monitoring into their empirical arts. On the other hand most professional ecologists, whether in universities or in applied institutes, were almost unshakeably attached to earlier and usually narrower approaches and did not want to know about ecology as a key to understanding and conserving man's environment. But for the firm base and understanding climate provided by the Nature Conservancy's scientists, research stations and field officers, the temptation to abandon the struggle might well have been too great.

Not only were conservationists and ecologists agreed in looking coldly on such a new initiative, but both groups showed strong suspicion that their own inadequate sources of support could be eroded by it, although the opposite soon proved to be the case. Leaders of the earth sciences were, with few exceptions, totally uninterested. Structurally, within IBP, the artificial exclusion of freshwater and marine conservation from the scope of CT, and the requirement to handle national liaison through the channel of often totally uncooperative Academies who had hardly heard of ecology and never of conservation, continually placed obstacles in the way. A further almost crippling handicap was the inadequacy of the available funds, which particularly hit a Section so concerned with remote, unpopulated and impoverished countries and regions, lacking in appropriate institutions to serve as bases and to develop studies. Many who would have helped within the Third World were so shocked and disappointed on learning the financial facts that they withdrew in disgust, excusably concluding that a world programme so impoverished could not be meant to be taken seriously. It is against this background that what was done within the CT sectional programme has to be viewed.

The programme was conceived as having a threefold task:

(i) to develop new techniques and systems for gathering, processing and analysing global ecological data, with special reference to the requirements of conservation;

(ii) to create the nucleus of a global network of field observers, analysts and users capable of further developing and applying the potential of these new techniques and systems, and

(iii) to demonstrate and disseminate the new methods, and to produce a definitive published explanation of a basis on which future work in this and related fields could be pursued.

The CT contribution, in common with that of other sections of IBP, was divided into a preparatory phase (1964–7), an operational phase (1967–72) and a winding-up and evaluation phase thereafter.

During the preparatory phase three requirements were taken as central. First came the problem of identifying rapidly, and with sufficient accuracy in field conditions, the different ecological types and relating them to their context. This was the subject of the first publication of CT and of IBP as a whole, *An Approach to the Rapid Description and Mapping of Biological Habitats* (Poore & Robertson, 1964). In the Convener's Foreword it was pointed out that desert and semi-desert areas exhibit in the simplest and most direct form the influence of edaphic, biotic and human factors on plant and animal associations. The distribution of these was often plainly traceable to obvious factors of rainfall, exposure, temperature, evaporation, soil, grazing, fuel-gathering and other such pressures. The coincidence in Jordan of strongly marked ecological zonation and well-developed aerial survey at large scales (already subjected to ecological interpretation) formed an attractive starting-point for extended studies in applied ecology. Accordingly, arrangements were made for a British Jordan Expedition which, in the spring of 1963, tested on the ground the feasibility of the approach and the problems which would need to be taken into account in giving guidelines to collaborators. In addition to the Convener, this expedition included Dr Poore, Sir Julian Huxley and a number of other experienced biologists and naturalists with a special interest in nature conservation.

The second central requirement was to express the detailed needs, in terms of information, as a series of points to be entered as answers to items in a clear and manageably compact questionnaire or check-sheet for each area or major part of an area. The check-sheet had to be so designed that results could be fed more or less directly into a computer with the minimum risk of distortion or error. It had also to be in the simplest possible form for widespread use in the field.

Lastly, the world-wide and comprehensive mission of the CT Section raised severe problems of comparability of data gathered by observers trained in many different schools of ecology, or in none. In the course of time, various classifications of vegetation or of ecological zones and types had been worked out in a number of centres, and had in some cases come into general use in certain circles or over certain regions. Having been designed to serve different functions, and relying upon different aspects and methods in relation to a particular regional vegetation and climate, they often gave results which were difficult to compare and interpret on a less parochial basis. The approach to a classification of classifications was hampered by the tendency to underrate the limitations of particular systems and to build up strong emotional attachments grafted on to the objective and practical merits of each. Some systems had become inseparably linked with cartographic modes of expression, while others had evoked large numbers of specialist research papers and surveys on an intensive localized basis.

CT originally looked for guidance in this area to the sister Section dealing with survey under Professor Heinz Ellenberg, but the evolution of that Section into PT, and its preoccupation with the productivity of ecosystems, undermined this partnership. Shortly after the consideration of the problem by an authoritative Working Group on Classification at the First General Assembly of IBP in Paris in 1964, UNESCO set up a working party which progressed on a much more leisurely time-table, irreconcilable with that prescribed for IBP, and CT was accordingly compelled to adopt its own working methods. Fortunately much help was available from Dr F. R. Fosberg, whose world-wide basis for compiling an inventory of plant formations had recently been published (Fosberg, 1961) and could readily be updated and revised to meet the need of a comprehensive objective inventory suited to the needs of a computerized data-bank.

The validity of this approach, and the successes and failures experienced in trying to apply it on a world-wide scale, are critically evaluated in this volume. On the whole it is evident that the scale of the global task mapped out, proved greatly to exceed the resources made available. Even had there been more adequate resources the effort was launched before its time. That is to say, far too few ecologists, conservationists and scientific institutions were as yet intellectually prepared to think and to work together in terms of a total ecological inventory of the planet, or even of a major part of it, however urgently such an inventory could be shown to be needed for the conservation of natural resources against modern human pressures. In this respect the CT Section failed to achieve its overall objectives, however many local and limited achievements may be credited to it.

It may be that, in realistic terms, such a world-wide effort was worth making, even if doomed to failure in the short run, as a pace-setter and forerunner of a more effective successor able to profit from its mistakes and its demonstrations. Only time can provide the answer, but meanwhile it has been judged essential, while the experience is fresh, to review it as objectively as possible. In the light of what we now know it seems questionable whether any alternative approach and programme could have achieved much more within the framework of IBP, with all its obstacles and limitations. The question in any case is probably incapable of being answered satisfactorily.

Much more significant and useful for the future is an analysis of the main features of the CT programme, with a view to revealing the extent to which they proved well conceived and capable of satisfactory fulfilment. Apart from a small number of globally concerned ecologists and conservationists, few have given serious thought to the principles and methods which must govern any world-wide strategy of ecologically-based land-use and conservation. Some individual criticisms and propositions simply indicate a failure to grasp the practical problem, and also may be open to question in regard to fundamentals. On the other hand it would be improbable, and indeed disappointing, if the result of so much varied effort did not enable plenty of misconceptions and defects to be detected on subsequent analysis, with a view to their avoidance or correction in any future efforts in this field. Already, since the CT operation ended, a number of publications on a national basis have demonstrated alternative possibilities for treatment of various aspects, and no doubt, others will follow.

It had been intended to complement this critical review with a detailed factual description and list of the sites and participants contributing to the international survey. Unfortunately international economic conditions have so restricted the probable demand for such an ancillary publication that the project has had to be dropped. This is the more regrettable as it would have provided the only satisfactory means of giving recognition by name to all the surveyors whose much appreciated efforts provide the basis for the present volume.

1. The choice of the check-sheet approach

Coordinator: G. L. RADFORD

1.1 Definition of the task

The aims and principles underlying the programme of the CT Section of IBP were set forth clearly and succinctly in Chapter 4 of the *Handbook to the Conservation Section of the International Biological Programme* (IBP Handbook No. 5, 1968), and the relevant section may be quoted in full:

From an early stage in the discussions leading to IBP it was recognized that any world-wide programme seeking to expand and to coordinate biological research would be incomplete if it did not also provide for the selection and protection of the sites and species which must be available for scientific study now and in the future. Moreover, human welfare in a broader sense depends upon the conservation of the natural resources of the world's environment, and the stimulation of conservation is thus an appropriate major aim of the Programme.

One of the first tasks which the CT Section has set itself is the examination of the range of ecosystems over the world and the assessment of the extent to which scientifically adequate samples of all the main types and their significant variants are already protected as national parks, reserves or research areas. Since it is well known that in many regions the existing series of protected areas is inadequate when judged against the background of environmental and community diversity, the assessment will also have as a main aim the selection of other areas which must also be preserved if the series is to be complete. The CT Section will also consider how best to provide for the protection of many scientifically interesting species whose survival cannot be ensured solely by the setting aside of reserve areas selected for their vegetation or general ecological characteristics.

The preservation of natural and semi-natural areas is important for the future of biology and for human welfare because they provide for:

(*a*) the maintenance of large heterogeneous gene-pools;

(*b*) the perpetuation of samples of the full diversity of the world's plant and animal communities in outdoor laboratories for a wide variety of research;

(*c*) the protection in particular of samples of natural and semi-natural ecosystems for comparison with managed, utilized and artificial ecosystems;

(*d*) outdoor museums and areas for study, especially in ecology;

(*e*) education in the understanding and enjoyment of the natural environment and for the intellectual and aesthetic satisfaction of mankind.

The CT Section cannot, however, confine itself to the selection of a series of reserves, however comprehensive these may be. It also seeks to ensure a scientific foundation for conservation throughout the world, and in particular for the assessment of conservation problems, for the management of plant and animal communities and species, and for the training of specialists and the education of a wider public in the aims and practice of scientific conservation.

The CT Section was constituted to concern itself primarily with environments on land. However, it is self-evident that no world-wide conservation effort can be narrowly restricted to one section of the total environment. Terrestrial, freshwater and marine ecosystems intermingle and interact in many situations where strict separation is entirely artificial. Therefore, while the CT Section concentrates primarily upon the land, it maintains contact with the PF and PM Sections so that the specialist knowledge of freshwater and marine biologists is associated with that of terrestrial conservationists in a comprehensive conservation effort.

CT has a role which links the various sections of IBP. Its work is vital to all sections because it seeks to select and preserve the essential material of sites and species upon which all biological work depends. Its selection of, and provision for, the management of these areas and species, and its conservation activities in a wide sphere, must draw upon scientific knowledge which has been accumulated inside and outside IBP. In this way the CT Section is a synthesizing agency, depending upon other specialist groups of biologists for information and at the same time serving them and the larger community by seeking to arrange the long-term protection of research areas and the conservation of the general environment.

The CT Handbook goes on to formulate criteria which should govern the selection of areas for inclusion in the final list of recognized and proposed reserves:

(*a*) The areas should, taken together, contain adequate and manageable samples of the entire range of major ecological formations or ecosystems in the world and illustrate the degree of variation within each.
The series should include sites which

(*b*) although they do not qualify for inclusion under the first criterion, support species of plants and animals of outstanding interest or great rarity;

(*c*) are of scientific interest because of the human management to which they have been subjected, even if this has in some cases led to more or less far-reaching modifications of the biota;

(*d*) are important because they have been the scene of detailed and well-documented research;

(*e*) contain, for example, deposits of peat, lignite or sediment from which information may be obtained about past vegetational and climatic changes, or are of special palaeontological importance;

(*f*) are of special physiographic or geomorphological interest and represent unusual habitats.

There followed certain considerations to be borne in mind in selecting the series of areas:

(*a*) areas should be included whether or not they appear to be under immediate threat. Experience has shown that no reliance can be placed on the survival by good fortune even of very remote areas, and a comprehensive series of the sites which are required for science and education must be selected and preserved if further irreplaceable material is not to be lost;

(*b*) other things being equal, preference should be given to sites that can conveniently be worked by existing or proposed research institutions, that can be supervised and managed effectively, and are least likely to be affected by adverse neighbouring development and by air and water pollution;

(*c*) areas must be of adequate size to support viable populations of the species which characterize them and for which they are established;

(*d*) research areas must also be large enough to allow for the increasing amount of land demanded by modern field experimental research.

(*e*) While consideration of amenities and attractiveness to tourists should not determine the selection of an area for protection for scientific purposes, there are sometimes advantages in locating scientific reserves near to or within larger areas of high landscape value, since it then becomes justifiable to protect a larger and more viable unit.

The policy thus defined in the CT Handbook required for its implementation a world-wide ecological survey, and this implied the collection of certain common items of information about all the sites to be examined. This information needed to be relevant to the policy, accurate and collected in such a form that useful comparisons could readily be made between one site and another. What was required, in short, was an appropriate recording programme together with an efficient system for handling and storing the data. It was important, moreover, that an initial survey, extensive rather than intensive, should be undertaken without delay, as a matter of urgency. These considerations determined the choice of a survey by 'check-sheet', the arguments for and against which are discussed below.

An essential feature of the CT programme was to be 'the examination of the range of ecosystems over the world and the assessment of the extent to which scientifically adequate samples of all the main types and their significant variants are already protected'. This implied some satisfactory system for recognizing and naming ecosystems so that information of the requisite precision could be conveyed unambiguously. The problem for the survey was that no generally agreed system was available, though the matter was under close consideration by ecologists and plant sociologists. Further problems of a comparable kind were raised by the need to convey information about soils and climates in the absence, here too, of generally accepted systems of specification and nomenclature. These problems are referred to in Sections 1.2 and 1.3 and are considered at greater length in Chapter 2.

1.2 Advantages of the check-sheet approach

Questionnaires are being used increasingly in the collection of information for a wide variety of purposes. Their use follows logically from the need to standardize data in a form suitable for analysis and comparison. They have the following main advantages over other means of collecting data from numerous contributors operating in widely scattered and often remote areas:

(i) requests for information have to be formulated simply and clearly, thus favouring speed and accuracy;

(ii) a firm basis is provided for the comparative treatment of the data;

(iii) data are collected in a form making for easy manipulation;

(iv) all these features facilitate the scientific interpretation of the data.

These considerations led to the choice of a questionnaire for use in the CT survey, and each of them is related to the design of the check-sheet and the organization of the survey in what follows.

1.2.1 Precise formulation favouring speed and accuracy

The most obvious advantage of a questionnaire would seem to be that questions can be so formulated that the answers to them supply required information at any chosen level of comprehensiveness and precision. This supposes, however, that the objectives of the enquiry can be and have been clearly defined, that requests for the necessary information can be conveyed intelligibly and unambiguously through the medium of a set of discrete questions, and that information in the desired form is sufficiently readily obtainable by those completing the questionnaire.

The most difficult task confronting the CT Section was not the definition of its basic objectives but the decision on how best to steer a middle course between an unrealistic demand for full ecological descriptions of all the areas already protected or proposed for future protection, and an excessive simplification of the real problem of securing useful preliminary information without undue delay. Two steps were involved in finding and embarking on this middle course: first, the adoption of a simplified concept of the ecosystem that is nevertheless scientifically adequate for the purposes of the survey; and, second, the precise formulation, in terms of this concept, of the questions to be included in the questionnaire. It was recognized that, if these two steps were taken successfully, the resulting questionnaire could be, in itself, a factor in encouraging the rapid collection of the requisite data. The total time required for a national contribution to the survey to become available to the organizers would depend on administrative procedures as well as on the time needed to complete the questionnaire. It was reasonable to suppose that the acceptability of the aims of the survey, and the amount of work judged to be involved, would tend to affect the priority assigned to the task. These considerations emphasized the need for a clear statement of objectives and for a precise formulation of all requests for information.

A questionnaire provides the most straightforward means of conveying to the surveyor exactly what is needed. It avoids unnecessary work on his part and, if well designed, will allow him the narrowest of margins for misinterpretation. The degree of success in this respect is related to the nature of the questions and the complexity of the subject. A questionnaire may consist of a single question at the head of a blank sheet of paper, requiring for answer a lengthy free-style description. As envisaged here it comprises a number of straightforward questions requiring for the most part simple yes/no choices or extremely short answers. The more nearly this is true, the more successful a questionnaire is likely to be.

Ecosystems, however, are highly complex and show continuous variation in many of the features by which they are characterized and distinguished. Questions for the survey had therefore to be so formulated as to be readily intelligible and answerable by the field-surveyor, while at the same time avoiding an excessive simplification that might lead to serious misinterpretation of the basic facts of the ecological situation. It was in search of this right balance that surveyors were invited to record, for each of the different types of vegetation and soil recognized in a survey-site, a code-designation indicating to which member of a series of simply defined discrete classes it could be allocated. The appropriate classifications were set out in the *Guide to the Check Sheet for IBP Areas* (Peterken, 1967), and their use obviated the need to make and record all the detailed observations and measurements that would be expected of specialists in the relevant scientific disciplines. Comparable guidance was given for answers to other questions involving the replacement of measurements of continuous variables by allocations to arbitrarily delimited classes in order to facilitate the work of non-specialist recorders. The considerable size of the *Guide* bears witness to the problems facing the designers of the questionnaire. It had to provide working definitions and recommendations for the guidance of recorders in all foreseeable circumstances, though it was intended more for consultation in cases of doubt than as a rule-book to be read assiduously from cover to cover before committing pen to paper. It must be conceded, however, that constant reference to the appendixes was essential for the correct recording of the codes for vegetation and soil.

Assuming familiarity with the body of the check-sheet and a need to refer to the *Guide* only for the codes for vegetation and soil, the time taken to complete a questionnaire in the field was found to range from half a day to a week. At the desk, with all the information immediately to hand, a questionnaire can be properly completed in about half a day, but extraction of information from published literature and files that are stored all in one place takes up to a day, varying with the level of organization of the information in the source documents. Items that are particularly time-demanding are on the one hand the latitude and longitude, altitudinal range and nearest climatological station, where these have to be taken from maps; and on the other hand information on types of soil and vegetation, where this involves abstraction from descriptive accounts, these rarely being in a form favouring easy allocation to classes of the systems adopted for the check-sheet. For very large sites which have never been fully surveyed it would obviously take a great deal longer than a week to complete a check-sheet, but it was never considered likely that any large amount of detailed surveying would be carried out solely or primarily for the CT survey.

The mode of presentation of the questionnaire has a considerable effect on the speed with which it can be completed, both speed and accuracy being increased if questions are readily understood and the required form of the answers made clear. When questions reach the level of requiring merely the choice of one or other of two simple alternative answers (two-state answers), comparisons between independently completed questionnaires are at their peak of reliability. As individual questions become simpler, however, the number required to collect the necessary amount of information becomes greater and the questionnaire tends to assume the appearance of complexity such as was evident in those earlier versions of the check-sheet when constraints were imposed by data-handling based on cards (p. 115).

1.2.2 A firm basis for comparison

A survey of the kind contemplated could be of little value unless strictly comparable data were collected from all survey-sites. This would be difficult to achieve by any other means than the use of a questionnaire, though information collected in other ways might, in certain instances, be transferable to check-sheets and thus made available for comparative purposes. This was true, for example, of that from the many sites and areas on which large bodies of data were accumulated in the course of research projects for the IBP biome studies. Information from sources such as published literature or private files was less likely to be in a suitable form – attempts at completing check-sheets from published descriptions of sites have proved of little value because what is irrelevant to the author's immediate purpose is so commonly omitted. In general it may be said that data need to be presented, and preferably collected, in accordance with a standardized recording technique for each relevant feature of the site. All items may be recorded independently, the whole approach being essentially objective; or, alternatively, groups of items may be linked by the adoption of appropriate systems of classification. The CT check-sheet, as already stated, asks for certain categories of information to be summarized in terms of suitably simple classifications of vegetation and of soils respectively. There are arguments in favour of both types of approach, but they have in common their reliance on a standardized procedure to ensure comparability of the recorded data. Much of the value of a questionnaire is lost unless there is a high level of standardization in the replies.

The mode of selection of sites for inclusion in the survey clearly will have a profound effect on the recorded range of ecological variation both within and between them, and therefore on the interpretation of any comparisons that may be made. National survey organizers were given freedom to select sites according to their own particular circumstances in relation to national as well as international requirements, and there were wide differences in approach (see pp. 195–6).

1.2.3 Ease of manipulation of the data

The task of data-manipulation is made easier the more nearly a questionnaire consists exclusively of questions demanding two-state answers involving a simple choice between two possibilities. The record of such a two-state choice is the unit of information termed a 'bit', and from the standpoint of data-handling the ideal questionnaire consists entirely of a series of bits. This consideration has deeply influenced the design of forms on which information is to be entered, ease of data-handling by mechanical or electronic means ranking next in importance to the relevance of the answers to the objectives of an enquiry.

The questionnaire is the ideal means for collecting precise information, its form reducing the amount of work required to prepare the data for processing. Punching can be carried out directly from a check-sheet to paper tape as a sequence of data-items matching the series of answers. It was found possible in this way to make the preparation and processing of the data a very straightforward operation and no constraints had to be imposed on the design of the check-sheet purely in the interest of ease of data-manipulation.

The following times were found to be taken for a single check-sheet of average length to pass through the various stages of data-preparation and data-processing, showing that it could normally be dealt with in two hours or less:

EDITING →	TAPE PUNCHING →	CARD PUNCHING →	VERIFICATION →	INPUT, PROCESSING AND STORAGE
25 minutes	25 minutes	20 minutes	30 minutes	15 minutes

This satisfactory speed of handling resulted from the decision to depart from the use of 40- or 80-column tabulator-cards for data-storage and as the basis for information searching and retrieval. The use of a computer considerably increases the flexibility of an information retrieval system and gives scope for more freedom in the style of questionnaire entries where this is desirable. The removal of these constraints radically altered the design of the check-sheet and, although further improvements in design might have provided a basis for a more closely integrated system, the overall efficiency is reasonably high. The retrieval system in its present form allows tabulation of the data and provision for analytical treatment.

1.2.4 A basis for scientific interpretation of the data

The collection of data through the use of standardized recording methods provides a firm basis for comparison and, given adequate site-selection and provided the data themselves are reliable, for their scientific interpretation.

A problem for the survey was that of adequate site-selection, and it must be conceded that only in countries with particularly well-organized participation could a really satisfactory solution be expected. The survey must, indeed, be regarded as a pioneer venture, and an important aim of the present study is to assess the extent to which synthetic interpretation is possible and scientifically valuable and then to recommend how future surveys of this kind might be improved.

1.3 Special problems of the check-sheet approach

The use of questionnaires is attended by a number of problems in both the organization of a survey and the collection, recording and interpretation of the data. The detailed organization of the CT survey is fully discussed elsewhere (pp. 194–5). The problems encountered were those common to any projects in which the activities of a large number of participants are co-ordinated by a single individual or a committee with varying degrees of enthusiasm and support. The concern in this section is to draw attention to problems arising as a direct result of the use of a questionnaire, though some matters of a broader interest are considered in connection with problems of sampling.

The principal disadvantages of the use of a questionnaire are the following.

(i) Some over-simplification in the collected data through the imposition of discrete classes where there is in fact continuous variation.

(ii) A danger of misinterpretation of the questions because of conceptual and other difficulties.

(iii) The introduction of bias into the recorded data in consequence of: (*a*) the selection and wording of the questions; (*b*) differences in training and outlook of recorders, and (*c*) inadequate control of the selection of sites for survey.

1.3.1 Over-simplification in the collected data

The use of questionnaires in situations involving continuous variables and complex interrelationships between variables leads inevitably to some degree of over-simplification in the recorded data. Completely objective independent recording is impossible in a comprehensive survey of this kind and suitably simple categorizations or methods of inventory are therefore unavoidable to provide a framework for recording. At best there will be some distortion of the true ecological situation. At worst the uniqueness of each site will not survive the insensitive pigeon-holing dictated by the need for a practical approach to the problems of specification.

At the level of detail required in the check-sheet almost all variables are to be regarded as continuous. The problem is, therefore, to define useful classes into which to group successive segments of the total range of variation. The *Guide* (Peterken, 1967) is in fact largely concerned with the explanation of such classes and with what the surveyor should record in certain foreseeable circumstances. The situation is simplest when a list of classes or alternatives can be made exclusive and exhaustive, or virtually so, at the chosen level of detail. The list of types of human impact is a good example, though pollution is an unfortunate omission: the low number of subsequent additions to the original list is testimony to its adequacy. There was little confusion between classes, what there was being concerned mainly with the distinction between selective tree-felling and logging to which explicit reference was made in the *Guide*. More complex situations posed problems either in answering the relevant questions or in the suitability of the data recorded. The distinctions between fresh, brackish and salt water are extremely difficult to make in working definitions, and completed check-sheets revealed that surveyors encountered problems in some instances. Definitions of different types of water-bodies, provided as guides for answering the questions in sections 11–13, proved inadequate because of their too exclusive reliance on dimensions while ignoring the origin and the basic physical and biological characteristics that may be critical for conservation-value. For this reason the questions in section 11 on fresh water, in particular, greatly simplify the real situation and provide an illustration of how the uniqueness of a site may be forgotten because of distortions arising from the recording procedure.

Our understanding of the true nature and causes of variation in vegetation is still very incomplete and the subject of much controversy, and therefore there has been no agreement, so far, on a general scheme of vegetational description and classification. It was, nevertheless, essential to adopt some scheme for use in the check-sheet survey. After prolonged consideration (pp. 91–3) it was decided to use the system proposed by F. R. Fosberg in 1961 and republished as an appendix to the *Guide* under the title *A Classification of Vegetation for General Purposes*. This has many features recommending it for the purposes of the survey (pp. 37, 66–7). On the other hand it was too recently devised to be fully comprehensive, and several proposals for modifications and additions had to be adopted during the survey. More important, its basically non-floristic character tends to make it rather coarse and insensitive compared with certain phytosociological systems. Surveyors were, therefore, invited to add some classes not included in the original list and provision was made in the data-handling procedures for dealing with additions, expansions of existing categories and also with intermediates and mosaics. They were also asked to list on the check-sheet, 'by their usual names, using Latin names for all species mentioned', the plant communities which occur within the surveyed site, and it was explained that the level of discrimination envisaged was that of the 'plant association'. This implied the use of any phytosociological classificatory system in common use in the nation or region in question, each plant community listed being allocated to its Fosberg formation by recording its code-designation.

Similar problems, and a comparable solution, figured in discussions on a suitable classification of soils, the system adopted being one proposed by D. F. Ball and also published as an appendix to the *Guide*. A more detailed consideration of the whole topic of methods of specification and classification to be employed in biological surveys comparable with the CT survey will be found in Chapter 2 of this volume.

1.3.2 Misinterpretation of questions through conceptual or other difficulties

A questionnaire enables requests for information to be presented in a simple yet precise manner which is likely to elicit the desired response. Imprecise framing of questions may lead to misunderstanding about what is required. So, too, may an over-complicated questionnaire, whether the complexity arises from the conceptual background to the questions, the manner of presentation, or both. In the survey there was no evidence of widespread misinterpretation through conceptual difficulties. Nonsense entries appeared a number of times in answers to the vegetational questions of section 7, chiefly through failure to understand the classification adopted for the survey, but this was by no means widespread. Section 14, dealing with 'outstanding features of the flora and fauna', provides an example of bad presentation leading to confusion. The definitions of 'rare', 'threatened', 'endemic' and 'relict' species proved to be inadequate and insufficiently distinct. This gave rise to inconsistencies in judgements as to the status of certain species, and these might have been avoided if a simpler terminology had been used.

Section 11 of the questionnaire, on fresh water, gave evidence of the operation of both main causes of misinterpretation. The first question of this section attempts to cover a complex situation but provides inadequate explanation and guidance in the Field Instructions. The result was the highest level of misinterpretation encountered in the whole of the questionnaire and a modification of the data-processing had to be devised to counteract it.

1.3.3 Introduction of bias into the recorded data

Bias arising from the selection and wording of questions

The questionnaire involved the selection and definition of a set of terms and categories. These were to enable a recorder to convey in an adequate manner the administrative status and the principal ecological features of a survey-site, and also the ways in which it has been modified by human activity. There is, clearly, scope for bias in the actual selection of topics to be covered in the questions or, alternatively, to be omitted from consideration. This is illustrated by the obviously too little attention given to animal communities and to faunistic features in general. This arose largely from the lack of an adequate system for the description and inventory of animal communities. It therefore became necessary to assume that habitat types could be used as a satisfactory basis for predicting the nature of animal communities to be found in a survey-area. This may be largely true of the lower trophic levels and of less mobile animals, but much more information is needed to predict with any confidence that particular large and mobile animals will occur within or near a specific site. The check-sheet invites reference to the presence of such species only if they are rare, threatened, relict or 'of biogeographical interest', and the interpretation of these terms is open to bias on the part of the surveyor.

In the check-sheet what were deemed the major environmental and biological features of any survey-area are explicitly asked for in the final questionnaire, and also provision is made for surveyors to mention additional features if thought appropriate. There is, nevertheless, the danger that important questions may be omitted and for that reason come to be regarded as unimportant or irrelevant to the survey. The primary advantage of a questionnaire, that the information secured can be precisely what is sought, therefore has the complementary disadvantage that in fact no more information is obtained than has been explicitly requested, apart from variable and unpredictable responses from certain surveyors to open invitations to add what they judge to be valuable. This applies to both intentional and unintentional omissions and underlines the need for very great care in the design of questionnaires. There must be a practical limit to the range of features about which relevant questions can be asked. With a subject as complex as an ecosystem or, still more, a group of ecosystems, it might seem that some overall bias must be introduced into the check-sheet, and therefore into the recorded data, as an inevitable consequence of whatever selection of features was in fact made. But it was envisaged from the start that features selected for inclusion might provide a basis for inferences at least about some of those omitted, the predictive capabilities of the survey then becoming one of the factors in judging its success. To this extent bias has been introduced by a deliberate selection. It must be conceded, however, that a more objective and independent coverage of a wider range of features would have been desirable and would have increased the reliability of any predictions.

Bias arising from differences in training and outlook of recorders

To whatever degree steps have been taken to ensure satisfactory site-selection and standardized entries in the check-sheet, it will remain possible for bias to be introduced into the recording by the surveyor himself. No amount of attention to detail in selection and presentation of the questions can allow for the varying interests and capabilities of the field staff. An examination of the background of the surveyors who completed a sample of check-sheets (p. 301) showed that they included leading ecologists, university professors, research scientists and other professional people, with varying amounts of training and experience, as well as students of biology and interested amateurs. At each level of general competence there was a range of specialized interest in various branches of the life sciences. The effect of this was particularly evident in answers dealing with 'outstanding floral and faunal features'. Here there was often a manifest bias in favour of particular groups of plants or animals. Frequently there was a note to the effect that no information at all was available on faunal features of special interest, even though the diversity of habitats or other characteristics of the site indicated that some were to be expected. The Czech and Slovak contributions clearly illustrate the problems in site-selection and in consistency of recording that may arise from the different specialist interests of those organizing work for the survey in neighbouring areas. After preliminary joint discussions aimed at harmonizing recording procedures, the two contributions were organized independently, in the Czech Socialist Republic (SR) by ecologists who selected for survey nature reserves that would include examples of the whole range of ecosystems occurring in that republic, but in the Slovak SR by foresters who sampled, in particular, all significant variants of forest ecosystems. Corresponding differences in emphasis are shown in the lists of rare and otherwise interesting species occurring in the survey-sites. The consequence is that comparisons between check-sheet data for the two republics tend to be misleading in several ways.

Bias through inadequate control of site-selection

National organizers for the CT survey were given a free hand to select sites for survey in a manner appropriate to the needs and resources of the country concerned, but this entailed the possibility of a lack of central direction that might seriously reduce the value of a national contribution. Whatever the method of site-selection adopted, there is need to relate the coverage of biological variation in surveyed sites to that in the country as a whole, and this can only be effected through guidance from a single authoritative source. The Country Check-Sheet was designed to assist in this way but met with no appreciable success. If representatives of the International Committee for IBP/CT could have been present at decisive stages of planning of all national contributions, some of the differences in approach to site-selection, which have precluded critical analysis both within and between contributions, might possibly have been avoided, a point to be borne in mind for future international surveys of comparable kind. Apart from site-selection there are other matters requiring attention at the planning stage of national contributions. These include procedures for the subdivision of sites and for the description of types of vegetation, and also the basis for judgements as to the rarity of species and the distribution of other examples of particular plant communities. At a more detailed level, and in the interest of standardized recording, consideration should also be given to the form in which notes on soils and other free-style entries should be recorded.

1.4 Design and contents of the check-sheet

1.4.1 Relationship of the questionnaire to the objectives of the survey

Having decided that a questionnaire provided the best means of collecting data of the requisite kind, attention was next directed to the type of questions to be asked and how they were to elicit information meeting the objectives of the survey. The first point to be made is that the decision had been taken to base the survey on existing individual *sites* having special interest for IBP/CT but varying in size and with definite boundaries which might enclose only a small part of the total area of interest, not on samples of uniform area nor on the whole of continuous areas covered by a feature of special interest. This being so, the basis for the selection of such sites for survey became important. Alternatives to this approach and the problems of selection are discussed elsewhere (p. 209). The present account deals with the relationship of the survey objectives to the final version of the check-sheet questionnaire.

The questionnaire covers eleven pages and is organized into nineteen sections (Appendix 1). The first two pages deal with basic information about the site, of the kind common to all comparable recording projects. Six sections are involved, each of two or more questions. The first five of these sections seek the minimum of information required as a background for the biological data. They ask for the surveyor's name and address and the source and date of the records; the designation of the site or subdivision concerned; the geographical location of the site; its national and international status, where appropriate; and its area and altitude. It was thought unnecessary to ask for administrative details on establishment and legislation of existing reserves. The sixth question concerns climate. Early trials of the check-sheet showed that much of the information sought was only patchily available and that when provided it often came from climatological stations remote from the sites and with very doubtfully similar weather. It was therefore decided to ask only for the location of the nearest meteorological recording station, from which data might be obtained later if thought desirable. This was recognized as less than satisfactory but was adopted as the best course available in existing circumstances.

The seventh section deals with vegetation and soils. It was realized that the extent to which the objectives of the survey could be met would depend heavily on how adequately the vegetation of the survey-sites could be characterized. Surveyors were asked to list the major plant communities occurring on a site, using any system of designation or classification in general use in the country or region. Each of the communities listed had then to be referred to its place in F. R. Fosberg's *Classification* (1961; also in Peterken, 1967), a non-floristic system depending solely on readily observable physiognomic and structural features of the vegetation and therefore suitable for world-wide comparative use. An estimate of the area occupied by each community was also requested, as a check on the completeness of the coverage and as indicating the scale of the inventory. For each community, too, information was sought about its soil in terms of Ball's *Classification of Soils* (in Peterken, 1967), a system comparable in ease of use and generality with Fosberg's system for vegetation, with the addition of descriptive notes covering features of colour, texture, drainage, acidity and depth. The eighth section was devoted to a question added to secure information on the extent to which each plant community listed for a site occurs elsewhere within the same country or region, and whether or not in protected areas. This information would be of considerable importance in planning for conservation. The Fosberg and Ball classifications are reproduced as Appendix 2 and Appendix 3 respectively.

The above three topics – vegetation, soils and the nation-wide or regional distribution of particular plant communities – constitute the heart of the survey for its purpose of ecological inventory. All the information collected for this section of the questionnaire relates to the individual communities recognized within a site, not to the site as a whole nor to any entire subdivision of it. It was thought unreasonable to treat other features on a plant community basis on the grounds that the relatively small increase in precision would not justify the greatly increased complexity of recording. It might, however, have been valuable to specify the altitudinal range of each plant community independently in sites with a large total range in altitude.

The next five sections, 9–13 inclusive, deal with the physiographic features of the site. The first, general landscape, was intended to provide a brief description of the main topographical features with a broad categorization of the degree and scale of dissection. The second concerns the length and physiography of any coastline present and whether its substratum is of rock, shingle, sand, mud, coral or ice; the presence and frequency of sheltered bays or inlets; and the tidal range. Three further sections cover the surface hydrology of the site. Section 11 asks for a categorization of superficial fresh water as permanent or intermittent, running or standing and, if standing, as swamps, ponds or lakes, characterizable as productive or unproductive. This information is sought as a guide to the spatial and seasonal distribution of open-water habitats and sources of drinking water. Section 12 deals with the presence of salt or brackish water bodies within the site and section 13 with the occurrence of fresh, brackish or salt water bodies, as lakes, rivers, estuaries, salt lakes, lagoons or the open sea, outside but immediately adjacent to the site.

Sections 7–13 deal with the general ecology of the site in terms of its main plant communities and physiographic features. The objective is an inventory of characteristics of the site relevant to conservation-planning and an indication of possible habitats for particular animal communities and individual species of plants and animals. Assuming an adequate coverage by completed check-sheets, the information obtained would allow the fulfilment of the declared aims of establishing what ecosystems are present in each participating country or region and how far each type is adequately represented in protected areas. This is the minimum prerequisite for framing a detailed plan for the conservation of all the main ecosystems occurring in a given area. The survey having shown which are the main ecosystems in the area, and which of them are already protected, means that it is possible to infer which have yet to be included in a series of reserves designed to be wholly representative. Some basis is required, however, for the recognition of particular examples as adequately *representative* of a *type* of ecosystem. This is a much more difficult matter than the comparison of an individual organism with the type specimen of a species, both because ecosystems are less readily delimited than are most species and because there has so far been no consistent practice of designating typical examples. The aim must be to choose an example from near the centre of what is judged the full range of expression of the ecosystem in question in the given area. Any particular example may be distant from the 'centre' because it is near the edge of its ecological range, because it has been significantly modified by human activity or for other historical reasons. It is such points that are covered by sections 14–17 of the questionnaire, dealing respectively with outstanding floral and faunal features, exceptional interest of the site, significant human impact and conservation status. The last item is concerned with the completeness of protection of the site, whether there is controlled or uncontrolled utilization, to what extent there is management for conservation and how far scientific research is permitted.

Besides constituting a series of representative ecosystems a reserve network should aim to include sites and features of some exceptional interest. Examples are those sites necessary for the continued natural movement of migratory animals and especially of birds and mammals, or including species and communities which live in unusual biological situations or are diminished in numbers or in territory occupied. Section 14 on 'outstanding floral and faunal features' was designed to elicit information of this kind, though the wording invites emphasis on species rather than communities. The next section, 15, asking for notes on the 'exceptional interest' of the site, does, however, provide an opportunity for recording communities with unusual features as well as other aspects of the site deemed worthy of particular mention.

An ideal conservation progrmme would doubtless aim to secure representative examples of ecosystems in a more or less unmodified state. This is now an impracticable aim for most if not all countries and it must be replaced by the objective of conserving those that still survive and of supplementing them with examples of their major man-modified variants. Information about the extent of human influence is invited in section 16, where the surveyor can indicate the history, kind and extent of human impacts and assess their current tendencies to increase, decrease or remain constant in intensity. Section 17 asks related questions about human activity forming part of a planned policy of site-management in the interests of conservation 'with the primary object of maintaining, restoring or creating an ecosystem of some special interest to biologists'.

One important purpose of nature reserves is to provide areas where ecological research and environmental and biological monitoring can be carried out undisturbed. It can be argued that the provision of sets of reserves which may include examples of unmodified ecosystems together with their most characteristic man-modified variants, will be of even greater value for furthering ecological understanding than if none of them had undergone change, especially if the changes result from disturbances of known kind, intensity and extent taking place over a known period.

Sections 18 and 19 ask for details of reference material, including maps and aerial photographs, relating to the survey-site and for any other relevant information.

The form and intention of each question is discussed in greater detail in Appendix 6, with consideration of the way in which it was answered and how it was processed for storage and retrieval of the information obtained.

1.4.2 Different versions of the final check-sheet

The Mark VII check-sheet was widely distributed as the final version of the questionnaire after successive modifications of earlier drafts. Apart from the standard form designated *Level 2*, on which the foregoing review of the questionnaire is based, three other versions eventually appeared. *Level 1* arose from a later decision to circulate a greatly simplified scheme for ecosystem inventory. Here the information requested about the vegetation of a site was reduced by retaining only the broadest of the categories in the Fosberg classification, and by omitting the area covered by each community recognized. The sections on soils and on the occurrence elsewhere of similar communities were also omitted. The American Institute of Biological Sciences prepared a much reduced version for use in Research Natural Areas and private reserves in the USA, and this was completed for seventy-eight sites.

Table 1.1. *Differences between Level 2 and other versions of Mark VII of the check-sheet: questions included or excluded*

Level 2	Level 1	French	USA	Common questions
1	P	P	P	C
2	P	P	R	
3	P	E	E	C
4	P	P	R	
5	P	P	P	C
6	P	P	P	C
7	R	M	R	
8	—	P	—	
9	P	P	P	C
10	P	P	—	
11	P	P	—	
12	P	P	—	
13	P	P	—	
14	P	P	—	
15	P	E	—	
16	P	E	—	
17	P	P	P	C
18	P	P	—	
19	P	E	—	

P, Present/unaltered; E, Present/expanded; R, Present/reduced; M, Present/modified in subject-matter; —, Absent; C, Common to each version.

Translations of the *Level 2* standard version were prepared in seven languages besides English: French, German, Polish, Russian, Serbo-Croat, Slovenian and Spanish. In six of these the original was translated without modification, but the French version was modified appreciably for use mainly in metropolitan France. It differed in a number of ways including the addition of requests for reference to the national series of maps and to systems of internal land administration. The most significant difference, however, was in the treatment of vegetation, lists of species being demanded rather than designations or descriptions of plant communities, and estimates of cover for each layer of the vegetation rather than the Fosberg coding of the community.

Table 1.1 summarizes the differences, question by question, between the standard *Level 2* and the other three versions of Mark VII of the check-sheet.

2. Problems of description and specification

Coordinators: A. R. CLAPHAM & G. L. RADFORD*

2.1 General consideration of the problems

Decisions on the claims of any area of land for protection as a national park or nature reserve must be based on a body of information about it whose scope and precision are such as to allow relevant comparisons with rival claimants. The most important features of the area for this purpose are its vegetation and the types of animals present throughout the year or for certain periods; but an adequate description for the guidance of those responsible for taking decisions on nature conservation must include information on the nature of the soils or other rooting media, on the topography and on other factors of the physical environment, in particular the climate. Most of these are features in which there can be wide and more or less continuous variation and which, therefore, pose difficult problems of description and designation. These problems are of two main kinds. In the first place there is need to secure general agreement on the form and content of records which will be of most value for comparison and selection, including ways of facilitating reference to different parts of a continuum. Secondly there is the practical consideration that any agreed scheme of description and specification must be within the competence of those undertaking the field surveys and must be adequately standardizable.

2.2 Vegetation-types

2.2.1 Historical background

The *Guide* (Peterken, 1967) points out that plant-assemblages differ from one another in their photosynthetic production and activity-supporting capacity as well as in intrinsic interest arising from the nature and variety of all the living organisms associated with them. It is, therefore, essential that any survey designed to provide basic information for the selection of sites for protection should give special attention to recording the vegetation of areas under consideration.

The *Guide* envisaged two different kinds of procedure for providing relevant vegetational information: by the objective recording of certain data judged to be of most value for the purposes of the survey, but without naming any

* *Contributors:* A. R. Clapham & G. L. Radford (2.1, 2.2.1, 2.2.4, 2.2.5, 2.3.3, 2.4); R. F. Isbell (2.3.2); D. Mueller-Dombois & H. Ellenberg (2.2.2); L. J. Webb (2.2.3).

vegetation-types; or by allocating the vegetation to its place or places in a previously agreed scheme or classification of vegetation-types. An early trial-version of the check-sheet asked recorders to note the presence or absence in the survey-area of, for example, dicotyledonous trees over 40 m and between 25 m and 40 m high, dicotyledonous herbs, grasses and bryophytes; all to be allocated to the appropriate cover-classes – 10% cover, 10–50%, over 50%, and clumped. This seems straightforward and objective but was found unsatisfactory in field trials for a variety of reasons. One of these was that recorders did not feel the recorded facts to be adequately informative about the vegetation of a survey-area, and felt it essential to make finer distinctions between vegetation-types though this entailed the adoption, even if unconsciously, of some kind of vegetational classification. There was the further difficulty that this kind of objective recording did not allow immediate comparisons between the newly surveyed areas and those already protected. It was, therefore, decided to look for a classificatory scheme that would be universally applicable and acceptable, easy to use in the field and possible to apply even with the small amount of information commonly available. There was no such system that was both universally applicable and also sufficiently detailed to make distinctions between vegetation-types at the level required for the comparison and selection of sites within a single country or region, and a compromise solution had to be sought. Sections 2.2.2 and 2.2.3 consider the available methods of vegetational classification and explain the choice made for the check-sheet survey.

2.2.2 Methods available for classification and their suitability for various purposes. D. Mueller-Dombois & H. Ellenberg

The following section is concerned with the general problem of classifying vegetation. A few proven methods will be discussed. It will be shown how these methods are related to the geographic scale, the objectives of the classification and the nature of the vegetation itself. An attempt will be made to assess the check-sheet survey in relation to these methods and its own specified objective, which was to provide a source of base-line data for conservation areas.

Vegetation can be defined as an assemblage of plants of from one to many species growing in areas of different sizes. Classification requires an identification of geographic segments of vegetation that show a certain degree of homogeneity within each segment. Degrees of homogeneity can be recognized at different levels of generalization in the spatial sense. For example, one may recognize small patches of grass cover (a few square metres in size) as homogeneous segments among a stand of irregularly dispersed trees. One may also recognize the trees together with the grass patches as an open forest or woodland, which may form a more broadly defined homogeneous segment that may be separated from a closed forest or scrub vegetation. Moreover, one may view the open forest, the closed forest and scrub of an area together as woody vegetation that can be separated from an adjoining herbaceous vegetation such as a grassland.

These are examples of separating vegetation segments by general life form and structural criteria. Vegetation segments may of course also be recognized by changes in species distribution, composition and quantities. For example, across a segment of a closed forest, one tree species may show a quantitative dominance in one area, while the rest of the forest may show a more uniform mixture of several tree species, permitting the recognition of two floristically defined closed forest segments, a mono-dominant forest and a mixed-species forest. Where such variations are gradual, the boundary allocation may cause some difficulties. Undergrowth species distributions may permit further subsegmentation of the mixed forest.

These few examples may suffice to illustrate that vegetation segmentation can be done at different degrees of homogeneity and different geographic scale (or detail). All levels of vegetation segmentation can be useful, but their usefulness varies with the objectives. It is, therefore, of utmost importance in vegetation classification to specify and understand at least the broader objectives.

General objectives

Since vegetation can be classified in so many ways, there is no single method of classification that can satisfy all purposes. The choice of a suitable method, however, is narrowed immediately by a statement of objectives. In this instance there are three general objectives.

Developing an inventory of existing types for conservation purposes

This can be described as the major goal of the IBP/CT check-sheet survey, and is a relatively new objective. It is, however, closely related to the long-standing aim of the plant geographer and vegetation scientist, which is to comprehend and order the vegetation diversity of certain territories. In this general objective, the main focus of attention is on the vegetation itself and not so much on the associated environments, which may be inferred from the vegetation or studied later in more detail.

An important question that may be asked immediately is, what is the size of the territory to be inventoried? Is it the world, a continent, an individual country or state, a county or a park? This leads to the specification of the geographic scale for classifying and mapping of vegetation.

Providing a framework of reference for other biological field studies and for local management

This general objective is rather pragmatic. It requires a relatively simple classification that can be agreed upon by non-specialists. The units must not be too many or too complex and yet must find considerable reality in the field, to be useful. The types must show integrity only for the local territory since the objective does not include the need to integrate the units with the vegetation of other territories. The classification must be more than a description of types. It must at least be developed into a key. Preferably it should be portrayed in the form of a map.

Understanding plant and community distribution and dynamics in relation to environment

This is a general and fundamental objective of the vegetation ecologist. It requires good floristic knowledge and a study of the environmental factors in different habitats or along gradients. An important methodological prerequisite is the laying out of sample plots or relevés,* because vegetation samples form the basis for arranging the data, which may result in either a vegetation classification or ordination, or in both. If such studies are considered to be useful for providing an ecological basis for other field research or for natural area management, the resulting arrangements should also be portrayed on a map. Mapping removes some of the complexities of geometric models, dendrographs, or tabular arrangements. The latter are, however, needed for the proper documentation of such community ordinations and classifications.

These three broad objectives are related to the levels of spatial generalization of vegetation.

Levels of spatial generalization

Vegetation is a geographical phenomenon. It cannot be understood without the space it occupies. The same applies to each vegetation segment or community. It must first be recognized in the field. Vegetation can be mapped at all geographic scales. But different scale ranges give different information sets, and these in turn provide for different general objectives. In the following paragraphs, scale ranges used for vegetation mapping will be discussed briefly: there are five levels, from the most general to the most detailed.

* relevés = vegetation samples in plots large enough to contain at least 90–95% of the species of a vegetation segment or community.

2.2.2 *Methods available and their suitability*

Level 1: very small-scale maps for global orientation, from 1:50 to 1:10 million

This scale range includes standard wall maps of the world and atlas maps of major world regions or continents. The most comprehensive recent world vegetation map in this category is that of Schmithüsen (1968) at the scale of 1:25 million (i.e., 1 cm on the map represents 250 km on the earth's surface). At this small scale, Schmithüsen was able to portray 143 world vegetation types on eleven pocket-sized map-sheets. The map units are shown by colour and symbol combinations. (An improvement for greater clarity would be to print the vegetation type numbers, shown in the legend, on their respective coloured fields.) Vegetation types depicted include such categories as tropical lowland rain forests, tropical evergreen dry forests, thorn scrub and succulent forests, summer-green conifer forests, alpine meadows, heath vegetation of the temperate zone, lichen- and moss-tundras, shrub deserts, etc. The vegetation categories coincide approximately with the 'biome' concept of IBP study groups, but many of Schmithüsen's units can even be considered as subdivisions of major biomes.

However, it is also apparent that Schmithüsen's vegetation units do not represent vegetation as it actually exists, because much of the earth's surface is variously modified, i.e., converted into agricultural lands, into industrial or urban areas. The map units merely outline certain ecological zones of assumed homogeneous growth-potential for the named vegetation types. There may be in these zones existing remnants of the kinds of vegetation for which the areas are named. For these remnant vegetations (provided they are still existing in their respective zones), the map establishes interesting differences and similarities (or equivalences) across our planet. Therefore, the map can fulfil a useful purpose in an inventory of world vegetation.

Level 2: small-scale maps covering individual continents or countries, from 1:10 to 1:1 million

This scale is generally used for mapping vegetation zones, i.e., areas characterized by certain key species or areas of so-called 'potential natural vegetation'. A good example of the latter is Küchler's (1964, 1965) vegetation map of the United States at the scale of 1:7.5 million. The map shows 106 vegetation types with such names as spruce–cedar–hemlock forest, redwood forest, Douglas fir forest, chaparral, sagebrush steppe, southern cordgrass prairie (*Spartina*), cypress savanna, Great Lakes spruce–fir forest, oak–hickory forest, etc. Thus, in contrast to Schmithüsen's world vegetation map which gives vegetation information only on the basis of physiognomy or general structure, Küchler's US vegetation map provides general floristic information by citing dominant species in the vegetation type names.

However, Küchler's map is similar to that of Schmithüsen's in one important aspect: it does not outline actually existing vegetation in the field. It only provides for the possibility that in the general outline of a map unit there may be one or more smaller tracts of vegetation remaining that may fit the type description given for the area. Apart from its general information value, such a map can become extremely useful for conservation purposes, because the actually existing remnants of the vegetation types can be located by points or asterisks directly on the map. In this way the representativeness of conservation areas or natural reserves can be evaluated across a large country. Wrongly handled, such a map can also be a dangerous tool, because it tends to overgeneralize vegetation type information. This overgeneralization can result in the omission of important variations in a regional vegetation for conservation purposes. Other good examples of vegetation maps in this category are the 1:1 million 'International Vegetation Maps' for South Asia, e.g., for Sri Lanka (Ceylon), prepared by Gaussen *et al.* (1964). In these maps, 1 cm represents 10 km in the field. A few actually existing but very broadly defined vegetation types are outlined or marked by symbols on Gaussen's maps. Moreover, these existing types are shown within their respective ecological zones. The 1:1 milion scale was chosen as a guide for the recently developed UNESCO classification of world vegetation types (UNESCO, 1973).

However, in most situations, existing vegetation types can only be located by a symbol and not mapped directly on such small-scale maps. It is therefore less confusing to refer to these maps as vegetation zone maps, or as ecological zonation maps, depending on whether the main source of information was obtained directly from the vegetation or from environmental parameters or from both. For example, Krajina's (1969, 1974) biogeoclimatic zonation map of British Columbia at 1:1.9 million, although couched in vegetation terms (by naming the map units after dominant plant species), is what it says, a map of ecological zones and not one of vegetation types. Each of his zones contains a number of significant vegetation types, which however cannot be mapped with sufficient clarity at that small scale.

Such small-scale ecological zonation maps are usually derived from already existing information and they may not require any additional field work. One way of derivation is to assemble existing vegetation maps prepared at larger scales and to reduce these by carefully reasoned boundary elimination into more generalized units. However, larger-scale vegetation maps covering contiguous large areas are rarely available (only for a few intensively studied countries), and ecological zonation maps may then be derived from environmental data. Most important at the small-scale geographic level are climatic data, and there are several proven methods that can be employed for the derivation of bioclimatic maps (e.g., the Köppen (1936), Thornthwaite (1948) or the Gaussen (1957) method). One of the simplest, easily understood and richly informative methods is Walter's climate diagram method (1957). At somewhat larger scales (approaching 1:1 million), topographic and soil data become important as additional environmental information for the delimitation of ecological zones. A good example of the latter kind of map, using climatic, topographic and soil data is the biogeoclimatic zonation map of Krajina mentioned previously. The distribution of certain key species, for example, dominant trees, was used as another guiding parameter for the drawing of zonal boundaries. However, the mention of key species in the unit name does not describe them as vegetation types.

Level 3: intermediate-scale maps for closer regional or subregional orientation at state or provincial level, from approximately 1:1 million to 1:100 000

Such maps may permit the mapping of structurally and broadly floristically defined vegetation units, such as alliances (*sensu* Braun-Blanquet, 1964) or dominance communities (*sensu* Whittaker, 1962). However, at this scale the vegetation is still often generalized into climax types or potential natural vegetation types. A good example of the latter is Küchler's (1974) map of the potential natural vegetation of Kansas at 1:800 000 (i.e., 1 cm on the map = 8 km in the field). It represents an enlargement of Kansas from Küchler's US vegetation map by an approximate factor of ten. This enlargement permitted Küchler to recognize twice the number (14) of potential vegetation types to the seven on his US map. In the text accompanying the 1:800 000 map, Küchler shows a photographic example of vegetation that still exists, for each of the fourteen mapped potential vegetation types. Yet it would probably be difficult to map these existing communities at this scale, because the landscape of Kansas is today dominated by man-introduced land modifications, to agricultural and other uses.

From this point of view, Gaussen's 1:200 000 climax vegetation map of France, a section of which is shown in Küchler (1967, p. 259) is particularly interesting. On this map, Gaussen shows several climax vegetation types by different colours. The climax types are not the existing vegetation types. They merely represent vegetation zones or belts established from existing remnant stands and from climatic and soil information.

For example, the beech climax zone may include meadows, apple orchards, heaths, tall shrub formations and beech forests. But Gaussen shows the type of land use of the area on the same map by a system of overprinted symbols denoting various forms of agricultural land use, such as corn, potato, rye, vineyards, plantation forests, etc. Totally man-modified or cultivated zones are shown without colour (white). Absence of land use symbols on the coloured fields of the climax zones indicates the presence of less modified or near-natural vegetations. The occasional overprinting of a climax type symbol may denote the presence of a more typical remnant stand. It is evident that such a map carries a significant information value for the objective at hand, i.e., an inventory of natural areas.

Vegetation maps at this intermediate scale range can show the outline of existing vegetation types only in relatively undisturbed or undeveloped landscapes, e.g., for national parks or remote areas with low human populations. An example is Mueller-Dombois' (1972) 'Generalized map of Ruhuna National Park, Ceylon', which was published at the scale of 1:140000. The map shows the actual outline of ten structural vegetation types, such as forest, scrub with scattered trees, scrub islands, short-grass cover, etc., which are based on Fosberg's (1967) system. This map was reduced from a large-scale map (at 1:31680) with twenty-eight structural vegetation types (Mueller-Dombois, 1969), which were generalized by elimination of those vegetation boundaries that were less important at the smaller scale.

Level 4: large-scale maps for research and management at county or national park level, from approximately 1:100000 to 1:10000 (i.e., 1 cm on map = 100 m in the field or 1 cm² = 1 ha)

This scale range is commonly used for mapping the outlines of existing vegetation types or communities. Such maps can be considered factual, since the hypothetical element of extending vegetation boundaries across man-modified terrain is avoidable at this level. Large-scale vegetation maps may require months or years for preparation, and when established may show the diversity and distribution of plant communities within specific nature reserves or natural areas.

The map units may be defined by all commonly used classification criteria, i.e., by structural vegetation criteria (*sensu* Dansereau, 1957; Küchler, 1967; Fosberg, 1967; and others), by species dominance criteria (*sensu* Whittaker, 1962; and other Anglo-American authors), by species association criteria (*sensu* Braun-Blanquet, 1964; and other European authors) or by combined vegetation and habitat criteria, *sensu* Kopp (Eberhardt, Kopp & Passarge, 1967; Mueller-Dombois, 1965; and others). These classification criteria will be discussed in more detail in the next section.

Structural and species dominance criteria are often more practical at the upper range of this geographic level, i.e., on vegetation maps with scales of 1:50000 to 1:100000, while floristic association criteria permit the portrayal of more detail, and thus are often applied at the larger scale range of 1:10000 to 1:50000. However, there is no absolute or direct relationship of classification criteria and map scales within this geographic scale range. The criteria used depend on the nature of the vegetation itself, on the more specific objectives and on the viewpoint of the mapper. For example, in the international comparison of forest site mapping methods in Switzerland (Ellenberg, 1967), which was carried out at the scale of 1:10 000, E. Schmid's species dominance and plant life form (= structural) criteria were applied to the same area as were the floristic association criteria of Braun-Blanquet. In this case, the authors' viewpoint entered strongly into the community classification and mapping schemes. On the other hand, areas with considerable species diversity may not permit recognition of species-dominance communities even at large geographic scales. In such cases, structural criteria may offer a better tool for subdividing a regional vegetation cover.

An example of a large-scale structural vegetation map is the map of Ruhuna National Park in south east Sri Lanka by Mueller-Dombois (1969), which was prepared from aerial photographic mosaics at the scale of 1:31 680. Fosberg's (1967) structural scheme was used in the preparation of this map. Yet the map scale was already so large that a number of additional units had to be established. This caused no difficulty in this tropical dryland region. Fosberg's scheme was flexible enough to permit establishment of additional map units along the same sort of criteria that are described in his system. For example, Fosberg (in Peterken, 1967, p. 104) recognizes four seasonal short-grass vegetation types. Only one of these applied to the Ruhuna Park grasslands, namely 1M21 = 'Seasonal orthophyll short-grass'. However, for mapping at this scale, six short-grass types were recognizable on the basis of structure: (1) short-grass cover without woody plants, (2) short-grass cover with scrub islands, (3) short-grass cover with forest-scrub islands, (4) short-grass cover with scattered trees, (5) short-grass cover with scrub islands and scattered trees, and (6) short-grass or graminoid cover with sections of sparse cover of barren areas near water.

Types 2 to 5 could, of course, also be treated as Fosberg's low savanna and shrub savanna types. They can be translated into the latter categories for regional comparisons between countries, but that was not the objective. Instead, the objective was to establish an integrity of map units for the dry zone in Sri Lanka. A similar structural vegetation map was prepared for Wilpattu National Park in the north eastern part of Sri Lanka's dry zone. The maps were used for periodic animal activity surveys (Mueller-Dombois, 1962; Eisenberg & Lockhart, 1972), for locating floristic and quantitative sample plots (Comanor, 1971), for soil surveys, etc. The more immediate purpose of the map was to establish a framework of reference for a number of research activities and wildlife management objectives.

Level 5: very large-scale maps for detailed local orientation, from approximately 1:10 000 to 1:1000 (i.e., 1 cm on map = 10 m in the field, or 1 ha = 1 dm^2 on the map)

Such detailed maps are not published very often, because they are usually prepared for rather special purposes of more local interest. A published example is the vegetation map of the Neeracher Riet by Ellenberg & Klötzli (1967), which covers a wetland bird sanctuary in Switzerland of about 100 ha in size. The map shows seventeen floristically defined communities, and the smallest variation recognized on the map is about 10 m^2 in the field. At this level it is sometimes possible to map individual species aggregations or even individual plants of larger life forms. The map was prepared for the conservation management of a unique ecological reserve, the largest existing fen habitat in northern Switzerland. Maps at such very large scales can be useful also for studies of vegetation development or succession, such as Pearsall's map of a river-mouth habitat in England (Tansley, 1939, pp. 604–5).

Maps at scales larger than 1:1000 are sometimes prepared for permanent research plots or sample quadrats. They are used as tools in the study of stand dynamics, for example the mapping of dynamic phases in a Yugoslavian climax forest (Mueller-Dombois & Ellenberg, 1974, p. 398), for the dynamic shifts of species in a community or for the location record of rare species and individuals. Such maps are usually not included in the concept of vegetation maps, but they are rather thought of as individual stand or community sample maps.

Systems for classifying vegetation

Structural vegetation type concepts and systems

The formation concept. Nearly all earlier attempts at classifying vegetation were based on physiognomic criteria that were more or less closely associated with features of the environment. Plant communities that are dominated by one particular life form, and which recur in similar habitats, are called 'formations' (in the physiognomic–ecological sense). Examples are the tropical rain forest, the mangrove swamp, the cactus desert, the grass steppe, the raised bog and the dwarf-shrub heath. Recognition of such types serves for initial orientation in setting subsequent studies into the proper perspective.

Originally, the term formation was defined physiognomically, that is through structural properties of the vegetation itself. Environmental attributes were added for closer description only. A later tendency has been to define the same concept climatically or as a geographic area, i.e., through properties outside the current vegetation cover. In the latter sense, the physiognomy of vegetation in certain areas of a macroclimatic or geographic zone was used only as a general indicator for the entire region. This has led to quite a different understanding of the same term. According to Clements (1928) a formation is the general plant cover of an area, which may include several physiognomic variations. These variations are inferred to belong to the prevailing, climatically controlled, physiognomic type. For example the prevailing physiognomic type may be a grassland, though the area may show stands of scrub and open forest. These would still be part of the grassland formation if occurring in the same so-called grassland climate. The same idea in the original concept is not a formation, but a vegetation zone or region. A vegetation region usually contains a mosaic of actual vegetation types. One of these vegetation types may prevail over larger areas in the zone, where it finds its most typical expression on non-extreme sites. Such vegetation was called zonal vegetation by Russian authors (Walter, 1971), which is similar to the climatic climax concept of Anglo-American authors, but less ambiguous. The Russian concept refers to a specific formation type of actually existing vegetation and not to potential vegetation, which may not really be present in an area. The world vegetation types mapped by Schmithüsen (1968), discussed previously, should be termed vegetation zones or vegetation regions. The boundaries of the major vegetation types actually occurring within such zones or regions might usefully be indicated by dotted lines.

Clements did recognize this zonal or regional vegetation mosaic, but he added to the confusion of the term formation by interpreting this vegetation mosaic as consisting of different developmental stages of the same formation. Clements converted the spatial side-by-side variation of vegetation, i.e., the different vegetation types or stages, into a postulated successional series, i.e., a time sequence. He believed that the regional side-by-side variations would develop into the same climax formation given enough time. This has led to some erroneous assumptions. Such a system that links all vegetation units to the final stage in succession constitutes a gigantic synthesis. However, it must resort to tenuous assumptions in many places. The system is inclined to force certain communities into preconceived positions. Such a system would be accompanied by many uncertainties in regions where the vegetation is almost everywhere modified by man. These uncertainties may be sufficient to make the system of little scientific value.

The synusia concept. A vegetation segment or plant community consists of plant species of different growth form and functions. For example, a grassland community may consist of scattered perennial bunch grasses, low-growing rhizomatous grasses and annuals. Each of these is a different type of life form and each life form type may be composed of several species. The assemblage of species of a life form type that grow together in the same habitat is referred to as a 'synusia' or 'union'. These synusiae have a certain individuality of their own in relation to the rest of the community, and for this reason were considered as the basic units of vegetation by some investigators, particularly by Gams (1918).

Very simple communities, such as annual grasslands may consist of only one plant synusia. More complex communities, such as forests may consist of ten or more synusiae. It is easiest to think of synusiae as layer communities (Lippmaa, 1939), such as moss, herb, shrub and tree layers in forests. But from a functional viewpoint, one may find more than one life form type in each layer, e.g., deciduous and evergreen trees in the upper tree layer, or geophytes, annuals and hemicryptophytes in the herb layer.

Synusiae may be identified with the help of a life form classification such as the well-known system of Raunkiaer (1937), which was developed into a key and further elaborated by Ellenberg & Mueller-Dombois (1967). Of course, synusiae may also be recognized less formally by broader life form classes. This depends on the specific objectives set for the investigation.

The advantages of the synusia concept are quite obvious: synusiae are easily recognized, even without knowledge of the species names. Descriptions of their combinations portray a clear picture of the communities and provide a certain idea of the habitat conditions. Synusial combinations can be traced even across the limits of different floristic regions and permit recognition of ecological relationships. Therefore, they are useful for world-wide comparisons as are the formations.

However, if they were used as basic units for classifying vegetation, one would arbitrarily separate the topographical and ecological unity of all those communities that consist of several synusiae, such as forest stands or heath communities. Synusiae should be treated as structurally definable subunits within a plant community.

Dansereau's profile diagram scheme. A well-known structural scheme is that of Dansereau (1951, 1957). His scheme employs six categories: plant life form, plant size, coverage, function (in the sense of deciduousness or evergreenness), leaf shape and size, and leaf texture. Each of these six categories contains a number of 'criteria' that can be used to characterize a vegetation segment in the field. For example, his plant life form category includes six general life form groups: trees, shrubs, herbs, bryoids, epiphytes and lianas; his size category includes three height classes: tall, medium, and low, which are defined quantitatively for certain life forms (e.g., low trees range from 8 to 10 m in height); his coverage category includes four criteria: barren or very sparse, discontinuous, in tufts or groups, and continuous. Each criterion is designated by a letter symbol. The letter symbols can be combined to describe and differentiate formations as units in the field, on aerial photographs or on a map. In addition, the map units can be interpreted further by schematic profile diagrams. These profile diagrams are established from a system of diagrammatic symbols, whereby each symbol denotes a structural criterion.

The method requires establishment of sample stands or relevés as is necessary for detailed floristic classification methods. Thus, Dansereau's method is more time-consuming than other structural schemes. The profile method is very formalized and the coded symbols have to be learned. In spite of its world-wide applicability, Dansereau's scheme is particularly useful for more specific purposes, such as the evaluation of military terrain or the study of structural detail in tropical rain forests, where taxonomic complexity presents itself as a barrier to studies in vegetation ecology (Holdridge *et al.*, 1971). The system may also be viewed as providing information complementary to the floristic association system of Braun-Blanquet (p. 48). Both systems work from below, i.e., from the detailed to the more general aspects.

Küchler's formula method. Another well-known structural system is that of Küchler (1967), which provides for a hierarchical approach. It begins with a separation into two broad vegetation categories of basically woody vegetation and basically herbaceous vegetation. Within the first category, Küchler distinguishes seven woody vegetation types: B, broadleaf evergreen; D, broadleaf deciduous; E, needleleaf evergreen; N, needleleaf deciduous; A, aphyllous; S, semideciduous (B+D); and M, mixed (D+E). In the second category he distinguishes three herbaceous vegetation types: G, graminoids; H, forbs; and L, lichens and mosses. These ten basic physiognomic categories can be further differentiated by whether they show a dominance of specialized life forms. The specialized life forms given in the system are five: C, climbers; K, stem-succulents; I, tuft plants; V, bamboos; and X, epiphytes. A third major distinction in Küchler's system is based on prevailing leaf characteristics in the vegetation segment: h, hard (sclerophyll); w, soft; k, succulent; l, large (> 400 cm^2); and s, small (< 4 cm^2). Further structural separations are made on height (stratification) and coverage of the vegetation. For height Küchler gives eight classes: 1, < 0.1 m; 2, 0.1–0.5 m; 3, 0.5–2 m; 4, 2–5 m; 5, 5–10 m; 6, 10–20 m; 7, 20–35 m; 8, > 35 m; and for coverage six: c, continuous ($> 75\%$); i, interrupted (50–75%); p, parklike or in patches (25–50%); r, rare (6–25%); b, barely present or sporadic (1–5%); and a, almost absent or extremely scarce ($< 1\%$).

With this set of categories and criteria, any vegetation segment may be characterized structurally by a formula composed of the letter and number symbols listed. Küchler gives various concrete examples and claims that the system can be applied to all map scales.

Fosberg's system. Fosberg presented a first (1961) and later a second (1967) approximation of a general structural classification of vegetation, which was adopted as a guide to classifying vegetation for the IBP and is reproduced in Appendix 2. One of the main features of Fosberg's system is that it is based – as are the schemes of Dansereau and Küchler – strictly on existing vegetation and purposely avoids incorporation of environmental criteria. This has the advantage that the vegetation units, when portrayed on a map, can be objectively correlated to independently established environmental patterns, because the vegetation boundaries are not in part delimited by environmental features. Where vegetation units are delimited in part by environmental features, correlation of such a vegetation map to environmental maps of the same area becomes problematic as this may result in circular reasoning.

The objective of Fosberg's scheme is to subdivide the vegetation cover of the earth into units that are meaningful for a large number of purposes by criteria that are applicable on a world-wide basis. These criteria cannot be floristic, because the distribution of plant species is geographically restricted. Therefore, they must be primarily structural.

Fosberg makes a distinction between physiognomy and structure. Physiognomy refers to the external appearance of vegetation and to its gross compositional features, implying such broad units as forests, grasslands, savannas and deserts, among others. Structure relates more specifically to the arrangement in space of the plant biomass. In addition, Fosberg uses function in the sense of seasonal leaf-shedding versus retention of leaves and specific aspects of growth or life form as important criteria for classifying the vegetation cover.

The vegetation is classified by use of keys. The first key begins with a breakdown into three possibilities – closed, open, or sparse vegetation. Thus, first consideration is given to spacing or cover of the plant biomass. Closed is defined as crowns or shoots interlocking, open as not touching, and sparse as separated by more than the plant's crown or shoot diameters on the average. Sparse vegetation is equated with the term desert, which is further defined as vegetation where plants are so scattered that the substratum dominates the landscape.

The first separation results in the first rank of vegetation units, which are called the 'primary structural groups' (namely closed, open and sparse). Within each of these, the second rank of vegetation units – called 'formation classes' – are separated.

For example, in the closed primary structural group, individual formation classes are distinguished as forest, tall savanna, low savanna (tall and low referring to height of grass layer), scrub, dwarf scrub, tall grass, short grass, broad-leaved herb vegetation, etc. Therefore, in the formation class breakdown, primary consideration is given to differences in the heights of vegetation layers and their continuity or discontinuity. But at least one of the layers in a vegetation unit must be continuous or closed to distinguish all of these formation classes from those in the open primary structural group.

Thirty-one formation classes are distinguishable in the first key. The individual formation classes are then further subdivided in separate keys. The first subdivision within each formation class key is by function, indicating whether the foliage is evergreen or whether there are leafless periods for the dominant layer. This functional separation distinguishes the third rank, called a 'formation group'. A further separation within the formation groups leads to the basic units, referred to as 'formations'. These are distinguished on the basis of dominant life form with emphasis on leaf texture (orthophyllous, ordinary leaf texture as opposed to sclerophyllous); leaf size (megaphyllous, at least 40 cm long and at least 5 cm wide; mesophyllous, of ordinary size; and microphyllous, for trees 2.5 cm greatest dimension, for shrubs, 1 cm or less), leaf shape (narrow versus broad); thorniness, and growth form (gnarled versus straight, succulence, graminoid, etc.).

Occasionally, the formations, which represent the fourth rank, are subdivided into 'subformations', the fifth and ultimate division. For example, the formation 'gnarled evergreen forest' is subdivided into two subformations, 'gnarled evergreen mossy forest' and 'gnarled evergreen sclerophyll forest'.

Each formation and subformation is supplied with at least one example of vegetation that fits the structural definition. A glossary defines all technical terms.

The classification system is necessarily artificial because, for example, the primary criterion of spacing may separate some environmentally or floristically closely similar vegetation into different primary structural groups. Yet, it may serve as a practical tool for mapping and organizing vegetation data for general purposes. Floristic associations can be studied within and across the structural frame given by the units. The structural vegetation units, when mapped, can be compared with climate, soil, historical and other environmental maps from which one can derive the major regional or zonal ecosystems.

The Australian system for IBP/CT. It seems significant that the Australian CT Committee of the IBP (Specht, Roe & Boughton, 1974) opted not to use Fosberg's system that was suggested for the IBP/CT survey. Instead the committee used a similar system developed by Specht (1970), which is reproduced here as Table 2.1.

As is evident from Table 2.1, the two primary criteria for classification are spacing and height of the vegetation. In this respect there is no basic difference from Fosberg's system. However, the Australian system uses one more category for spacing, and its height divisions are more detailed for trees. Three divisions are given for the height of trees, where Fosberg's system gives only one, namely forest. The other main life forms, shrubs and herbaceous plants, are each separated into two height classes as in Fosberg's system. The Australian divisions are defined in quantitative terms, i.e., spacing in per cent

Table 2.1. *Structural formations in Australia*

(From Specht *et al.* (1974, p. 6)

Life form and height of tallest stratum[a]	Projective foliage cover of tallest stratum[a]			
	Dense (70–100%)	Mid-dense (30–70%)	Sparse (10–30%)	Very sparse[b] (< 10%)
Trees > 30 m[c]	Tall closed-forest[a]	Tall open-forest	Tall woodland	Tall open-woodland
Trees 10–30 m[c]	Closed-forest[a]	Open-forest	Woodland	Open-woodland
Trees 5–10 m[c]	Low closed-forest[a]	Low open-forest	Low woodland	Low open-woodland
Shrubs 2–8 m[c]	Closed-scrub	Open-scrub	Tall shrubland	Tall open-shrubland
Shrubs 0–2 m[c]	Closed-heath	Open-heath	Low shrubland	Low open-shrubland
Hummock grasses 0–2 m	—	—	Hummock grassland	Open-hummock grassland
Herbs (including moss, ferns, hemicryptophytes, geophytes, therophytes, hydrophytes, helophytes)	Closed-herbland[d] (1) Closed-tussock grassland (2) Closed-grassland (3) Closed-herbfield (4) Closed-sedgeland (5) Closed-fernland (6) Closed-mossland	Herbland[d] (1) Tussock grassland (2) Grassland (3) Herbfield (4) Sedgeland (5) Fernland (6) Mossland	Open-herbland[d] (1) Open-tussock grassland (2) Open-grassland (3) Open-herbfield (4) Open-sedgeland (5) Open-fernland (6) Open-mossland	Ephemeral herbland

[a] Isolated trees (emergents) may project from the canopy of some communities. In some closed-forests, emergent *Araucaria*, *Acacia*, or *Eucalyptus* species may be so frequent that the resultant structural form may be classified better as an open-forest.

[b] Some ecologists prefer to ignore scattered trees and shrubs, equivalent to emergents in a predominantly grassland, heath, or shrubland formation.

[c] A tree is defined as a woody plant more than 5 m tall, usually with a single stem. A shrub is a woody plant less than 8 m tall, frequently with many stems arising at or near the base.

[d] Appropriate names for the community will depend on the nature of the dominant herb.

foliage cover and height in metres. Life form characteristics are used as an additional criterion in the height separations. For example, tall shrubs are defined as reaching from 2 m to 8 m in height, while low stature trees are defined as reaching from 5 m to 10 m in height. In their overlapping height ranges the two kinds of woody plants are separated by presence or absence of basitonic branching. While it may work in Australia, this separation would probably cause difficulties on a world-wide vegetation basis.

The two-way breakdown by spacing (i.e., percentage cover) and height in Table 2.1 results in twenty-six structural 'formations' for Australia, which can be compared with the thirty-one formation classes of Fosberg that were intended for world-wide application.

From this it can be inferred that a separation into low, intermediate and tall forests is necessary for Australia, if a structural system is to make any sense there. Fosberg suspected the need for this height distinction. In his suggested refinement on p. 81 in Peterken (1967), he admits that his one height classification of forest may result in lumping rather unlike types, such as subarctic spruce taiga with giant Douglas fir forests in northwestern America. But Fosberg points out that his system is flexible enough to accommodate such refinement. Another point is the four-way breakdown of spacing by the Australian classification into, for example, closed-forest, open-forest, woodland and open-woodland as opposed to Fosberg's three-way breakdown into closed-forest, open-forest and savanna. It would seem that Fosberg's system provides the flexibility for refinement in recognizing, for example, two spacing-subclasses in his open-forest category, so that the two systems are not incompatible. This means that the Australian spacing categories could be translatable into the Fosberg system if the need arises. It is interesting that the Australian system includes structural 'subformations' under its three spatially defined herbland formations, whereas none are shown for the woody formations. Most, if not all, of these herbland subformations can also be translated into Fosberg's herbaceous subformations. The absence of structural subformations for the Australian woody formations indicates a greater need for using floristic subdivisions, although L. Webb shows (in Section 2.2.3) that a number of structural criteria were found useful in distinguishing tropical rain forest types in N. Queensland. No mention is made of seasonality in the Australian system – certainly a reflection of the near-absence of deciduous forests in Australia.

The formation system of UNESCO. This classification system was established by the UNESCO Committee on Classification and Mapping of the World's Vegetation based on a list supplied by Schmithüsen & Ellenberg. The system was published initially by Ellenberg & Mueller-Dombois (1967) and then, in slightly modified form, by UNESCO (1973). The latter version includes a colour and symbol scheme for 225 vegetation types compiled by Gaussen. The purpose of this classification is to serve as a basis for mapping world vegetation at a scale of 1:1 million in terms of vegetation units that indicate parallel environments or habitats in different parts of the globe. Existing classifications were reviewed and these have influenced the thinking of the committee (notably Rübel's system). But none of the existing systems were found entirely suitable for the intended purpose. As in Fosberg's system, vegetation structure forms the main separating criterion. However, terms referring to climate, soil and landforms were included in the vegetation names and definitions, wherever they aided in the identification of the units. The reason for this is that significant ecological differences in habitat are not always reflected by easily definable structural or physiognomic vegetation responses. For example, tropical lowland rain forests differ ecologically from tropical montane rain forests. Yet, their structural differences are apparent only in certain regions and not on a world-wide scale.

The vegetation units are listed in hierarchical order under each of five 'formation classes'. The five formation classes are: I, closed forests; II, woodlands or open forests; III, scrub or shrubland; IV, dwarf-scrub and related units; and V, herbaceous communities. Thus, spacing and height of dominant growth forms are treated as parallel criteria in distinguishing formation classes. Each woody formation class is subdivided into 'formation subclasses' on the basis of whether the vegetation is mainly evergreen, mainly deciduous, or xeromorphic. These are then further separated into 'formation groups' by the macroclimate in which they occur. For example, distinguished among closed forests that are mainly evergreen are tropical ombrophilous (or rain) forests, tropical and subtropical seasonal forests, tropical and subtropical semi-deciduous forests, temperate rain forests, etc. The next lower subdivision is the 'formation'. Formations in tropical rain forests are tropical lowland rain forests, submontane and montane rain forests, tropical cloud forests, tropical subalpine rain forests (usually transitional to woodlands), tropical alluvial forests, tropical swamp rain forests and tropical bog forests. The next lower level represents the 'subformation', which together with the formation is considered the main map unit. For example, the tropical cloud forest is subdivided into a broad-leaved subformation (the most common form) and a needle-leaved or microphyllous subformation.

The classification gives an outline of all better known formations of the earth. The system is flexible and allows inclusion of additional units if this should become necessary. It provides a framework that permits the accommodation of an unlimited number of floristically quite different units (that occur in various localities scattered over the earth's surface) into physiognomically and ecologically equivalent abstract categories.

Both the UNESCO classification and Fosberg's scheme can be applied to categorize vegetation in the field and on maps in comparative terms within each scheme and also between the two schemes.

Fosberg's scheme provides a ready field tool for mapping at large and intermediate map scales. It allows one to establish pure vegetation units for correlations with environmental units mapped independently at the same scale. Because of its strictly structural orientation it may group ecologically quite different vegetation to the same unit. For example, tropical lowland and montane rain forests may form one vegetation unit. However, the ecological difference would become apparent upon comparing the vegetation units with environmental maps of an area, and there would be no danger of circular reasoning.

The UNESCO scheme gives some environmental–geographic information at the start and therefore conveys an immediate orientation that appears useful for a world-wide inventory. It provides for an outline of major vegetation types and a general overview that can serve for immediate statistical purposes. For example, endangered vegetation in different parts of the world may be singled out for conservation. Specific mapping criteria may have to be worked out regionally within this framework. These can then be based conveniently on a combination of regionally significant structural and floristic criteria.

All structural systems are artificial. For example, an open-forest or woodland may differ from a closed forest only because of some disturbance. However, the primary objective of these schemes is identification of given vegetation. An arrangement according to ecological, sociological or historical relationships would handicap the diagnostic value of such a classification. Moreover, it would hardly ever be completed, since ecological, sociological and historical relationships are the objects of continuing research and readjustment.

Ellenberg's classification of world ecosystems: a functional scheme

Ellenberg (1973) has presented a scheme for classifying the world into a hierarchy of ecosystems from a functional viewpoint. The largest and all-encompassing ecosystem is the 'biosphere', i.e., the outer skin of our planet (soil, water and atmosphere), in that it is the life-medium of organisms. It includes the oceans to their maximum depths. The biosphere is subdivided into two main groups according to type of energy source: 'natural or predominantly natural ecosystems', i.e., those whose functions depend directly on the sun as energy source, and 'urban – industrial ecosystems', whose functions depend on reconstituted energy (fossil fuel and recently, also atomic energy).

Six main separating criteria are used at different levels in the hierarchy:

(*a*) prevailing life-medium (air, water, soil, buildings),

(*b*) biomass and productivity of the primary producers,

(*c*) factors limiting the activity of primary producers, consumers and decomposers,

(*d*) mechanisms regulating the gain or loss of matter or of nutrients, (*e*) relative role of secondary producers (i.e., of the herbivores, carnivores, parasites and other mineralizers), and

(*f*) the role of man in the ecosystem (i.e., his role in the origin, development, energy flow and mineral cycling of the ecosystem, particularly his function in supplementing energy sources).

A hierarchical order is obtained by defining successively smaller ecosystems within larger ecosystems. Starting with the biosphere, the next lower size-level is referred to as the 'mega-ecosystem'. Five mega-ecosystems are recognized by the life-media (criterion *a*) that they represent (capital letters as used in Ellenberg's key): M, marine ecosystems (saline water); L, limnic ecosystems (fresh water); S, semi-terrestrial ecosystems (wet-soil and air); T, terrestrial ecosystems (aerated soil and air); and U, urban–industrial ecosystems (the creations of man). The first four are predominantly natural while the last is artificial.

'Macro-ecosystems' are the next lower size-level within each mega-ecosystem. The macro-ecosystems are still very broad or inclusive units that are separated mainly by the criteria (*b*) to (*d*) (e.g., forests). 'Meso-ecosystems' are considered the basic units of this scheme. They are the 'ecosystems' in the most commonly understood sense. A meso-ecosystem is considered a relatively uniform or homogeneous system with respect to the abiotic conditions as well as the life forms of the prevailing primary and secondary producers (e.g., a cold-deciduous broad-leaved forest with its animal life). 'Micro-ecosystems' are subdivision of meso-ecosystems, which depart with respect to a certain component (e.g., a lowland, montane, or subalpine cold-deciduous broad-leaved forest with its animal life). 'Nano-ecosystems' are considered to be small ecosystems that are spatially contained within larger ecosystems and that exhibit a certain individuality of their own (e.g., a wet depression in a montane deciduous broadleaf forest).

Within almost all ecosystems one can recognize strata or other 'partial systems', which can be analysed individually. At least three partial systems can be recognized generally: 'Topo-partial system', i.e., a layer or other topographically stratified segment within an ecosystem (e.g., the topsoil in a forest). 'Substrate-partial system', i.e., a small island-like community within an ecosystem (e.g., a moss-covered log in a forest). 'Pheno-partial system', i.e., a partial system that appears only during a certain time of the year (e.g., an algal bloom at the surface of a lake).

The classification scheme includes a special scale for defining the kind and degree of human influences for each ecosystem to be classified. Four kinds of human interference are recognized: 'harvesting' of organic materials and minerals, which are significant for the metabolism of an ecosystem; 'adding' of mineral or organic materials or organisms; 'toxification', i.e., adding of substances which are abnormal for the metabolism of the ecosystem and which are detrimental to important organisms or organism groups; 'changing of the species composition', i.e., by suppressing existing species or by introducing alien species into the ecosystem. The degree for each of the types of human interference is expressed by a scale of increasing severity from 1 (e.g., no harvesting) to 9 (e.g., destructive harvesting).

For world-wide comparisons of ecosystems the scheme also includes a biogeographic separation into nine regions, such as tropo-American, tropo-African, tropo-Asian, Australian, etc. Each of these biogeographic regions can be further subdivided into biogeographic subregions or provinces. All criteria are identified in the scheme by letter symbols and a decimal system. These provide for classifying any ecosystem by a short formula on a world-wide basis. An overview, in form of a key, shows the four predominantly natural mega-ecosystem types (M, L, S and T) subdivided to meso-ecosystems and in some examples to the level of nano-ecosystem and partial system (where well known). The scheme can be completed as further knowledge becomes available.

The key makes a major division between aquatic (M+L) and land ecosystems (S+T) on the basis of structure. The vertical extent of predominantly natural land ecosystems (in contrast to aquatic ecosystems) is not determined by their life-medium (soil and air) and the availability of light, but by the height to which the dominant vascular plants grow. It follows that the terrestrial ecosystems are defined primarily by vegetation structural criteria, and their classification is based on the UNESCO formation system. Therefore, meso- and micro-ecosystems are divisions somewhat parallel to formation and subformation types, but they are described in functional terms (criteria (*b*) to (*f*) as far as these are known). It may also be noted that the second structural unit concept of synusia has given rise to the functional concept of partial system as used in this ecosystem scheme.

While the scheme is based entirely on structural–functional criteria, it is also clear that any exact investigation of ecosystems cannot ignore the species composition that forms the living matrix of the system. On the contrary, for any detailed investigation of ecosystems it is desirable to derive as complete as possible a species list of the participating plants and animals. Moreover, abundance determinations should be made for at least those species of plants and animals that are significant for the productivity and maintenance of the ecosystem. These lists are then usefully ordered or classified according to animal- and plant-sociological viewpoints.

Floristic vegetation units and systems

Species dominance community-type concepts: the sociation and consociation. Single, easily noticed plant species provide the simplest floristic tool for attaining, relatively fast, a certain order in the great variability of plant communities. These have always been used even by untrained persons, for example, in differentiating forest stands (beech forest, pine forest, etc.). Such a simple classification can also be very satisfactory for scientific purposes, if the area is floristically poor. In Scandinavian countries the most abundant or the most dominant species are used for distinguishing the so-called sociations.

Du Rietz (1921) considered the 'sociation' the basic unit of vegetation classification and defined it as a recurring plant community of essentially homogenous species composition with at least certain dominant species in *each* layer. For example, the East German pine-heath communities form a *Pinus sylvestris–Calluna vulgaris–Cladonia* sociation, certain beech forests a *Fagus sylvatica–Allium ursinum* sociation, etc.

Du Rietz speaks of a 'consociation' if only the *upper stratum* of a several-layered community is dominated by one species. As a type concept, a consociation can also be understood as a class composed of individual concrete sociations, whose upper strata are dominated by the same species, while the lower strata may be dominated by different species in each vegetation sample. The term consociation was used also by Clements, Tansley and Rübel in a very similar way. Consociations are more common than sociations particularly in species-rich areas. An example is the oak forests of England, which, according to Tansley, represent a consociation with very variable undergrowth. Few oak forests have the same dominants in the herb layer; one example is the *Vaccinium*–oak forests on acid soils.

Petersen (1927) tried to apply the consociation concept to the classification of meadow communities in central Europe. He distinguished meadow types by the dominance of certain grass species, one dominant species characterizing a meadow type. However, because of the great number of species in central and southern Europe, there are rarely meadows with only one dominant species. Therefore, it would be necessary to consider most communities as mixed types or they would not fit into Petersen's system at all.

This difficulty with regard to the sociation and consociation concepts exists in all regions with large numbers of species, where many species compete for the same habitat. A good example is the tropical rain forest in continental lowland areas. Therefore, sociations and consociations have no universal applicability as units in vegetation classification.

However, even in such communities where single plant species have become dominant it is often not satisfactory to classify them as belonging to a certain consociation type. It was found that the same species may become dominant under different habitat conditions, whereby the associated flora may differ considerably in response to the differences in environment. For example, the tall reed grass *Phragmites communis* may grow in pure stands at the margin of larger lakes with occasional admixture of *Scirpus lacustris* or other tall semi-aquatic plants. *Phragmites* is found also to form vigorous stands at river margins in the tidal ranges, in habitats with considerable daily and annual fluctuations in water level. The associated plants named above cannot grow under these conditions. Instead, a more or less rich geophyte-flora is found growing there in the spring, especially the yellow-flowering *Ranunculus ficaria* and *Caltha palustris*. It is obvious that the two *Phragmites* consociations can be considered one unit only for very superficial reasons.

Thus, community types defined by a single dominant species (consociations) may lump together very different habitats. Moreover, the single dominant species concept cannot be applied in many regions. It is better to use a more flexible concept of floristic dominance types, where community types can be recognized by one or more dominant species in the prevailing synusia. This, in fact, is the most widely used community-type concept in North American vegetation studies (Whittaker, 1962), and in Australian studies (Specht, Roe & Boughton, 1974). It lends itself to a relatively easy and informal system of classifying communities in many parts outside the continental tropics. In such floristically simpler areas, dominance community-types may be used effectively as the first floristic subdivisions of formations. They correspond approximately to the European type concept of alliance (Ellenberg, 1959) which, in the more formalized system of Braun-Blanquet, forms the floristic unit above the level of association.

Because more than one dominant species is often used to designate these dominance-community types, they have been called 'associations' by Clements. These so-called 'associations' are usually very large and heterogeneous in habitat conditions and they differ entirely from the European association concept, which is discussed in the next section.

The association concept. It is quite possible in the examples cited above to differentiate several vegetation units if one considers the associated as well as the dominant species. Units that are floristically defined in this manner are called associations. In contrast to a sociation, an association does not have to show a single dominant species in each layer. Instead more than one species per layer may be used to define an association.

Following a resolution of the International Botanical Congress in Brussels in 1910, it was agreed to apply the term association only to communities 'of definite floristic composition, uniform physiognomy and when occurring in uniform habitat conditions'. As understood on the continent of Europe, an association refers to a relatively small vegetation unit, a unit below the level of consociation. The 1910 International definition of the term association was rather strictly interpreted in continental Europe. However, an exact fulfilment of the three requirements (definite flora, uniform habitat, and physiognomy) is not always possible.

The requirement referring to a 'uniform' habitat is particularly difficult to fulfil. A uniform habitat may be found in several field situations, but the vegetation samples to be grouped into an association-type can never have identical habitats, because no two places on the earth's surface have exactly the same site factors. Likewise, the criterion of definite floristic composition needs closer definition. In classifying, it is impossible, even though ideal, to consider all species to be of equal significance. Because of the great variability of communities, one would have to distinguish as many 'units' as there are plant communities. Even two closely similar vegetation samples will not have identical species lists. Yet, closely similar vegetation samples will have a certain proportion of species in common. Therefore, it is possible only to emphasize certain groups of species, namely those that recur commonly in different locations of a region. Only those communities are put into a type that show the same groups of species. Such groups can be distinguished either by comparing a large number of vegetation samples (i.e., by tabular comparison) or in other ways. An association-type, therefore, can be defined as a unit of vegetation derived from a number of vegetation samples or relevés that have a certain number of their total species in common. An individual association member, i.e., a concrete community, can be recognized in the field by the presence of certain species of a diagnostic group.

Unfortunately, the Brussels definition does not really specify the criteria that were meant to be applied in distinguishing an association. As a result, two entirely different association concepts evolved in continental Europe and North America. The only criterion common to both these different interpretations is that an association name is made up of a combination of species names. In North America, Clements (1928) interpreted the term association very broadly to refer to the first subdivision of a formation. This broad association concept is still widely used in the United States and in Australia. Since Clements' 'formation' was actually the general plant cover in a given macroclimatic region (i.e., a vegetation mosaic), his association concept was more or less a climatic subregion of which a selected vegetation cover was used as an indicator. For the whole of North America Clements recognized three so-called climaxes – a grassland, a scrub, and a forest climax. Each climax was subdivided into a few 'formations' (= regions) and each 'formation' was subdivided into two or more 'associations'. For example, in the forest climax, the Pacific coastal forest (region) was called the *Thuja–Tsuga* association and the *Larix–Pinus* association. Clements defined an association floristically by joining the names of two regionally dominant species and then implied that an association was a grouping of two or more consociations. The term consociation was understood *sensu* Du Rietz. Thus, Clements' association concept was even more inclusive than the consociation concept, which defines community types by single dominant species.

Braun-Blanquet's floristic association system. In brief, the system consists of preparing species lists in relevés and then processing these lists in synthesis tables. In these tables, the species common to several relevés are identified and emphasized. This process has recently been automated by computer programs (Spatz, 1972; Ceska & Roemer, 1971). The species unique to each relevé are not ignored, but they are not given the same value as the species that recur together in a number of relevés. These common species groups are the key to the identification and mapping of vegetation units.

The association, as previously defined in the continental European sense, is considered the basic unit in Braun-Blanquet's (1928, 1951, 1964) system. Therefore, his system can be called a floristic association system. Other vegetation units are recognized by the same tabulation technique, but as units above or below the rank of association. In this way all units are interconnected in form of a hierarchy, but each unit is identified by certain common groups of species.

The different ranks are usually designated by a particular ending added to the root of the scientific genus name of an especially characteristic species. The following summary gives a general outline:

Rank	Ending	Example
Class	-etea	Molinio–Arrhenatheretea
Order	-etalia	Arrhenatheretalia
Alliance	-ion	Arrhenatherion
Association	-etum	Arrhenatheretum
Subassociation	-etosum	Arrhenatheretum brizetosum
Variant	No ending	*Salvia* variant of the Arrhenatheretum brizetosum
Facies	-osum	Arrhenatheretum brizetosum bromosum erecti

The lowest unit in this system, the facies, is not characterized any longer by exclusive species (i.e., 'character species'), but merely by dominance of a certain (or several) species. Therefore, it corresponds in some respects to the consociation or sociation. However, it is viewed here in relation to the other ranked units, whose geographic coverage is progressively larger.

Recently, the tendency has developed to distinguish associations merely by differentiating species. This implies dispensing with the requirement of character species for an association. This development results from the experience that there are only few character species in the strict sense. However, the alliances retain their own character species, while orders and classes usually show numerous character species.

The segregation of different vegetation units by differential species is based on tabular comparison of vegetation relevés. Therefore, it is based on a purely inductive method. However, ranking of the units into the previously discussed system, that is, in particular the solving of the question as to which of the units can be considered associations, depends on the personal judgement of the investigator.

Evaluation

The check-sheet survey in relation to the stated methods and its specified objectives

The general objective of the CT check-sheet survey was to obtain in a relatively short period a description of natural areas and research sites with their vegetation types in internationally comparable terms. The specific purposes were to find out which major vegetation types or ecosystems are receiving conservation status, and to determine how representative these vegetation types are on a world-wide basis. It is clear that this survey had to be based on a structural system, because species distributions are by nature provincial, i.e., confined to floristic provinces.

It may be said that the general objective has been met most efficiently by the adoption of Fosberg's system. The reason is that of the structural systems this method is the easiest to apply; and it carries the most universal meaning. Dansereau's and Küchler's systems appear to be equally universal in application, but they require more detailed observations in sample plots. Küchler's formula method may take an intermediate position as far as time investment is concerned. Its unit hierarchy is not so formalized and resulting structural formulae may permit too many combinations to achieve a ready overview of parallel types. But this is again a matter entirely of purpose. Dansereau's profile method is the most detailed and time-consuming. It is probably most useful at the large to very large scales, and is almost comparable in detail of unit-separation to the floristic association system of Braun-Blanquet.

The Australian system can be considered a regional refinement of Fosberg's system in so far as it leaves out (for Australia) unnecessary units (for example, all deciduous woody vegetation types), while it incorporates refinement in forest height classes and plant cover density units. The latter are quantified. This makes the Australian types more objectively assessable. But the Australian system is not different in principle from Fosberg's scheme, and the Australian formations and subformations can probably all be translated into Fosberg types, should the need arise.

The UNESCO system is more specialized than Fosberg's in the sense that its application requires more experience to yield good results. A person with a good knowledge of world vegetation types may be able to translate all of Fosberg's types into the UNESCO system, should that become necessary during the planned UNESCO mapping project.

A word of caution may be added regarding the proposed 1:1 million map scale for the UNESCO units. The scale rarely allows mapping the outline of existing structurally defined formation types. Therefore, it would be more appropriate to call the UNESCO units 'formation zones'. For an inventory of conservation areas it would be necessary to locate the existing remnant formation types in their respective zones and to mark these by shading or with asterisks on the proposed 1:1 million international vegetation maps.

The first of the two specific CT check-sheet objectives – to find out which of the major vegetation types or ecosystems are receiving conservation status and which not – poses a more complex problem. Of course, the conservation status itself may be readily established from the check-sheets. But the problem of vegetation type diversity of an area can only be resolved in reference to a specified geographic scale range, because Fosberg's structural categories can be identified at different levels of homogeneity. For example, the concept of a 'gnarled evergreen mossy forest' (Fosberg unit 1A1 3) may be interpreted as a cloud forest belt (at the scale of 1 : 1 million) in one surveyor's mind, while it may be rather exactingly interpreted (at the scale of 1:10 000) by another.

It must be remembered that classification of vegetation usually involves two levels of abstraction. The first level is introduced in the segmentation or subdividing of a vegetation cover by the homogeneity concept of the investigator. One must decide what range of variation in the vegetation cover one can reasonably allow within a vegetation segment or unit. The second level of abstraction is introduced in the grouping of similar segments into vegetation categories or classes. This part of the classification process depends on the similarity concept of the investigator.

When vegetation has been abstracted twice in this fashion the established categories may or may not have much reality in nature. The classification itself gives very little information on this question unless it is well documented by vegetation samples (i.e., relevés), but even then the validity of the classification is not always easy to assess. The test of validity of a vegetation classification comes when the established categories are depicted on a map. Therefore a vegetation classification cannot really be considered complete until it is supported by a vegetation map.

The check-sheet survey can thus show only to a first level of approximation the true diversity of structural types in the areas surveyed. A more definite answer can only be obtained through a map showing the Fosberg units present in the check-sheet sites.

The second specific purpose of the IBP/CT survey – to establish what sort of representation the check-sheet areas give on a world-wide basis – can hardly be answered with this survey. This is so because there is generally not enough knowledge on the vegetation types outside the surveyed areas. This information can only be achieved through a world-wide effort to map vegetation types. This leads to the following recommendation.

Hierarchical mapping for conservation purposes: a recommendation

The insufficient cover of the IBP/CT check-sheet survey in terms of world distribution of available areas has been brought out in Chapter 1. But even if the distribution of check-sheet returns had approached a more complete global coverage, the survey can only be viewed as a first approximation. As such it has established a momentum that should be utilized and developed further through the next internationally co-ordinated research programme, notably by the UNESCO MAB 8 Project.

One cannot reasonably expect that a complete survey of globally available vegetation types at the previously discussed large geographic scale of Level 4 can be made within the time constraints of even a 10-year programme. However, it is at this large map scale of 1:10 000 to 1:100 000 that ecological field research and natural area management have to operate in order to be locally significant and effective. For one thing, conservation of biological resources becomes meaningful only when we begin to take stock of the species in the communities and ecosystems. We must further understand their quantitative relationships, their dynamic tendencies, and the ecological roles of at least the more important species, whether they be dominant or rare and endangered.

A global survey for conservation of species and ecosystems would best be approached through a programme of hierarchical mapping. A world inventory of conservation sites or ecological reserves should relate all sites to a 'system of ecological zones'. Furthermore, because of the provincial nature of species distributions, it is of utmost importance that superimposed on the system of ecological zones is a 'system of biogeographical provinces'. These provinces will serve to emphasize the uniqueness of the biological populations that comprise ecosystems occurring in otherwise structurally and environmentally similar vegetations in different parts of the world.

It is important that a global survey of this sort uses all existing information. At the broadest level of generalization (Level 1), Schmithüsen's 1:25 million world vegetation map can be adapted with relatively little extra work. Firstly, the world vegetation formation zones indicated on that map may be used to search for existing remnants of real vegetation types of world formations (to be located by dots on the same map). The same map should also be supplied with the boundaries of world biogeographic provinces of such categories as suggested by Ellenberg (1973), i.e., tropo-American, tropo-Asian, Australian, etc. Secondly, a search should be made for Level 2 vegetation and ecological zonation maps, e.g., Küchler's map of the USA or Krajina's map of British Columbia. On these maps also the still existing remnant vegetation types of zonal significance should be located by asterisks. Wherever possible, biogeographic boundaries should be indicated. The UNESCO plan of generating a comparable set of international vegetation maps (UNESCO, 1973) at the scale of 1 : 1 million deserves the greatest support in this respect: once put into action, the mapping project can supply reliable information on the status of world ecosystems. The 1 : 1 million mapping project should make full use of the tremendous advances recently made in remote sensing technology. In this way a real breakthrough could be achieved by mapping world ecosystems in considerable detail.

Any important individual area should then be enlarged to Level 3 map scales (ranging from 1 : 100 000 to 1 : 1 million) for an inventory of major ecosystems within states, provinces, or island groups. The next enlargement to Level 5 maps can then be related to the global network of ecological reserves.

The large geographic scale range at Level 4 (1 : 10 000 to 1 : 100 000) is the one that forms the underpinning of the various vegetation type concepts and classification systems discussed above, because all of these are based on experience gained in the field with real (in contrast to potential) vegetation.

Recall the previously mentioned ambiguity of the homogeneity concept relating to segmentation of a vegetation cover into communities. This ambiguity can be minimized by specifying the geographic scale. For example, if an area is to be classified into communities at the scale of 1 : 10 000, different investigators are likely to stress similar subunits, particularly if the classification system is specified also. Ambiguity at the second level of abstraction in classification – the similarity concept of the investigator – would likewise be minimized by specifying geographic scale and classification system. The similarity concept can even be made objective to some extent by using similarity indices.

It must be understood clearly that the task of surveying the biological resources on our planet is not complete until we produce an inventory of species populations with information on their grouping, quantities, and dynamic status in their respective communities. This task can only be accomplished through intensive local area studies involving the establishment of a large number of sample stands or relevés. There is little doubt that the most successful method for this purpose is the relevé method of Braun-Blanquet. It must be emphasized that the method cannot serve to establish a world-wide hierarchy of floristically defined communities, because species ranges differ from area to area. However, the sampling of species lists in the field, with indications of their quantities in a series of relevés or sample stands, can be done in all vegetation areas of the world. It is the most thorough and the most rapid community analysis method for this specific purpose. Moreover, there are a number of relatively simple, rapid, and useful data processing methods available, ranging from the two-way table technique to the dendrogram method of cluster analysis (Mueller-Dombois & Ellenberg, 1974).

In the establishment of a global network of ecological reserves, it would seem appropriate that first urgency is put on the development of the 1:1 million international vegetation maps. Second priority should be given to the more detailed floristic and faunistic local area surveys. For the specified purpose at hand it would seem appropriate to promote the relevé method as the best formal inventory technique for local area research and management. Depending on the nature of the regional vegetation, the relevé method can be employed in connection with large-scale structural vegetation maps established through Fosberg's classification criteria, or it can be used in connection with any other large-scale existing vegetation map (e.g., one based on species-dominance criteria). Moreover, the method itself may supply the criteria for mapping floristically defined finer subdivisions by yielding – in most situations – diagnostic or key species through the two-way table technique. These key species in turn can serve as the basis for mapping floristically defined community types which are useful as a framework for natural area research and management.

2.2.3 *The special problem of tropical forest classification:* L. J. Webb

The problems of vegetational classification have been well summarized in recent times (e.g., Whittaker, 1962) and, as Shimwell (1971) remarks, it no longer seems profitable to continue arguments about which is the 'best method'. Whether the classification and description remains simply an exercise in indexing and pigeon-holing data by a set of more or less arbitrary criteria, or becomes an extended study of the syntaxonomy of plant communities, will depend on the aim of the work, the facilities available and the urgency with which conclusions – at least tentative ones – are required.

For IBP, with the object of the Conservation of Terrestrial Communities (CT), the decision is clear: it is necessary to provide an objective scientific basis for the conservation of habitats and species so that maximum biological diversity and a range of representative ecosystems are maintained. In addition, as Peterken (1968) points out, it may be necessary to reserve special sites with some outstanding feature which can only be subjectively assessed as of wide scientific interest.

A standardized approach which uses a key based on a combination of physiognomic, structural and functional features was provided by Fosberg (1967). The selection of criteria is admittedly arbitrary, and the categories are fixed at the very general level of formation or subformation. They may be refined by adding some 'significant information' including quantitative data and floristics, but excluding environmental characteristics. The advantage of such a comprehensive and generalized system is to make vegetation data comparable on a world scale. However, while this may be satisfactory for the selection of 'international' reserves for biological research, it may not be adequate to identify the range of ecosystems of a single country or region, for which lower-level vegetation units are required. This inadequacy is intensified in countries, especially tropical ones, with relatively large floras and high diversity of communities. In these countries, where the flora is generally not well-known, there is no doubt that floristic classification, at least on a comprehensive and rapid scale, is impossible. There seems no alternative in such complex situations to using some kind of physiognomic–structural classification of vegetation, and the question then becomes whether adequate refinement (which must avoid floristics) of Fosberg's key is possible. If it is not possible, what other method could be used which would preserve the simplicity, rapidity and value for broad comparisons of the Fosberg classification, while yielding a more detailed coverage of important vegetation types on a reasonably broad scale?

Before attempting to answer this question, it is necessary to refer to the general problems of structural classification. The question will then be examined in relation to the tropical vegetation of North Queensland for which a detailed map and vegetation classification are now available, and for which a series of trials of different structural and floristic classifications has recently been made.

Problems of structural classification in relation to IBP purposes

The validity of a classification of vegetation is usually established by the closeness with which the vegetation units are correlated with physical environmental factors such as limits of rainfall, altitude or soil drainage, due allowance being made for the effects of disturbance. The vegetation units thus become indicators of environmental types or of patterns of 'site potential'. These environmental correlations may not, however, be relevant to the aim of IBP/CT, which is to conserve 'special areas for permanent scientific study' (Peterken, 1967), and for which the primary role of reserves selected from an international standpoint is as 'sources of materials and facilities for biological research' (Peterken, 1968). To satisfy this aim it would be desirable to select representative sites by reference to a classification of ecosystems, but this is impracticable. It has, therefore, been found most convenient to deal with the vegetation component, a procedure dignified by the assertion that vegetation occupies a basic position as the primary producer in an ecosystem and is therefore somehow an integrated expression of it.

The emphasis on vegetation, and especially the restriction to physiognomic–structural attributes in the classification of complex tropical vegetation, raises the following problems:

Level of specification of the type

It is clear that the level at which structural features cease to be informative or useful in typology has to be determined for the particular study area. It has been shown that the more complex the flora, and hence the greater the variety of life forms and structural attributes, the more effective is structural classification. Conversely, floristic classification is superior in regions with simple and usually well-known floras (Webb, Tracey & Williams, 1974). A structural classification proposed by Webb *et al.* (1970) produced a more detailed subdivision of the rain forests of eastern Australia than was possible by the systems of Ellenberg & Mueller-Dombois (1967) and Fosberg (1967). The Australian structural types of rain forest were correlated not only with macroclimatic factors but also with major differences in soil texture, soil aeration and soil nutrient availability, notably of phosphorus and calcium (Webb *et al.*, 1970; Webb, Tracey & Williams, 1976). This suggests that the structural typology was able to identify vegetation units below the level of subformation, i.e., that the structural types approximated the level of floristic 'alliances' (groups of related 'associations'). As recent numerical analysis of floristic data from selected North Queensland sites shows, several rain forest associations are represented within one structural type in the same region (Dale & Webb, 1975; L. J. Webb, unpublished data).

Historical aspects

Similar structural patterns of vegetation tend to recur throughout large regions on a continental and intercontinental scale, irrespective of major floristic differences. Thus structural classification, except in a most rudimentary way (e.g., by involving special life forms such as cacti, bromeliads, sclerophylls, etc.) is not able to extend above climate in the hierarchy of environmental factors, and misses the role of historical factors which ultimately determine the variety of taxa available in a given area (van Steenis, 1953). Yet it is these differences in flora and associated fauna from region to region, as well as the spectrum of their variations within each region, which is of interest to IBP/CT. This deficiency may be circumvented by introducing the concept of 'biotic provinces' (Dasmann, 1972, 1973*a*), i.e., by identifying biogeographic regions which are distinguished by vegetation, flora or fauna. In highly diversified places, within which there are major environmental changes over a small area, it is necessary to include several different vegetational formations within the same biotic province, otherwise provincial boundaries could not be shown except on large-scale maps.

By applying structural classification of the vegetation within biotic provinces rather than across an arbitrary geographical region, the taxonomic characteristics of particular areas resulting from differences in biological history are preserved. Dasmann (1973*a*) has already provided a tentative classification and maps of the biotic provinces of the world, and the application of this method for the definition of conservation areas in tropical north-eastern Australia will be described below.

Recourse to floristics (and faunistics) is inevitable in the case of the 'Special sites' defined by Peterken (1968). Only in this way can criteria such as rarity of species, unusual scientific interest (such as relict or primitive species), biological diversity, etc. (Holdgate, 1970) be satisfied.

Probabilistic factors in the pattern

In complex tropical and subtropical forests, determinate and probabilistic ecological factors underlie the pattern of regeneration of plant species (Webb, Tracey & Williams, 1972). Many species are ecologically vicarious within a more or less uniform determinate physical environment, so that a sample plot located within a given floristic association (Dale & Webb, 1975) will not catch the total number of species available and which are characteristic of the forest type. Richards (1971) has discussed this problem in species-rich tropical communities, noting that it would be impossible to define a 'minimum conservable area' unless rare species present in low densities were able to replace one another without affecting the integration of the community (cf. Poore, 1964). The most realistic solution, as Richards (1971) suggests, seems to be to include in the sample plot as many as possible of the units of the patchwork or mosaic of the forest which are recognizably different in structure. Further, if species recruitment for all the different niches in complex tropical forests requires a 'vast hinterland of habitat' (Elton, 1973), then the larger the reserve the better.

Seral structure

The mosaic structure of 'climax' rain forest is associated with patches of regeneration of different age, height, and composition, which have resulted from wind-throws and other forms of disturbance, i.e., the forest is a mixture of primary and secondary patches. This secondary growth raises another problem for structural classification, which is very sensitive to the effects of recent historical events (such as cyclones or logging) which drastically modify height and density of canopy and the distribution of life forms such as tall pioneer forbs, lianes and saplings with uniform stem diameters (Webb, Tracey & Williams, 1976). Should the sample area to which the structural classification is applied comprise only forest which is apparently mature, integrated and stable, or should it include seral patches? The general implications of this question of successional status as a valid basis for classification were discussed by Peterken (1968), who took the pragmatic view that vegetation to be reserved should be sampled as it exists, irrespective of its relationship (if that is known) to the 'potential natural vegetation' of the site.

Secondary growth forests tend to simulate simple mature forests, i.e., simplification of physiognomic–structural features associated with certain limiting environmental factors for a mature forest type results in a structural convergence with early seral stages of ultimately complex communities. Since the latter pioneer communities tend to be dominated patchily by a few characteristic tree species, termed 'natural monocultures' by Richards (1971), it seems that, as with simple forests, structural classification should be abandoned in favour of nominating a limited number of floristic types characteristic of disturbance of a given area. Accordingly structural classification should be reserved for apparently mature forest.

Criteria for choice of reserves

Although the purpose of the structural typology is essentially biological, i.e., to identify representative ecosystems for reservation in the tropics, it cannot be isolated from a selected number of environmental features which are relevant to the feasibility of reservation. Different associations occur on a large or small scale depending on the scale of variation of the physical environment which determines them. Thus while a small area may be adequate to preserve a given structural type, a relatively large area would be necessary to cover floristic variations at the association level. These determinate variations in the vegetation type are additional to probabilistic variations within each association. Once the environmental correlations of the associations are established in the field, estimates of the area of reserve required are assisted by a knowledge of the variation in soils, topography, etc., in the region, and this is a further argument for the recording of relevant environmental data. Thus extensive, more or less uniform, terrain with one major vegetation type will present fewer problems of location of a reserve than a mosaic of different community-types on different soils, topography, etc.

It is also relevant to note the importance of size and shape of area, form and length of boundary, and ecological context (van Leeuwen, 1966; Webb, 1966; Margalef, 1968) in deciding the location of reserves of representative ecosystems. Social and political factors, such as land tenure, proposals for clear-felling for pulpwood extraction, etc., are also highly relevant but are outside the scope of this paper.

Application to North Queensland forests

Design of the structural pro forma

An 'open-ended' structural typology which provides for the addition of attributes which may be ecologically significant in newly surveyed areas, or relevant to a special purpose (such as wildlife habitat) provides the opportunity for collecting more biological information than can be supplied by a key using fixed categories. Thus a list of structural attributes can be drawn up for a particular region after a preliminary survey has shown which attributes are necessary to provide a satisfactory and rapid ecological classification using the methods of numerical analysis (Webb *et al.*, 1970). This list can be further refined to yield a working *pro forma* for this and other regions, and provision is made to add 'prominent features not recorded'. A *pro forma* listing various properties of physiognomy, struction and function, with values scored either by presence or absence or on a 0–3 scale, was designed on the basis of experience in north-eastern Australia, and New Guinea (Webb, Tracey & Williams, 1976). The *pro forma* was later successfully used, with some on-the-spot modifications, to classify *Nothofagus* forests in Central New Guinea (Hynes, 1973, 1974), in an ecological survey of vegetation in the New Hebrides (A. N. Gillison, personal communication), and to make a structural comparison of New Zealand and south-eastern Australian rain forests and their tropical affinities (L. J. Webb, 1976). The *pro forma* provides a rapid and compact method for collecting relevant data, and the resulting structural description and classification of the vegetation in the different regions yielded a framework of reference which would not have been possible in the absence of floristic data or of very detailed information about relevant factors of the physical environment.

A *pro forma*-derived typology has the advantage, especially if it is applied within biotic provinces and so preserves regional taxonomic differences, that the field data are allowed 'to sort themselves'. Additional field data can be added as they become available, and are either accommodated within existing categories or form new ones. There is plenty of room for innovation in a biological sense, and the danger is reduced of using an elaborate analysis to confirm what, substantially, was already known, i.e., as predicted and constrained by a schematic key.

For example, the basic structural *pro forma* which was developed to enable the classification of tropical forest vegetation (Webb *et al.*, 1970; Webb, Tracey & Williams, 1976) has been modified by the addition of special properties relevant to niche occupation and food resources for birds. A multi-disciplinary study is under way in North Queensland to determine plant and bird populations in particular situations involving mosaics of sclerophyll forest and rain forest on the humid tropical lowlands, and some progress reports have been made (Kikkawa & Webb, 1967; Webb *et al.*, 1973; Kikkawa, Webb & Tracey, 1974). However, discussion here will be restricted to the use of the structural *pro forma* for classification of North Queensland forest sites, and comparison of the results with the Fosberg (1967) system.

Results of the comparison

Area studied. The North Queensland humid tropical zone is part of the very broad 'Queensland coastal' biotic province suggested by Dasmann (1973*a*), and which requires subdivision. It is convenient to recognize far northern, northern and southern coastal provinces, which approximate the floristic zones based on Burbidge (1960) which were used, together with Köppen climatic types, to define bioclimatic regions in the Queensland IBP survey by Specht, Roe & Boughton (1974). The northern province was also separated as one of seven floristic zones in a floristic classification, using numerical analysis and a modification of the Jaccard measure, of 1147 tree species from 265 rain forest sites in eastern Australia (L. J. Webb, unpublished data).

The northern humid tropical province near Cairns (lat. 15° 30′ to 18° 30′ S) comprises a complex pattern of different rain forest and sclerophyll forest types whose distribution is controlled by rainfall (range 1250–4500 mm mean annual, decreasing westwards), altitude (from sea-level to 1600 m), soil parent material (ranging from acid schists and granites to andesites and Recent basalts), soil depth and drainage, and the influence of wildfires.

The Quaternary history of the region is complex, and there have been extensive readjustments of the limits of distribution of rain forest and sclerophyll forest over the past 30 000 years, according to recent palynological studies (Kershaw, 1970, 1971, 1973). The region contains a remarkable array of narrow endemics isolated in refugia such as deep gorges and cloudy wet slopes of the coastal granitic mountains, as well as a high proportion of monotypic taxa and primitive angiosperms. The distribution and fragmentation of these are evidently related to Recent climatic changes, and provide special problems for locating areas for biological conservation.

Results of structural analysis. Physiognomic–structural data were collected on *pro formas* for thirty rain forest sites and fifteen sclerophyll forest sites in the Bloomfield River area, north of Cairns (lat. 15° 30′ to 16° 15′ S, long, *c*. 145° 20′ E). The characteristics of the sites and the methods of numerical analysis of the data are fully described by Webb, Tracey & Williams (1976). The first division of the analysis cleanly separated the simpler sclerophyll forests from the more complex rain forests. At the 11-group level there were five groups of sclerophyll forest sites, each of which was well correlated with differences in soil depth, texture and drainage. Each group was dominated by a tree species with a characteristic physiognomy, notably bark texture. The classification of the sclerophyll forests, together with their structural equivalents in the system of Fosberg (1967) and Specht (1970) are given in Table 2.2.

The five groups revealed by numerical analysis evidently reflect the physiognomy of the dominant tree species, and in structurally and physiognomically simplified situations such as this floristic classification based on the dominant, would seem both feasible and preferable. Four of the groups are placed under only one category in the Fosberg key.

The thirty rain forest sites separated into six groups. The classification of the groups in relation to the intuitive structural classification of Webb (1959, 1968) and Fosberg's key is shown in Table 2.3. The simpler secondary forests are clearly separated from the mature forests. The three large groups of mature

Table 2.2. *Structural classification and equivalents of fifteen sclerophyll forest sites in tropical Queensland*

Group from analysis of *pro forma* data	Structural type	
	Fosberg (1967)	Specht (1970)
Seven sites dominated by *Eucalyptus polycarpa* F. Muell., with rough fibrous bark	1J1 2 (a low savanna)	Grassy woodland
Three sites dominated by *E. alba* Reinw. or *E. tereticornis* Sm. with glassy-smooth bark	1J1 2 (a low savanna)	Grassy woodland
Two sites dominated by *Tristania suaveolens* Sm. or *Melaleuca quinquenervia* (Cav.) S. T. Blake with papery bark	1J1 2 (a low savanna)	Sedgy open forest
Two sites dominated by *Casuarina* spp. with rough corky bark and needle-scale leaves	1J1 2 (a low savanna)	Grassy open forest
One site with *E. cullenii* Cambage with deeply furrowed 'iron bark'	2A1 2b (a steppe forest)	Low grassy open woodland

Table 2.3. *Structural classification and equivalents of thirty rain forest sites in tropical Queensland*

Groups from analysis of *pro forma* data	Structural type	
	Webb (1968)	Fosberg (1967)
Three secondary forests, av. height 10 m, with emergent sclerophyllous *Acacia aulacocarpa* A. Cunn. ex Benth.	Low mesophyll vine forest + sclerophylls	1A1 5(?) or near 1D1 1–1D1 3b
Three secondary forests av. height 18 m, with emergent sclerophyllous *Eucalyptus* spp., *Acacia polystachya* A. Cunn. ex Benth.	Low mesophyll vine forest + sclerophylls	1A1 5(?) or near 1D1 1–1D1 3b
Seven mature forests, mostly in wet gullies	Six complex mesophyll vine forests, one notophyll vine forest	1A1 1
Eight mature forests, mostly on moist lower slopes and in gullies	Eight complex mesophyll vine forests	1A1 1
Eight mature forests, generally in drier situations on exposed ridge tops or gravelly and stony slopes	Two complex notophyll vine forests, two notophyll vine forests, four mesophyll vine forests	1A1 1
One mature forest on wet cloudy mountain summit at higher altitude than other sites	Notophyll–microphyll vine-fern forest	1A1 1

forests appear to reflect a gradient from moist sheltered gullies and lower slopes to drier more exposed ridges or shallower soils. The groups of mature rain forests are fairly well separated by the intuitive structural classification, but fall within only one of Fosberg's classes. The mixture of sclerophylls and orthophylls presented by the secondary forests also seems to require an additional class.

It is relevant to note that in a parallel study, using the structural *pro forma*, of forty-one mature rain forest sites in Papua New Guinea (Webb, Tracey & Williams, 1976), eighteen site-groups were revealed by structural analysis and were mostly well correlated with major environmental differences. The sites were distributed among seventeen structural types according to the intuitive classification of Webb (1968). However, in the Fosberg system thirty-three sites belonged to 1A1 1, four sites to 1A23, three sites to 1A1 3a, and one site to 1A1 2a. Considerable refinement of the first category for multistratal evergreen forest (rain forest) in Fosberg's key seems necessary.

Other comparisons. In the floristic classification of 265 rain forest sites in eastern Australia (L. J. Webb, unpublished data), sixty sites in the humid tropical province of North Queensland yielded twenty-four distinct site-groupings, all well separated by major environmental differences. The sites belonged to nine structural types according to Webb (1968), but with the exception of some types of gnarled evergreen mossy forest (1A1 3a) and swampy palm forest with orthophylls (1A1 2), the rain forests belonged to only one class (1A1 1) in Fosberg's key.

Further data for testing different kinds of classification are provided by a map of vegetation types of the humid tropical province of North Queensland, recently prepared at a scale of 1 : 100 000 from high altitude aerial photographs (L. J. Webb & J. G. Tracey, 1975). This map also enables some discussion of the problem of scale, which always underlies the selection of representative areas. The scale of 1 : 100 000 is considered to be optimal for the relatively rapid mapping of 'land units', i.e., areas in which there is a limited number of land components occurring in a constant sequence and forming a characteristic landscape pattern (Gibbons & Downes, 1964; Downes, 1969). The 'land system', which consists of a combination of land units, and which can often be associated with geomorphological features on aerial photographs, is too general for purposes of detailed land-use planning, including nature conservation.

On the other hand, the 'land component', a part of the land unit which is uniform in relation to potentials and problems for a particular form of land use, is too detailed for the general survey required for IBP/CT. The ease or difficulty of drawing boundaries between vegetation types at a scale of 1:100 000 is relevant also to the sensitivity of the structural classification of vegetation. Thus Fosberg's formation classes, derived from the spacing and height of plants, are equivalent to the primary patterns which are easily recognized on the aerial photographs. At the scale of 1 : 100 000 it was possible to draw boundaries (excluding cleared and grossly disturbed areas) between eleven different types of vegetation extensive enough to map. Where the pattern was too intricate to map, a further seven vegetation complexes or mosaics were recognized and separated. The validity of the boundaries was then checked by ground survey or, in relatively inaccessible areas, by helicopter survey. However, large areas of tall, close-canopied and uniform-looking rain forest, which transgress wide limits of altitude, rainfall and soils, and which were found on the ground to embrace numerous structural and floristic types, could not be differentiated on the aerial photographs. This rain forest corresponds largely with Fosberg's category 1A1 1 of tall, multistratal, orthophyllous and evergreen forest. It was, however, possible to map seven different structural types of rain forest, based on smoothness, brokenness and height of canopy, and presence of characteristic emergents and co-dominants such as eucalypts, acacias and palms. Only the gnarled mossy forest and palm forests are catered for additionally in Fosberg's key. Refinement of his 1A1 1 to include sclerophyll emergents of different kinds, and degree of evenness of canopy, would be helpful.

Within the seven major structural types of rain forest, which approximate to land systems, some forty different rain forest types, corresponding to land units, could be separated on the ground on the basis of major environmental differences, such as altitude, rainfall and cloudiness, soil parent material, exposure, etc. All these different rain forest types, which are exhibited well above the floristic association level, and which may conveniently be regarded as alliances or similar groupings below the subformation level, are subsumed by only four categories in Fosberg's key: 1A1 1, 1A1 2c, 1A1 3a and 1A2 3(?).

In the vegetation complexes or mosaics, in which a wide variety of forest, scrub, swamp, marsh, meadow, savanna and grass communities are intermingled, the Fosberg key was much more useful, and some twenty-two of Fosberg's subformations could be identified. Although these communities have only a fragmentary representation in the humid tropical province of North Queensland, they are widespread elsewhere, e.g., savannas in northern Australia, and swamps and marshes in Papua New Guinea.

Finally, the North Queensland map includes some seventeen structural–floristic types of sclerophyll woodland, which can be distinguished only on the ground but which all belong to Fosberg's 1I1 2 ('evergreen broad sclerophyll tall savanna') and 1J1 2 ('evergreen broad sclerophyll low savanna').

Conclusions

The fact that the large number of vegetation types which occupy extensive areas in the North Queensland biotic province, and which can be differentiated by floristic classification, intuitive structural classification, or by numerical analysis of structural data collected on a *pro forma*, and would all be placed within only few of Fosberg's classes, indicates that refinement of the Fosberg categories is needed. This refinement is actually provided by the additional properties listed in the structural *pro forma*, which were derived essentially from the structural features found to be relevant by other workers, and which were found to yield a meaningful classification in North Queensland (Webb *et al.*, 1970).

The Fosberg classification was also found to be unsatisfactory for the sclerophyll forests, not only in North Queensland but also throughout Australia (Specht, Roe & Boughton, 1974). A major difficulty is the absence of a category of vegetation spacing, additional to closed, open and sparse as defined by Fosberg. Thus the Australian survey found it necessary to use the classification of Specht (1970) which supplies this extra category.

The insensitivity of the Fosberg classification at a level relevant to the identification of conservation areas in complex tropical forests has theoretical and practical implications. Two quite distinct but complementary purposes are involved in the recording of vegetation data: standardized description and adaptive classification. Fosberg's system is a classification which is necessarily generalized in an attempt to compare vegetations on a world scale. His key to 'formation classes', based simply on density of crowns and height of plants, is logical and provides a primitive description, similar to an aerial view, of a given stand of vegetation. However, when his classification extends to more detailed classes, i.e., to the level of 'formation group' (characterized by seasonality of the dominant layer), and to 'formation' (based on dominant growth form, leaf texture and features such as thorniness), it is inevitable that local 'context-dependent' vegetation situations become involved. These cannot all be provided for in a rigid key of any manageable size, even though it may be possible, as Fosberg suggests, to subdivide a formation by characters of growth form and other features which are not, however, further defined in his scheme.

A universal system for the description and recording of the structure of vegetation was proposed and successfully used by Dansereau (1958), and more recently by Mills & Clagg and the US Army Corps of Engineers Waterways Experiment Station (WES), described by Dansereau, Buell & Dagon (1966). These descriptions employ a shorthand system, with symbols for standardized profile diagrams, and aim at recording 'all the essential variables for structural comparison with an acceptable degree of repeatability' (Dansereau, Buell & Dagon, 1966). The structural attributes which are recorded partly depend on the purpose for which the resulting classification and terrain evaluation (Grabau, 1968) are applied: thus survey for military purposes is more interested in various kinds of penetrability of the vegetation. The WES method for this purpose is, however, very time-consuming, both in the field and in the subsequent analysis, even though it is not concerned with floristics. Dansereau, Buell & Dagon (1966) conclude that the general acceptance of a universal system of vegetation recording which is adequate and practicable is 'far in the future', and that numerous modifications of any proposed system are inevitable. For this, a continually evolving version of IBP/CT based on international co-operation and the central collation and analysis of descriptive data is vital.

It seems, then, that there are at least two methodological requirements if an acceptable system is to be evolved:

(i) What is needed is not a standardized classification based on the definition *a priori* of vegetational classes as in Fosberg (1967), but the definition of rules to determine such classes, i.e., a standardized description which is extendable and 'open-ended'. This should be derived from previously developed methods so that earlier results can, to some extent, be translated into the new evolving format, and so that comparisons are possible with earlier work. (The properties of vegetation used in the standardized *pro forma*, and the values by which they are assessed, should be described fully and if necessary illustrated in a practical handbook for data recorded in the field.)

(ii) A computational system is required to produce a classification at a suitable group level, once a reasonable number of sites, whatever their distribution regionally or internationally, has been sampled. The computational system should then be able to allocate additional sites to known classes or to determine that a new class be established. It should also be able to deal with new vegetation properties which may be added to the *pro forma* in the field and accepted as necessary for classification.

Finally, it seems inescapable that any attempt to identify areas of genetic resources for conservation must proceed beyond structural classification to taxonomic groupings. The structural classification proposed here would be valid only within a framework of biotic provinces as proposed by Dasmann (1972, 1973). Collation of relevant biological data to erect reliable boundaries for biotic provinces in each country involved in the conservation of terrestrial ecosystems is therefore an urgent prerequisite, especially in regions with complex biotic communities which exhibit fine-grained patterns of variation as the result of historical influences.

2.2.4 Further consideration of the Fosberg system and its value for biological surveys

Section 2.2.2 has discussed general problems of vegetational description and classification, pointing out that no single system is likely to serve all the various purposes for which vegetational information may be required. When the check-sheet survey was initiated it was decided that Fosberg's classification should be adopted as a basis for collecting and storing vegetational data for general purposes, both on account of its intrinsic merits and because it was already available for use: the UNESCO classification (pp. 41–3) appeared first in 1967 (Ellenberg & Mueller-Dombois, 1967) and in its final form only in 1973. The Fosberg system is explained in some detail on pp. 37–9 and the key to its formation-classes, with descriptive designations of the component formations and subformations, is reproduced in Appendix 2. It was selected for use in the survey because it depended on readily observable features of the vegetation, irrespective of floristic composition, so that it did not call for expert taxonomic knowledge and could be expected to yield comparable information from all parts of the world.

The *Guide to the Check-Sheet for IBP Areas* (Peterken, 1967) has an Appendix by Fosberg himself on what he has termed a *Classification of Vegetation for General Purposes.* He notes the importance of descriptions of vegetation-types, and of maps showing their distribution, not only for students of ecology and plant sociology but also for workers in many related scientific disciplines and for those in agriculture, forestry and the numerous other practical fields 'where vegetation has a bearing'. He goes on to stress the present 'difficulty in relating information from one area with that in another, and in making comparisons', and concludes that such comparisons would be greatly facilitated if a generally acceptable classification could be devised on a world basis. 'It would then be possible to organize the vegetation information available for the whole world into a co-ordinated body so that any part of it would be readily available in usable and comparable form.'

The most obvious similarities and differences between areas of vegetation are those in general aspect and structure. They include the presence or absence of sizeable woody plants and, if they are present, the height, crown-form and spacing of the dominant trees or shrubs, the size, shape and texture of their leaves and whether evergreen or deciduous, and also characteristics of the subordinate vegetation, in particular the distinctness and completeness of its layering. If woody plants are sparse or absent the aspect of the vegetation is determined by the life-form, stature, density and seasonal changes of its most prominent components, whether grasses or other grass-like plants, broad-leaved herbs, bryophytes or lichens. These are all readily observable features for the accurate recording of which no expert taxonomic knowledge is required, and these are the features used by Fosberg in the key to his formation-classes and formation-groups and in the brief definitions of his formations and subformations.

It is a familiar fact of plant geography that vegetation-types occurring in similar environments but in widely distant parts of the world commonly agree closely in these features of general aspect and structure even when they are very different in floristic composition. In some instances, indeed, such homologous communities are characterized by members of the same families or even genera of plants, though rarely of the same species unless as a result of established introductions, deliberate or accidental. Physiognomically and structurally similar vegetation-types are also known to result from similar histories of management or mismanagement in widely separated areas. All this being so, knowledge of the Fosberg formation to which a piece of vegetation belongs can convey a good deal of additional information about it, so that it has considerable predictive value. It allows inferences as to the general character of the vegetation and its environment such as its suitability for various kinds of animals, its stability, its vulnerability to fire or soil-erosion, its exploitability by man and its past exploitation and other matters of importance for workers in those various fields 'where vegetation has a bearing', including nature conservation. Also the information is not only easy to record and convenient to store but is readily intelligible, and the inferences from it are of world-wide applicability. This is in sharp contrast with floristic information, the collection and interpretation of which depend upon expert taxonomic knowledge and ecological understanding of *local* vegetation.

It was recognized from the first, however, that the information derived from the use of the Fosberg classification would be insufficient for the full purposes of the check-sheet survey, with its emphasis on nature conservation. Conservationists are certainly interested in the permanent protection of representative examples of physiognomically and structurally different types of vegetation and their associated types of animals, but they are no less concerned to conserve particular assemblages of plant and animal species, and especially to save from extinction many rare and threatened species all over the world. This entails protecting the communities of which they are components, as well as the habitats essential for the persistence of these communities. This special interest of conservationists in the floristic composition of different vegetation-types cannot be met adequately by the use of a system of vegetational recording taking little or no account of the individual species present. For this reason recorders for the check-sheet survey were asked not merely to allocate the vegetation-types of surveyed sites to their Fosberg class but also to list the plant communities present, giving for each its 'usual name' and the Latin names of dominant or otherwise significant species and also of rare or specially interesting species. There being no floristic classification of world-wide applicability, recorders were invited to use any system prevalent in their own country or region, supplying references to published descriptions of the communities named.

Problems in the use of the Fosberg system

Field trials and analyses of check-sheets completed for the survey have revealed some difficulties in using the key to formation-classes and in interpreting the brief descriptions of formations and subformations. It has also become clear that there are some omissions that prevent the assignment of a satisfactory Fosberg coding to certain types of vegetation.

Difficulties in the interpretation of key and definitions

Down to the level of the formation-class the Fosberg classification is essentially hierarchical, proceeding by successive dichotomies and readily intelligible from the simple key provided. Beyond this level it is non-hierarchical, no further keys are provided and allocation to formations and subformations involves the correct interpretation of short descriptive phrases (p. 252 *et seq.*). This is, on the whole, satisfactory in practice but it has certain consequences best understood by reference to that set of formation-classes characterized by features of their chief woody elements: 1A–1K, 2A–2F and 3A–3B (Appendix 2, p. 248 *et seq.*). There are no serious problems in arriving at these formation-classes by use of the key, and the two formation-groups into which each is divided are in general readily separable from a knowledge of the aspect of the vegetation in different seasons of the year. The real problems arise in distinguishing formations within the same formation-group. As an illustration we may consider scrub dominated by the broad-leaved, mesophyllous, deciduous, thorny shrub *Crataegus monogyna*. It may be placed either in 1B21a, 'Mesophyllous deciduous orthophyll scrub', or in 1B24, 'Deciduous thorn scrub', according to whether more importance is attached to the nature of the leaves of the dominant shrub or to its possession of thorns. The addition, in brackets, to the descriptive phrase for 1B24 of the word 'Thornbush' is made of less value by its omission from the Glossary. Also, vegetation with scattered 'trees' of one or more species of *Crataegus* falls under 1J21, 'Deciduous orthophyll savanna'; but if with scattered 'shrubs' of *Crataegus*, under 1K24, 'Mesophyllous deciduous thorn shrub savanna', even though the formation 1K21 is defined as 'Deciduous orthophyll shrub savanna' and is apparently the shrubby counterpart of 1J21. Here the surveyor needs some guidance as to the relative importance he must attach to leaf characters on the one hand and to the presence or absence of thorns on the other.

This is not to overlook the real difficulty of deciding in certain instances whether a species with a few weak thorns, or one bearing thorns in some circumstances but not in others, should or should not be regarded as thorny, but such problems are unavoidable for the working biologist and he must overcome them, for himself and for others, as best he can. The more general question of how to reach correct decisions on the interpretation of the descriptive phrases might be answered in either of two main ways. The phrases might be modified or expanded so as to remove difficulties such as are exemplified by the allocation of *Crataegus* scrub to its appropriate formation. It may ordinarily be possible to achieve this, but the inevitable lengthening of the descriptive phrases may operate against any real gain in clarity. The alternative would be to replace the phrases by an extension of the key down to formation level.

In the set of formation-classes and formation-groups under consideration there is a basically similar pattern of division into formations, and it would therefore be possible to make use of an identical series of dichotomies for each, so as to extend the hierarchical classification down to that level. Fig. 2.1, (*a*)–(*c*) shows how this might be done for woody plants, but there are many other possibilities. The question of priorities such as that between the nature of the leaves and the presence or absence of thorns would then be resolved by the deviser of the key and the surveyor would be relieved of the onus of that decision. It would still be desirable to include a brief description of each formation and subformation, since the presence or absence of any given feature is made explicit in a dichotomous key only for the single feature used in the initial dichotomy. If the first division is made on whether the chief woody plants are or are not thorny, then it is known that all formations in the one half of the classification have thorns and all those in the other half lack them. But if the same question is asked at a later point in the key, then all formations split off in earlier dichotomies may or may not be thorny, the key giving no information on the point. An independent check on the correctness of an allocation would then be possible only by having a set of descriptions of each of the formations in the formation-group, with mention of all the features used in the key and perhaps some others as well. That being so it would seem to lighten the task of the surveyor to have the dichotomous key and the hierarchical structure of the classification extended throughout.

Another type of problem for the surveyor arises from the use in the key, or in descriptive phrases, of terms which are either ambiguous or whose meanings overlap. The terms 'microphyllous' and 'sclerophyllous', both used in definitions of formations, exemplify the problem. The former is defined in the Glossary as 'with very small leaves; as used here, for trees, leaves 25 mm in greatest diameter and less, for shrubs 10 mm and less, for dwarf shrubs 5 mm and less'; and the latter as 'texture of leaves hard, stiff or coriaceous, as opposed to orthophyllous', while 'narrow sclerophyll' vegetation has 'needle-like or narrowly linear hard or stiff leaves'. These Glossary definitions cannot be made mutually exclusive and it is therefore important to avoid using

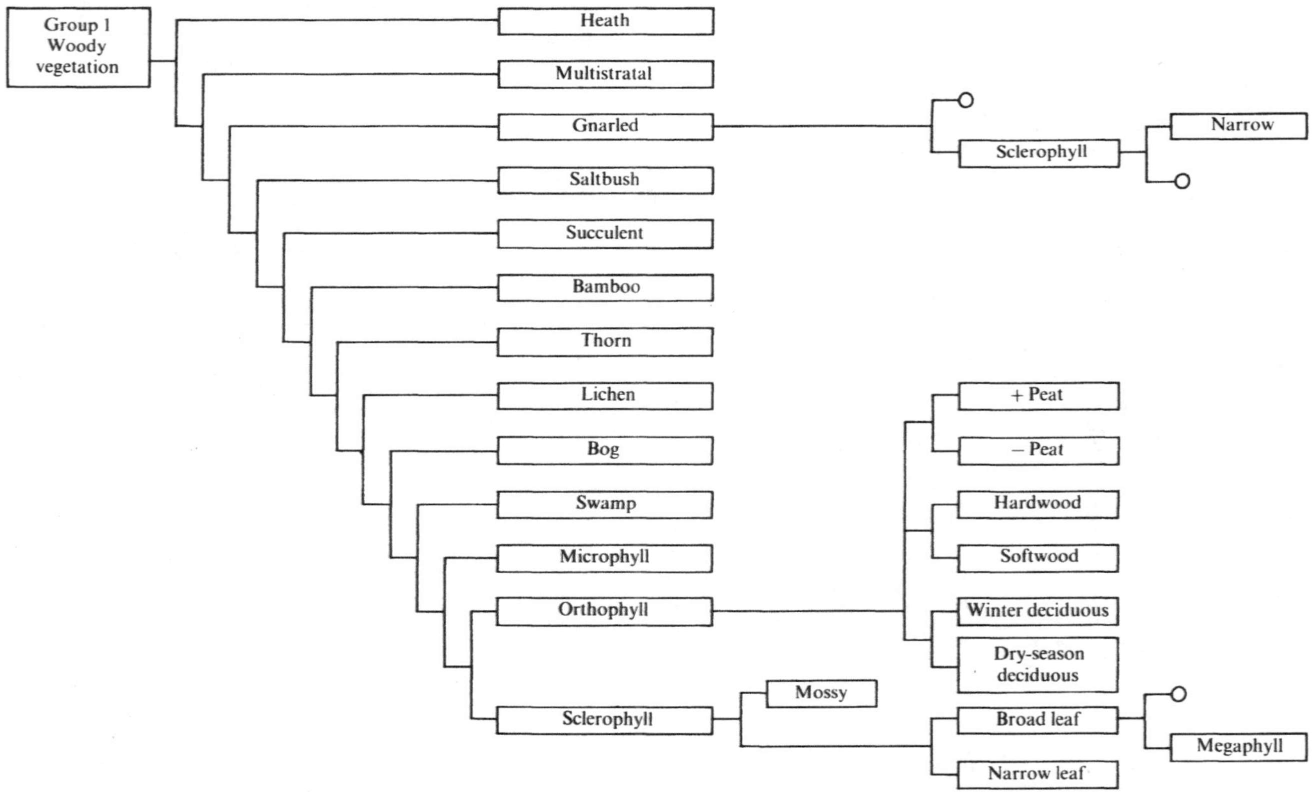

Fig. 2.1(*a*). For legend see p. 71.

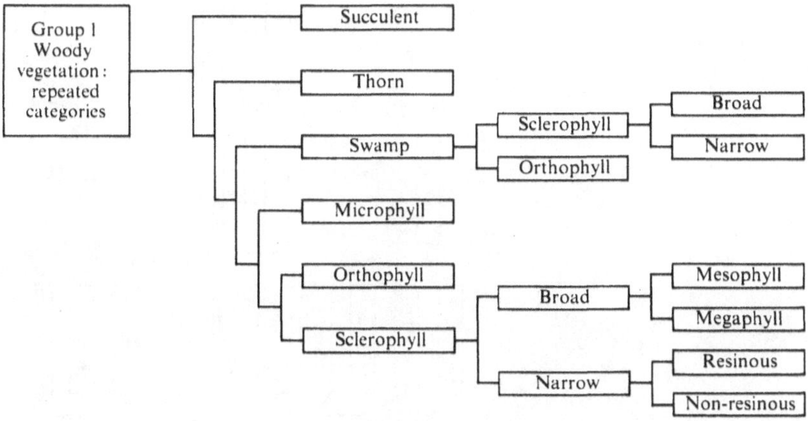

Fig. 2.1(*a*)–(*c*). Example of a series of dichotomies extending the Fosberg key down to formations for vegetation dominated by woody plants.

the terms as though they were alternatives. The phrases defining formation 1A17, which includes coniferous forests, describes it as 'Evergreen narrow sclerophyll forest (needle leafed forest)', with 'Leaves (or equivalent) linear or scale-like, hard'. Forests dominated by conifers with scale-leaves or short needles might seem to qualify equally well for the 'Microphyllous evergreen forest' of 1A19, the additional description 'Frequently but not always of compound-leafed trees, sometimes thorny' not being definite enough to clinch matters. Heaths dominated by the scale-leaved *Calluna*, moreover, are treated as 'Microphyllous evergreen dwarf scrub' in a subformation of 1C12, 'Evergreen broad sclerophyll dwarf scrub'. The terms in question seem, therefore, to be treated sometimes as mutually exclusive alternatives and sometimes as relating to different features of the same leaf. The consequent problems for the surveyor, and others of the same kind, could here again be avoided by extending the dichotomous key down to formations or even subformations and by more careful definition of terms in the Glossary.

Omissions from the original list of formations

The reprint, issued in 1972, of Fosberg's *A Classification of Vegetation for General Purposes*, which appeared as an Appendix to the *Guide to the Check Sheet for IBP Areas* (Peterken, 1967), included a list of 'proposed additional formations and subformations to the Fosberg classification' arising from field trials of the check-sheet in various parts of the world. Early in 1974, G. L. Radford and R. J. de Boer drew up an extended list of proposed additions arising from their examination of completed check-sheets and from correspondence with surveyors and others. This list is reproduced on pp. 285–6.

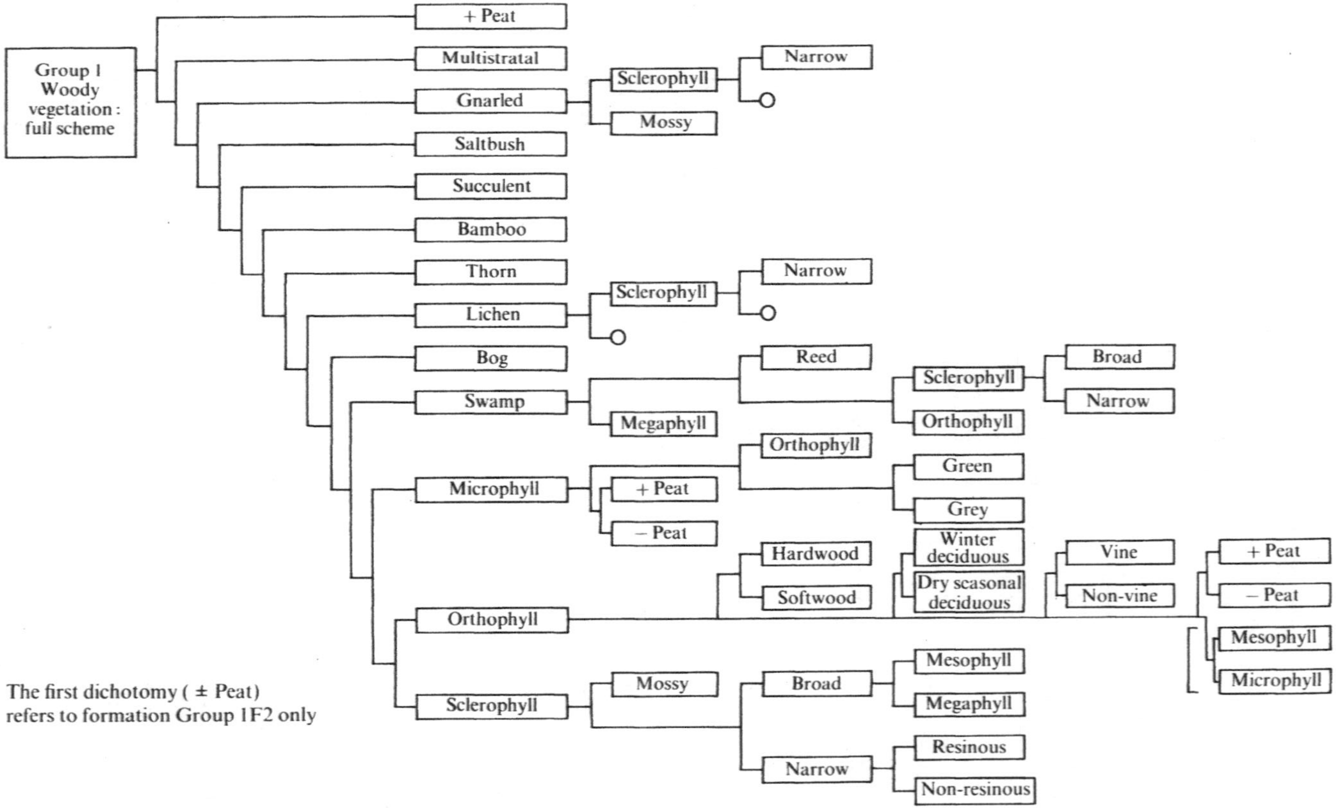

Fig. 2.1 (c). For legend see p. 71.

Table 2.4. *Forest communities of the Westhoff–Den Held classification of Netherlands vegetation, for the most part 'associations', all of which correspond with the Fosberg formation 1A21*

Westhoff–Den Held coding	Syntaxon (vegetation-type)
36Ab1	Betuletum pubescentis
37Aa1	Querco roboris-Betuletum; deciduous phase
2	Fago-Quercetum, excl. *Quercus–Ilex* community
3	Convallario-Quercetum dunense
38Aa1	Carici remotae-Fraxinetum
2	Consortium of *Carex remota* and *Populus nigra*
3	Pruno-Fraxinetum
4	Macrophorbio-Alnetum
5	Violo odoratae-Ulmetum
6	Fraxino-Ulmetum
7	Anthrisco-Fraxinetum
8	Crataego-Betuletum
Ab1	Stellario-Carpinetum

Comparisons, based on check-sheet returns, between the Fosberg system and floristic classifications in local use

Completed check-sheets from countries where a floristic classification of vegetation is in general use provide material for comparisons with the Fosberg system. Such comparisons have particular importance for deciding whether the Fosberg or some other non-floristic system should continue to be used in future surveys, even if generally acceptable floristic systems became available for the whole of extensive regions.

Netherlands

A country for which this comparison may readily be made is the Netherlands, which has a generally accepted, comprehensive and detailed classification of plant communities based mainly on floristic criteria and developed from the Zurich–Montpellier system. This is the classification due to Westhoff & Den Held and used in their *Planten Gemeenschappen in Nederland* (1969). This was used consistently both in the check-sheet survey and also in a recent independent examination of samples of natural vegetation throughout the country. All the associations of the Westhoff–Den Held classification found in the Netherlands have been allocated to their Fosberg formations (see Appendix 4), and the comparison is of considerable interest, particularly as revealing the difference between the systems in level of discrimination. Thus the single Fosberg formation 1A21, 'Winter-deciduous orthophyll forest (hardwood forest)', covers thirteen different Westhoff–Den Held associations (Table 2.4); and a still more striking example is provided by the Fosberg formation 1M21, 'Seasonal orthophyll meadow (short grass)', which covers no fewer than sixty-three associations. On the other hand some Fosberg formations are represented in the Netherlands by only part of a single floristic association, an example being 1B18, 'Microphyllous evergreen scrub', which comprises only the taller phase of the healthy Genisto pilosae-Callunetum. On average there are about 5.5 different Westhoff–Den Held associations to each Fosberg formation.

Japan

The main islands of Japan lie between latitudes 31° N and 46° N, the smaller islands of the Ryukyu and of the Bonin and Volcano arcs extending further southwards almost to the Tropic of Cancer. Over this distance of about 3000 km, ranging from subtropical ro subarctic latitudes, there is great diversity in geology and topography as well as in climate, with a corresponding diversity in the appearance and floristic composition of the vegetation. Losses consequent upon the climatic fluctuations of the Pleistocene period have been relatively small, in contrast, for example, with those of the British Isles, and the flora remains very species-rich, with many endemics and relicts of great phytogeographical interest. Intensive studies of the flora have been undertaken over a long period and, in more recent times, considerable progress has been made towards the recognition and description of all the major terrestrial plant communities of the country with a view to ensuring the conservation of representative samples of as many as possible of them. This has entailed the adoption for national use of a system of specification and classification which, like that used in the Netherlands, is primarily floristic and based on the Zurich–Montpellier system. After the decision by the Central Committee for IBP/CT to use the Fosberg system as the general basis for recording vegetation-types encountered in the check-sheet survey, the Japanese CT Committee drew up a list of equivalents in the two systems for the assistance of their survey-teams. 'Most sites of biologically important vegetation, flora and fauna have been recorded on check-sheets', and in most instances vegetation-types are classified in terms of both the Japanese and Fosberg systems.

A list of equivalents, such as that received in the Central IBP/CT Office in 1972, shows much the same lack of detained correspondence as is evident in the comparable list for the Netherlands (p. 291). On the one hand a single Fosberg formation or subformation may be represented by many different communities of the Japanese classification. Thus the subformation 1A16a, 'Mesophyllous evergreen broad sclerophyll forest', is shown in the 1972 list as the Fosberg equivalent of the twenty 'plant communities' given in Table 2.5. Most of these rank as 'alliances' or 'suballiances' in the floristic classification (see Numata, Yoshioka & Kato, 1975) and have usually been

Table 2.5. '*Plant communities*' (*comm.*) *of the Japanese system listed as equivalents of the Fosberg formation 1A16a*, '*Mesophyllous evergreen broad sclerophyll forest*', *in whole or in part*

Fosberg formation	Plant communities of Japanese system
1A1 6a	*Ficus wightiana* comm.
	F. microcarpa comm.
	Elaeocarpus sylvestris var. *ellipticus* comm.
	Castanopsis cuspidata var. *cuspidata* comm.
	C. cuspidata var. *sieboldii* comm.
	C. cuspidata–Machilus thunbergii comm.
	Machilus thunbergii comm.
	Quercus (*Cyclobalanopsis*) *glauca* comm.
	Quercus (*Cyclobalanopsis*) *salicina* comm.
	Quercus (*Cyclobalanopsis*) *acuta* comm.
	Quercus (*Cyclobalanopsis*) *gilva* comm.
	Quercus (*Cyclobalanopsis*) *myrsinaefolia* comm.
	Distylium racemosum comm.
	Cinnamomum camphora plantation
1A1 6a or 1B1 4a	Mixed forests of evergreen *Quercus* (*Cyclobalanopsis*)
1A2a and 1A1 6a	*Fagus crenata* – evergreen *Quercus* (*Cyclobalanopsis*) comm.
1A2a and 1A1 6a	*Quercus acutissima–Q. serrata* – evergreen broad-leaved tree comm.
or	
1B2 1a and 1B1 4a	*Q. acutissima* (*Q. variabilis*) – evergreen broad-leaved tree comm.
	Q. serrata–Castanea crenata – evergreen broad-leaved tree comm.
1B1 4a or 1A1 6a	*Q. phillyraeoides* comm.

divided further into two or more 'associations'. On the other hand a single floristic community may cover so wide a range in structure and physiognomy as to be referable to more than one Fosberg formation, as is true of the community dominated by *Picea glehnii* in Hokkaido, shown as equivalent either to 1A1 7a, 'Resinous evergreen narrow sclerophyll forest', to the shrubby 1B1 3d, 'Evergreen narrow sclerophyll swamp', or to 1D1 2a, 'Open narrow sclerophyll swamp', according to the height and openness of the spruce canopy. In general, however, the Japanese floristic classification, like that used in the Netherlands, leads to a much finer degree of discrimination than does the Fosberg system.

Scandinavia

In the Scandinavian countries a Joint IBP Committee encouraged the preparation of agreed schemes for mapping the plant communities of forest, fjeld and sea-shore, these being later combined in a 'Uniform Classification' for use not only in mapping but also in recording vegetation for the check-sheet survey, in conjunction with the Fosberg system. This was essentially a phytosociological system, based on methods developed by Scandinavian

Table 2.6. *Forest associations of the 'Uniform classification' of Scandinavian vegetation all of which correspond with the Fosberg formation 1A21*

Association	Ref. no.	Explanatory notes and descriptions
Populo-Quercetum	Forest (12)	Poor oak forest with *Populus tremula*, grasses, Ericaceae and mosses
Melico-Quercetum	Forest (13)	Richer oak forest with grasses and herbs
Deschampsio-Fagetum	Forest (14)	Poor beech forest
Alno(incanae)-Prunetum	Forest (15)	Forests of *Alnus incana* and *Prunus padus* along streams
Alno(incanae)-Fraxinetum	Forest (16)	Rich *Fraxinus* forests in favourable conditions
Equiseto-Fraxinetum	Forest (17)	*Equisetum*-rich forest occurring infrequently along spring-lines
Ulmo-Tilietum	Forest (18)	Forest of *Ulmus glabra* and *Tilia cordata*, rich herb-layer
Ulmo-Quercetum	Forest (19)	Vigorous *Quercus–Carpinus* forest with *Fraxinus* and *Ulmus*. Much affected by agriculture
Fraxino-Fagetum	Forest (20)	Rich beech forest on wet flushed soils. Dominated by *Fraxinus* and *Acer* in some instances
Melico-Fagetum	Forest (21)	Pure beech forest, species-poor

ecologists over a long period. In Appendix 5 (p. 296) the allocation of the associations to their Fosberg formations shows relationships of much the same kind as between the Westhoff–Den Held and Fosberg system in the Netherlands. The Fosberg formation 1A21, 'Winter-deciduous orthophyll forest (hardwood forest)', for example, covers ten different forest associations of the Uniform Classification (Table 2.6), the corresponding figure for the Netherlands being thirteen. The average number of forest associations in each Fosberg formation represented is 2.4. Comparable figures for fjeld and sea-shore are not always available because the phytosociological classification may here extend only to alliances.

These simple numerical comparisons show that the floristic systems in question lead to a more detailed classification of vegetation-types than does the Fosberg system, and go much further in meeting the special requirements of conservationists. Another kind of difference is illustrated by the fact that one of the Westhoff–Den Held forest associations found in the Netherlands, normally referable to Fosberg formation 1A21 as a winter-deciduous hardwood forest, contains very variable amounts of the evergreen *Ilex aquifolium* (holly), although the great majority of the component woody plants are deciduous. When there is a great deal of holly present the appropriate Fosberg formation becomes 1A16 (Evergreen broad sclerophyll forest), despite the fact that Westhoff & Den Held find it impossible to make any clear distinctions on floristic grounds. There are also several Westhoff–Den Held associations that have to be placed in different Fosberg formations at different stages of development of the communities. Thus the early or scrub phase of developing Salicetum arenario-purpureae falls into formation 1B22, 'Deciduous swamp scrub', when the larger willows (*S. purpurea*) are no more than 5 m high, but the mature phase with taller willows has to be placed in 1A22, 'Deciduous swamp forest'. This allocation of parts of a single floristic association to different Fosberg formations would be equally necessary if the height of the dominant woody plants were limited by environmental factors rather than stage of development, with the difference that the division would then be permanent. The Fosberg allocations would disregard the very high degree of floristic similarity and real affinity between the two stands of vegetation, while the purely floristic classification would fail to take account of ecologically significant differences in physiognomy. That the two types of vegetational classification provide different and complementary information about vegetation emerges also from comparisons between the Scandinavian Uniform Classification and the Fosberg system. Whereas Fosberg formation 1O12, 'Moss meadow', is represented by communities placed in four different alliances of the Uniform Classification, the single alliance Arctostaphyleto-Cetrarion nivalis is in seven different Fosberg formations (p. 296–9) because a substantial floristic similarity persists over a considerable range in physiognomy and structure.

In those derivatives of the Uppsala and Braun-Blanquet systems of vegetational classification used in Scandinavia, the Netherlands and Japan, and many other countries, the higher classes are commonly defined partly in physiognomic or physiognomic–ecological and partly in floristic terms. Thus Formation I of the Westhoff–Den Held system in use in the Netherlands comprises vegetation of aquatic and amphibious plants and includes the Classes Lemnetea, Zosteretea, Ruppietea, Charetea, Potametea and Littorelletea; Formation VI has the single class Phragmitetea of reed-swamps dominated by tall grasses and sedges; Formation XII consists of shrubby communities and includes the Classes Franguletea, Salicetea purpureae and Rhamno-Prunetea, and Formation XIII comprises all forests in its three Classes Alnetea glutinosae, Quercetea robori-petraeae and Querco-Fagetea. Each Class is further divided into Alliances and Associations, mainly on floristic grounds. Despite the largely physiognomic basis for the Formations they do not correspond very closely with Fosberg's formation-classes. The Classes of Formation I, for example, fall some into Fosberg's 1P, 'Submerged meadows', and some into 1Q, 'Floating meadows'; and the boundary between scrub and forest is differently placed. The more important difference is that the *association* of the Netherlands system is a much finer unit than the formation or even the subformation of the Fosberg scheme.

2 *Problems of description and specification*

Some countries have adopted vegetational classifications based on different principles from those underlying the basically floristic systems considered above. Tunisia and Australia have been selected as examples.

Tunisia

The check-sheet survey of proposed conservation-sites in Tunisia, carried out in 1969 by Radford & Peterken with the help of local scientists (Radford & Peterken, 1971), gives rise to some interesting comparisons. The intention of the visit was to survey IBP areas and at the same time to test the check-sheet in the field and gain experience of the use of the Fosberg system. A set of sixth-three sites was selected from lists previously prepared from various sources (p. 179), but there was time during the short stay in Tunisia to visit rather fewer than half of them and check-sheets could be completed only for these, even with the assistance of local scientists who accompanied Radford & Peterken in the field. An attempt was nevertheless made to record some relevant information about the remaining sites, and it was here that difficulties arose with the Fosberg coding of the vegetation-types reported as present in them.

Comprehensive ecological and floristic studies extending over some years had made it possible to publish in 1966–7 a set of five vegetational maps covering northern Tunisia at a scale of 1:200 000 (*Carte Phyto-Ecologique de la Tunisie Septentrionale*, M. Gounot & A. Schoenenberger) to be followed by two further maps of central and southern Tunisia at a scale of 1:500 000 (H. N. Le Houerou, 1969). In this way the whole country had been mapped, though not in great detail. This being so, it might be thought that sufficient information was already available for completing the vegetational sections of check-sheets for sites anywhere in the country and certainly for that majority of the sites (forty-six out of sixty-three) located in northern Tunisia and covered by maps at a scale of 1:200 000.

The colours and symbols used in the maps are based on the concepts of the Toulouse school of plant geography and represent units in a hierarchy extending downwards from zones (*étages*) define bioclimatically as pre-humid, humid, sub-humid and semi-arid (Emberger, 1960). These are divided into subzones (*sous-étages*), upper and lower, and further into *variantes* distinguished by the severity of the winter. At the next lower level are the *séries*, each comprising an inferred 'climax' type of vegetation together with successional and degradational stages. The more distinctive variants within each series are termed *groupements forestiers*, in contrast with the *groupements cultigènes* of long-cultivated areas whose climax vegetation cannot readily be inferred from the existing plant communities. Each *groupement*, finally, may comprise a number of *faciès*. The *Quercus suber* Series (Série du Chêne Liège) will serve as an example. This is characteristic of the Lower Subzone of the

Humid Zone west of Bizerta and is on more or less strongly leached brown forest soils over non-calcareous substrata or occasionally on surface-leached soils over limestone. It thrives best in an annual rainfall of 800–1000 mm. There are five main *groupements* in the area. The first (CA: *Groupement à Quercus suber, Cytisus triflorus*) is found for the most part at altitudes above 450 m, while CB (*Groupement à Quercus suber, Pistacia lentiscus, Erica arborea*) is restricted to warmer localities, mostly below 450 m except on south-facing slopes, and CZ (*Groupement à Quercus suber, Pistacia lentiscus, Quercus coccifera*) comprises the littoral cork oak forests found in the *Variante* of the Lower Subzone of the Humid Zone characterized by milder winters. In the remaining two *groupements Quercus suber* occurs either not at all or only very sparsely. CC (*Groupement à Arbutus unedo, Erica arborea, Cistus monspeliensis, Pistacia lentiscus*) comprises fire-induced modifications of various faciès of CB, and CK (*Groupement à Arbutus unedo, Quercus coccifera, Erica multiflora*) consists of various stages of degradation of CZ: in both the vegetation is usually a *maquis* of varying stature and varying degrees of openness but locally may consist of little besides the fern *Pteridium aquilinum* or the grass *Ampelodesma mauretanicum*. Within each of the five *groupements*, indeed, there are edaphically-determined faciès and others representing stages in degradation of both vegetation and soil as a consequence of fire, over-grazing, erosion, etc. Each faciès is characterized floristically as well as ecologically, but they are placed in the same groupement because they are envisaged as ecologically interrelated stages in one or more sequences of vegetational change within the domain of a single climax, not because of their floristic affinities.

These faciès are shown on the maps by printing the symbol of a characteristic species in appropriate parts of the area carrying the colour and symbols denoting the *groupement*. At a mapping scale of 1:200 000 this allows only extensive stands of a particular faciès to be mapped distinctively. For the Djebel Rorra site in the woodland region of Kroumirie near the Algerian frontier, for instance, the map shows no faciès of the *groupement* CA, but field-work within the site showed six different communities constituting a series of increasingly degraded derivatives of the climax cork oak forest. Significantly for Radford & Peterken each of these is assignable to a different Fosberg formation, though all are mapped as *Groupement à Quercus suber, Cytisus triflorus*. The climax forest falls in Fosberg formation 1A16a, the progressively lower-growing shrubby communities are allocated to formations 1B14a, 1B18a, 1H24 and 1K25 respectively, and finally a grassy phase in heavily grazed clearings has to be placed in 1M21. Consequences of this bioclimatic and dynamic–ecological basis for the classification are, then, that structurally and floristically different communities may not be distinguished when, as often, the finest mapping-unit is the *groupement forestier*; and that structurally and floristically near-identical communities may be mapped as different if they are corresponding faciès of two or more different *groupements* or *séries*, as is frequently the case.

Problems have thus been seen to arise for the check-sheet survey where a vegetational classification, used locally for mapping and other purposes, is based on quite different principles from those underlying either the Fosberg or the mainly floristic systems. This is not to deny that it is of great scientific and practical importance to be aware of that network of basic relationships, spatial and temporal, between different vegetation-types which is made explicit in Emberger's classificatory scheme. The concept of mutually replaceable plant communities, varying widely in inherent stability, is an indispensable consideration for those faced with the organization of future biological surveys in the interest of land-use planning in general, as well as of nature conservation in particular. This point is discussed further on pp. 93–4.

Australia

A question arising from the consideration on pp. 73–80 of completed check-sheet returns from the Netherlands, Japan and Scandinavia is whether the Fosberg system could be so modified as to meet the serious criticism that it fails in its present form to discriminate between vegetation-types which are very different from the standpoints of both the land-use planner and the conservationist. Recent studies in Australia throw some important light on this question.

In 1968 the Australian IBP/CT Committee decided to extend investigations already in progress in some of the States and to undertake a conservation survey of the whole continent and of Papua New Guinea. The first step was to develop a vegetational classification which would be applicable throughout the area to be surveyed and would be simpler to understand and use in the field, under Australian conditions, than the Fosberg scheme. This Australian classification is briefly explained and discussed in Section 2.2.2. The highest classificatory unit, the 'structural formation', was defined as a 'series of climax plant communities which have essentially the same structural characteristics though possibly differing in floristic composition in all strata'. Table 2.1 (p. 40) shows how the structural formations so far recognized in the survey area are defined and differentiated.

Up to this point the Australian classification has features in common with the Fosberg scheme, the structural formations, like the Fosberg 'formation-classes', being defined in terms of the openness of the vegetation and the life-form and stature of the top-storey plants. The primary groupings of the Fosberg formation-classes are based on the extent to which the vegetation *as a whole* is closed. Primary structural group 1, 'closed vegetation', with 'crowns or peripheries of plants touching or overlapping', for example, includes amongst its eighteen formation-classes 1D, 'open forest with closed lower layers', and 1H, 'open dwarf scrub with closed ground cover'. Group 2, 'open vegetation', with 'plants or tufts of plants not touching but crowns not separated by more than their diameters; plants, not substratum, dominating landscape', includes 2A, 'steppe forest' (tree layer and lower layers open, lower layers may be open or sparse) and 2C, 'dwarf steppe scrub' (open predominantly woody vegetation less than 0.5 m tall). Primary structural group 3, 'sparse vegetation or desert', has 'plants so scattered that substratum dominates landscape'. In the Australian scheme one basis for the primary two-way division is the 'projective foliage cover' of the 'tallest stratum', irrespective of the degree of closure of lower layers of the vegetation. The limits are drawn at 70% (between 'dense' and 'mid-dense'), 30% (between 'mid-dense' and 'sparse') and 10% (between 'sparse' and 'very sparse'). In the Fosberg scheme, moreover, 0.5 m is the top-storey height separating 'dwarf scrub' from 'scrub' and 5 m that separating 'scrub' from 'forest', there being no further division by height and no other distinction than in stature between shrubs and trees. The dividing-lines between Australian structural formations are at top-storey heights of 2 m and 8 m for shrubs and at 5 m, 10 m and 30 m for trees (Table 2.1, p. 40). The correspondence is seen to be far from close, the Australian scheme being distinctly more precise in defining the boundaries between classificatory units. It pays little attention at this level of the hierarchy, on the other hand, to plants of other layers than the uppermost. It is therefore of interest to find that, if account is taken only of vegetation dominated by woody plants, the numbers of Australian structural formations and of Fosberg formation-classes are identical at twenty of each. This arises because the finer division by heights of the trees and shrubs of the tallest stratum in the Australian system is balanced by the absence of distinctions based on the nature and degree of closure of layers other than the uppermost.

Effects of these two major points of difference are made evident in Table 2.7 which shows for vegetation-types dominated by woody plants that a single Fosberg formation-class such as 1A or 1E may include several different structural formations and, conversely, a single structural formation, such as 'Tall woodland' or 'Low open-woodland', may comprise several different formation-classes. It will be seen, too, that the Australian scheme does, but the Fosberg scheme does not, distinguish between shrubs and trees with heights between 5 m and 8 m. The Fosberg formation-class 1A therefore covers both 'Low closed-forest' and 'Closed-scrub' in this height-range, and similarly for vegetation-types with more widely spaced woody plants. It becomes clear, then, that even at this level of classification the information implied by the allocation of a vegetation-type to a class in one of the two schemes must commonly be insufficient to allow an unambiguous conversion to its appropriate class in the other.

The Fosberg and Australian classifications diverge more radically at the

Table 2.7. *Comparison of Australian with Fosberg classification*

Percentage cover of tallest distinct stratum

Height of tallest distinct stratum (m)	100–70		70–30		30–10		10–0	
100–30	Tall closed-forest	1A, 1D, 2A	Tall open-forest	1D, 2A	Tall woodland	1E, 1F, 2A, 2D	Tall open-woodland	1E, 1F, 1I, 1J, 2D, 3A
30–10	Closed-forest		Open-forest		Woodland		Open-woodland	
10–8	Low closed-forest		Low open-forest		Low woodland		Low open-woodland	
8–5	Same, or Closed-scrub		Same, or Open-scrub		Same, or Tall shrubland		Same, or Tall open-shrubland	
5–2	Closed-scrub	1B	Open-scrub	1G, 2B	Tall shrubland	1G, 1K, 2B	Tall open-shrubland	1G, 1K, 2E, 3B
2–0.5	Closed-heath		Open-heath		Low shrubland		Low open-shrubland	
0.5–0.1		1C		1H, 2C		1H, 2C		1H, 2C, 2F, 3B

next lower levels of the classificatory hierarchy. In the Fosberg scheme the pairs of formation-groups, into which the formation-classes are divided according to the seasonality or otherwise of the vegetation (e.g., evergreen or deciduous tree-foliate), are further subdivided on non-floristic features into 'formations' and 'subformations'. The differentiating features mainly concern the size, shape and texture of the leaves of the most prominent components of the vegetation. In the Australian system the structural formations are divided, where the necessary information is available, into 'alliances', 'associations' and 'societies', defined as series of climax plant communities, within the same structural formation, having progressively increasing degrees of *floristic* similarity:

(1) 'Alliance.' Plant communities having related species as dominants of the uppermost stratum, and possibly the same or related species in the understorey, may be grouped together under one alliance.

(2) 'Association.' Plant communities which have the same species as dominants of the uppermost stratum, though possibly different floristically in lower strata, may be grouped together as an association.

(3) 'Society.' Plant communities which have the same species as dominants of the uppermost stratum, and the same species prominent in the lower strata, may be grouped together as a society.

The 'alliances' of the Australian system are thus seen to be defined, in general, in floristic terms: vegetation-types are placed in the same alliance if their top-layer dominants are of the same or related species. 'Tall shrubland' in Queensland, for example, is divided into five alliances dominated respectively by *Eucalyptus signata* and *E. intermedia*, *Melaleuca symphyocarpa*, *Bauhinia* spp., *Acacia aneura*, and *A. cambagei* (Specht, Roe & Boughton, 1974). In the treatment of closed rain-forests in Queensland assignable to structural formations 'Tall closed-forest' and 'Closed-forest', however, there is a departure from this practice. The procedure devised by Webb and his co-workers (pp. 56 *et seq.*) is followed and the alliances are defined for the most part in structural and physiognomic terms (Specht, Roe & Boughton, 1974, p. 138). The seventeen alliances are grouped into Tropical, Sub-tropical, Submontane and Coastal, and the eight alliances of the Tropical group are designated:

(1) Complex mesophyll vine forest.
(2) Mixed mesophyll vine forest.
(3) Semi-deciduous mesophyll vine forest.
(4) Semi-deciduous notophyll vine forest (±araucarian emergents).
(5) Mesophyll feather-palm (*Archontophoenix*) vine forest.
(6) Mesophyll fan-palm (*Licuala*) vine forest.
(7) Simple mesophyll–notophyll vine forest (±sclerophyll emergents).
(8) *Calophyllum inophyllum* (strand vegetation).

Here advantage is taken of the structural complexity and physiognomic diversity of these rain-forests to extend a basically non-floristic classification below the level of the structural formation. In the descriptive names of the alliances the term 'vine forest' refers to the relative abundance of lianas; and 'complex', 'mixed' and 'simple' indicate degrees of conspicuousness of such structural features as robust lianas, vascular epiphytes, plank buttresses and large compound leaves. Reference is also made in the names to differences in the duration, size, shape and texture of the leaves of prominent components of the vegetation. Ostensibly floristic terms such as 'fan-palm', 'feather-palm' and 'tree-fern' are used because they have very familiar physiognomic implications. Only the last of these alliances has a strictly floristic designation, and even that is qualified by the addition of 'Strand vegetation'.

In the Fosberg system a non-floristic classification is extended from the level of the formation-class to that of the formation and subformation mainly by the use of leaf-characters, those other structural and physiognomic features found valuable in differentiating forest-types in Queensland having a very restricted application. Fosberg has indeed stated (1967) that his classification of vegetation-types by structural and functional features 'has been carried down to the lowest levels at which these categories of information provide clear separations of units. The ultimate units, below this, must clearly be floristic, and will correspond generally to the associations recognized by most systems'. The Australian alliance is an appreciably finer unit than the Fosberg formation – check-sheet returns from certain States show between two and three times as many alliances as Fosberg formations (Table 4.14, p. 175). The effective use of a structural–physiognomic system down to alliance level must therefore be regarded as an important achievement and one inviting careful consideration in the planning for future biological surveys.

Future surveys, comparable in kind and scale with the check-sheet survey, are likely to aim at providing the data needed for taking decisions on a wide range of aspects of land-use, including nature conservation in the broadest sense. The collection of data on a regional or world scale, followed by its storage and analysis in a centre serving the whole area surveyed, will make it more readily possible than at present to co-ordinate programmes of action where, as in nature conservation, this is agreed to be desirable. This leads to the conclusion that the data to be stored and interpreted must be readily and certainly obtainable by surveyors having no more specialized training and experience than can reasonably be expected of the general field-worker. They must be collected in a uniform manner over the whole of the area surveyed and, as far as possible, their significance for assessing land-use potential must be generally intelligible and not entirely dependent on the knowledge of local specialists.

A non-floristic classification of vegetation may be expected to be based on features which can be observed and recorded in the field without dependence on expert local taxonomic knowledge. This is a matter of general importance but is of particular relevance to surveys of species-rich vegetation in the wet tropics, much of it still very imperfectly known from the taxonomic standpoint. There is the further consideration that an assessment of the significance of taxonomic lists demands familiarity with the general implications of the presence or absence of particular species, many of which may be of quite restricted geographical distribution. The outcome of a strictly floristic survey will therefore be effectively interpretable only by scientists combining the requisite local taxonomic and ecological knowledge.

The first question to be answered is whether this degree of parochiality of information on the abundance, presence or absence of individual species can be overcome by substituting some kind of non-floristic information about vegetation-types for the floristic information commonly regarded as essential. It must be conceded that, where the main concern is with the conservation of individual species of plants and animals, a detailed knowledge of species present in given areas is indispensable. The second question is, therefore, whether information about the species present in a survey-site, sufficient for at least the initial purposes of a large-scale survey, can in any way be inferred without the need to draw up detailed lists on the spot.

The Queensland workers have for many years been investigating the use of structural characters in vegetational classification and their general conclusions are stated and discussed by Webb in Section 2.2.3. Of particular relevance is the comparison made recently of the properties of floristic and physiognomic–structural data concerning complex rain-forest communities (Webb *et al.*, 1970). This study was based on data collected from seventy rain-forest sites, mostly in the coastal belt from Cape York southwards to the neighbourhood of Port Macquarie in New South Wales, and covering substantially the whole range of rain-forest types in eastern Australia. Environmental data collected for each site included 'the broad features of surface geology, topography, slope, and aspect; soil texture, depth and drainage; latitude, altitude, seasonal distribution of rainfall, temperature (including mean temperature of the coldest month and frost incidence); and, for about half the sites, a limited amount of soil–chemical data (total and available P, exchangeable cations and pH)'. The meteorological data either came from the nearest comparable climatic station or were estimated from regional rainfall maps and calculated from lapse-rates. In all, twenty-two environmental attributes were recorded. The floristic records were spot-lists of the common canopy-trees, and the structural data were obtained by asking the field-observers to complete a questionnaire on twenty-four characters of the vegetation, including those used in the classification of Queensland rain-forests for the Conservation Survey and few others characteristic of transitional and temperate forests further south. There were also included three features previously shown to be valuable environmental indicators: type of branching (unbranched, branched low down or shrub-like branching), bark characters and presence of erect thorny stems and branches. The observers were asked to estimate canopy height and to score the remaining characters on a 3-point scale.

The floristic and structural data were classified independently by the use of both agglomerative and divisive computer programs. There was little to choose between the two floristic analyses, but the divisive structural analysis proved unsatisfactory, in part, it seems, because of the particular numerical model used but also because 'a single structural feature may be associated, to a greater extent than is likely with a single species, with any one of a number of alternative environmental situations', so that division on a single structural feature tends to produce site-groups representing mixtures of different environments. For these reasons the comparison between floristic and structural classifications was restricted to the results of the agglomerative–polythetic analyses by the Canberra programs CENTGLAS and MULTBET for qualitative floristic and quasi-continuous structural data respectively.

This comparison shows that, down to the 10-group level, the floristic and structural analyses produced site-groupings about equally informative as to climatic and edaphic features of the sites: there was no support for the view that structural features are less efficient than floristic at lower levels. The floristic analysis led to a greater geographical localization of site-groupings than did the structural: all the New South Wales sites fell into only three different groups at the 10-group level of the floristic analysis but were scattered over eight groups in the structural analysis. This suggests that floristic composition was more influenced by historically-determined facts of species-distribution, and especially by past changes in temperature, than were structural features, and therefore these tended to diagnose a particular environment irrespective of its geographical location. Within a restricted region, on the other hand, the floristic analysis provided a more precise definition of habitat through species-assemblages with strong site-preferences. The structural characterization of environment tended to depend more on moisture than on temperature and to be less clear-cut, but was for that reason more informative about the intergrading of habitats. It also revealed that different combinations of environmental factors can give rise to structurally similar forests so that, for example, climatic information may need to be available before inferences can be drawn about levels of soil fertility.

In the final discussion of their results the authors of this important paper conclude that a vegetational classification based on a set of structural–physiognomic features recorded by untrained observers has proved as useful as one based on floristic composition and has provided a more detailed subdivision of rain-forests in eastern Australia than does the world-classification proposed by Fosberg. In particular it has provided a more detailed separation of relevant environmental types. They suggest that with a more comprehensive set of structural–physiognomic features, previously shown empirically to respond to climate and soil, a site-assessment could be made that would be sufficiently detailed for land-use planning.

The authors warn that it must not be concluded either that there are no problems in the recording of suitable physiognomic features or that a wholly satisfactory world-wide physiognomic classification of vegetation is readily feasible. Apart from matters of sampling procedure there is the general problem of how to record such features, whether merely as present or absent or, if quantitatively, in what way. While there seems no reason to doubt that a suitable physiognomic classification could be devised in any one part of the world, the restricted geographical distribution of certain well-marked life-forms, such a those of the bromeliads and cacti of parts of the western hemisphere, might present difficulties for the construction of a single world-wide system operating down to the lowest levels of the classificatory hierarchy. What does seem likely, however, is that the present Fosberg scheme might usefully be modified and extended in its scope, perhaps along such lines as those explored by the Australian workers.

This consideration of the Australian survey-procedure provides some support for the decision of the Central IBP/CT Committee to adopt, in the Fosberg scheme, a mode of recording vegetational information having advantages over purely floristic listing in ease of use in the field and in relevance to general land-use planning irrespective of geographical location. It emphasizes, however, that the Fosberg scheme has shortcomings and that further research is necessary to see how far it is possible to devise a recording procedure combining a high site-assessment capability with reasonable demands on field-observers and at the same time having something approaching world-wide application. The Australian effort, and especially that of the Queensland workers, gives some encouragement for the belief that this may not be an unrealistic endeavour.

None of this amounts to a claim that physiognomic–structural features of vegetation should be used to the exclusion of floristic composition. Detailed floristic data are ultimately indispensable for certain aspects of nature conservation and are always likely to be helpful in site-assessment at the finer levels of discrimination, and the data-store must surely aim at the eventual inclusion of both types of information. The second question raised on p. 85 was whether there might be some way of gaining information about the species present in a particular survey-site, sufficient for at least a preliminary consideration of the significance of the site for nature conservation, even if no lists were made on the spot. It must be said at once that nothing can satisfactorily replace *in situ* listing by competent taxonomists. It is nevertheless true that, within quite large areas having more or less uniform climate and a similar range of topography and soil-type, the plant communities which characterize certain combinations of environmental features have much the same species-lists throughout the area. If such areas of floristic similarity were recognized and delimited, lists could be drawn up for each of the vegetation-types encountered within them. With knowledge of the relevant environmental data, and of the physiognomy and structure of the vegetation, it should then be possible to predict with some confidence what species would be found in any site within the area. The resulting lists would necessarily be tentative and provisional, but could nevertheless be of some value in early stages of the analysis of survey findings during the present period of shortage of trained taxonomists.

The establishment of 'biotic provinces' of the kind envisaged above has been recognized in recent years as being of great potential value not only for students of the geographical distribution of plant and animal species but for biologists in general. Much more work will be necessary, however, before their boundaries can be defined satisfactorily and community lists drawn up. The most recently published maps covering the whole of the earth's surface are those showing the 'biogeographical provinces' of Udvardy (1975), prepared as a contribution to UNESCO's 'Man and the Biosphere' programme in continuation of the work of Dasmann (1972, 1973*a*; IUCN, 1974). Dasmann aimed at providing a framework within which to assess how far the world's ecosystems were already represented within protected areas, taking account of the fact that ecosystems more or less identical in physiognomy and structure might differ to widely varying extents in species-composition. Udvardy admits that it would be desirable, before attempting to demarcate biogeographical units for conservational purposes, to prepare a comprehensive catalogue of the distribution of both ecosystems and species. Meanwhile he proposes a hierarchy of units the highest of which is the 'Biogeographical realm', a 'continent or subcontinent-sized area with unifying features of geography and fauna/flora/vegetation'. They are eight in number and comparable with Wallace's 'regions' but with the addition of Antarctic and Oceanian realms. Each is divided into a number of 'Biogeographical provinces', equivalent to Dasmann's biotic provinces. These vary widely in size. In the Nearctic Realm, for example, Udvardy lists twenty-two provinces of which '2. Oregonian' occupies only the coastal belt of the States of Washington and Oregon, while '4. Canadian taiga' comprises most of Canada south of the tundra and east of the Rockies. The provinces are based on the characteristic climatic climax vegetation, their boundaries being those of the plant formations of Weaver & Clements (1938) except where the inclusion of more than one distinctive fauna or flora calls for the subdivision of the area over which a single formation extends: areas having fewer than 65% of their animal species in common (usually, in practice, of their mammals and birds) are regarded as belonging to different faunistic provinces. The provinces of Dasmann and Udvardy correspond fairly closely with those of Good (1947), which are based on floristic resemblances though with no clear indication of the criteria adopted.

The delimitation of these provinces is based on information which must vary widely in completeness from one part of the world to another and must often be inadequate for its purpose. The placing of boundaries, other than those determined by the physiognomy of the inferred climax vegetation, seems to have depended more on the distribution of species of mammals and birds than of higher plants, presumably because more complete information was available for these animal groups. Udvardy envisages the eventual delimitation of subprovinces, districts and subdistricts so as to arrive at units of much greater floristic and faunistic uniformity than the present provinces, but sees this as a task for regional experts. There will then be a need for systematic listing, district by district, of the species present in plant communities specified in terms of the climatic, topographical and soil data normally collected in the course of biological surveys, together with information on non-floristic features of the vegetation, preferably going beyond what is necessary for a Fosberg coding. It is arguable that the delimitation of these smaller biogeographical units and the provision of representative species-lists for their component communities would be a valuable priority task for ecologists and taxonomists in the immediate future, and one that might well be thought appropriate for adoption by the organizers of Project 8 of the 'Man and the Biosphere' programme. Only with this task well under way can a judgement be made on the feasibility of predicting the species-composition of survey-sites for which lists cannot be made on the spot for lack of taxonomic experts.

2.2.5 Concluding observations on vegetational recording

Records of the vegetation must always be an essential part of the information to be collected and stored in biological surveys, whether undertaken to provide a basis for a policy of nature conservation or for land-use planning in general. What vegetational data should be recorded, and in what way, will depend on the scale and scope of the survey. The main concern must be that records shall provide the maximum of relevant information about the vegetation itself and about the potentialities of the site. Within a single country this is likely to be achieved most readily by the adoption of procedures for recognizing, describing and classifying different types of vegetation, and for elucidating and reporting their relationships with each other and with their environment, that are generally used and understood in the country in question and that will enable full advantage to be taken of the results of earlier investigations employing the same procedures. The success of an international project like the check-sheet survey, on the other hand, will depend on information allowing comparisons between widely separated areas and planning on a world-wide scale.

Throughout the nineteenth century botanical explorers and others interested in the geographical distribution of different types of natural vegetation reported their observations much as Alexander von Humboldt proposed in his *Ideen zu einer Physiognomik der Gewächse* (1806). Humboldt selected sixteen basic physiognomic types of plants as a basis for describing vegetation without naming the component plant species. This would enable useful information to be recorded for areas where the native flora had not yet been studied and, more important, would permit comparisons to be made between areas having few or no species in common. The subsequent impressive increase in knowledge and understanding of natural vegetation on a world scale became evident with the publication at the end of the century of A. P. F. Schimper's great *Pflanzengeographie auf physiologischer Grundlage* (1898). Schimper stressed the importance primarily of climate, but also of soil-type and other factors, in determining the general aspect of the undisturbed vegetation of an area and described those major types of natural vegetation which extend across continents in broad and more or less latitudinal belts and which derive their physiognomic uniformity from being dominated throughout by plants of similar life-form. Schimper showed thirteen of these 'formation-types' in his vegetational map of the world, six of them different kinds of forest. Later workers increased this number, further subdividing the basic forest, scrub, grassland and desert formations so as to effect a finer classification. Thus Rübel (1936), modifying proposals by Brockmann & Rübel (1912), recognized twenty-eight formation-classes, of which thirteen were dominated by woody plants, and claimed that 'the entire vegetation of the world can be included in these comprehensive formation-classes'.

Towards the end of the nineteenth century, with a rapidly growing knowledge, especially in parts of Europe, of the individual plant species characterizing local types of vegetation, the description and classification of vegetation was developing along quite different lines. There was now a strong emphasis on the plant community as an assemblage of named plant species and on the value of particular species as indicators of ecologically significant features of present and past environments. Attention was concentrated on vegetational units of far smaller size than the formations of the plant geographers, these 'associations' being defined in floristic rather than physiognomic terms. Already in 1915 Braun-Blanquet, for long leader of the main Central European school of phytosociologists, was stating that a classification based on physiognomy was 'trop vague pour être employé avec fruit dans une étude approfondie', and this view has been repeated frequently up to the present day. There have been differences of opinion about what features of floristic composition should be used in distinguishing one association from another, but the concepts and practice of the Braun-Blanquet (or Zurich–Montpellier) school have been adopted, with or without some modification, in many parts of the world. Following the standard European procedure records are being made not only of the species present in sample-areas of specified size, and of their relative abundance and gregariousness, but also of a number of environmental features such as soil-type and moisture regime. In this way a body of information is being accumulated which relates each association to its characteristic environment and thus affords a steadily strengthening basis for inference from floristic composition to other matters of scientific interest and to aspects of potential land-use.

The organizers of the check-sheet survey were thus faced with difficult decisions. A world-wide survey, and one needing to be initiated with the minimum of delay, must necessarily involve vegetational recording in many regions where the flora was as yet little known and where the surveyors could not be expected to have received much local taxonomic training. This would point to the desirability of adopting a non-floristic recording system capable of yielding information of real value for the purposes of the survey, easy to use anywhere in the world and permitting comparisons between one surveyed site and any other. Against this was the evidence that floristic classificatory units were, in general, much more sensitive to small differences in features of the controlling environment than were the larger physiognomic units of the plant geographers. Most conservationists, moreover, were interested in the protection of particular rare and threatened species and of the plant communities in which they occur. It was these opposing sets of considerations that led the organizers to recommend the adoption of some single non-floristic system for use everywhere, and its reinforcement wherever possible by naming the plant communities encountered in a particular site in accordance with usual national or regional practice. This would commonly entail allocation to floristic units of known local significance.

Two non-floristic systems, both of them much more detailed than the earlier physiognomic classifications, had strong claims for selection, but the UNESCO system (pp. 41–2) was not yet available in its final form and accordingly that of Fosberg was adopted. Its keys, brief descriptions and glossary appeared to make it a satisfactorily simple matter to assign a vegetation-type to its appropriate formation or subformation, though in the event some difficulties arose (pp. 68–71). Since data on the location, altitude, climate, topography, geological substratum and soil, and also on past and present human impacts, were also to be recorded for each site, it was expected that the value of the Fosberg coding for assessing the potentialities of a site would be steadily enhanced.

The Fosberg classification in its present form may seem to have proved rather less useful than was hoped. But it is important to remember that the collected data are being stored so as to remain conveniently accessible, and that the check-sheet survey was always envisaged as a testing-ground for future biological surveys, the data from which would be added to the existing store. It may be some decades before there is adequate taxonomic knowledge, especially of equatorial regions, for full floristic recording to be possible everywhere. It is becoming clear, moreover, that it is in particular in the species-rich tropics that, for many purposes, description and classification in physiognomic-structural terms promises to be sufficiently informative to provide an adequate alternative to floristic recording (pp. 54–65). Even where full floristic information is available, additional inferences can be drawn from the physiognomy of the vegetation of an area, given certain environmental and historical information, that may supplement in a valuable way the conclusions to be drawn concerning nature conservation and potential land-use in general. What is required is the collection, as far as possible, of all data of known or strongly presumed relevance to biological site-assessment and the interests of conservationists.

Further and much more extensive investigations of the kind initiated for rainforests by the Queensland workers (pp. 59–63) will have to be undertaken before the list of requisite basic data can be finalized, but already enough is known to enable a useful start to be made at any time. What is becoming clear is the overriding need for the *basic facts* about each major vegetation-type and its environment. Informed views as to the allocation of each to the codings or classes of any particular systems of vegetational classification are of undoubted value and should form part of the data to be collected, but for the time being they cannot, and should not, replace the basic data. This conclusion arises in part from our lack of detailed knowledge of the natural vegetation in many parts of the world. More important is the consideration that every system of classification depends on a selection of certain features of the vegetation, or its environment, to provide a basis for the delimitation of classes. These features may or may not be of real value for decisions on land-use, including the desirability of establishing nature reserves, and in any case are unlikely to cover all the features of major significance for such purposes. There is the further point that any classification based mainly or entirely on floristic criteria must necessarily encounter problems arising from differences in the geographical range of its component species. Normally a plant community will be found to change progressively in species-composition in all directions from the area where it was initially studied and specified. Not only does its species-composition change but its special ecological requirements will also vary to a more or less significant extent with changes in general environmental conditions, so that the ecological implications of its presence or absence may be affected in important ways.

The alternative to asking field-surveyors to allocate each of the vegetation-types they encounter to its coding or class in a recognized classificatory system, is to elicit from them the basic data deemed most relevant to the aims of the survey. This is most readily achieved, as suggested by the Queensland workers, by the provision of a *pro forma* to be completed for each vegetation-type (pp. 58–9). This would invite information on the physiognomy and structure of the vegetation, including estimates of the height and density of the different layers, the presence or absence of special life-forms such as palms, ferns, lianas, higher-plant epiphytes, stem-succulents, etc. It would also ask for such facts of floristic composition as can be provided. A full list of species might be invited but must be recognized as an unattainable ideal at present for the many areas where the flora has been explored incompletely or is still virtually unknown. Even where adequate knowledge is available the compilation of complete lists in species-rich vegetation will require at least one expert taxonomist in each survey-team. There should be less difficulty in supplying names only of the abundant and conspicuous species, and in particular the dominants of each layer, since these are the plants most likely to be familiar to the non-specialist field-worker and could most easily be collected for subsequent identification. Also substantial progress in the near future towards delimiting biogeographical units of sufficient uniformity, with the preparation of representative species-lists for a range of plant-communities included in each, might, as suggested above, enable data-centres to overcome some of the difficulties arising from the present shortage of taxonomists.

The storage of both floristic and non-floristic data in the way suggested would provide a good basis for inferring the nature of the vegetation of a survey-site and for interpreting it in ecological terms. To what extent a single *pro forma* could serve for vegetational recording in all parts of the world is highly problematical. It would almost certainly be desirable to prepare different versions at least for each major physiognomic type of vegetation, so as to avoid burdening the field-surveyor with large numbers of irrelevant questions. But, even if different, these versions would need to be so closely interrelated that all the data could be processed and stored as a single homogeneous set so as to serve as a sound basis for comparisons between any of the surveyed sites.

Apart from basic vegetational information a comprehensive *pro forma*, like the present check-sheet, would need to invite information about the geographical location and legal status of each site and about climatic, edaphic and biotic aspects of the environment it affords for living organisms of all kinds. The opportunity should also be taken for recording views as to the allocation of its geographical and topographical situation, climate, soil, vegetation, etc., to their classes in appropriate systems of classification. The statement of such views should not be regarded as indispensable parts of a *pro forma* completed by the field-surveyors: the basic data provided should enable the participating national organization or the data-centre to arrive at satisfactory allocations. But the views of local experts, familiar both with the area in which the survey-site falls and with systems of classification in general national or regional use, could be additional information of considerable value. This would be particularly true of expert assessment of the local significance of the presence or absence of indicator species such as the *groupes socioécologiques* of Duvigneaud (1946) or the *ökologische Gruppen* of Ellenberg (1963) and others.

Of comparable value would be the interpretation of the existing vegetation and soil in dynamic-ecological terms. Most systems of vegetational classification are characterized by hierarchies of classes based on degrees of either physiognomic or floristic similarity. An alternative basis is provided by the fact that a number of different types of vegetation are able to replace each other in any particular area. What is there at a given time, is a consequence of past as well as present factors of the total plant environment, and may be either stable (in the sense of undergoing no general directional change) or in process of more or less rapid change, with related changes in microclimate, fauna and other features. In many instances it has been possible to understand in detail why and how one vegetation-type replaces another in the course of successional or degradational change, the former occurring spontaneously, the latter usually a consequence of some form of human activity. Stages in these sequences of change may differ as profoundly, both in appearance and in floristic composition, as do stable closed forest at one extreme and poor weedy grassland, continuing to deteriorate through overgrazing, at the other. Yet it makes theoretical as well as practical sense to envisage a form of classification in which all these mutually replaceable vegetation-types are brought together in a single class, as has in fact been done by the American school of ecologists founded by H. C. Cowles and especially associated with the name of F. E. Clements, and by the group centred on Toulouse in France, with H. Gaussen and L. Emberger prominent amongst its leaders. The Toulouse school has concerned itself in particular with providing a scientific basis for general-purpose vegetational mapping in the Mediterranean area (pp. 78–80). No opportunity should be lost for recording the views of local experts in the dynamic ecology of the vegetation and soil of any site within the area covered by such a classification: they could be of great value in predicting the outcome of any contemplated system of management or of alternative land-use.

In conclusion it must be frankly conceded that it is not yet possible to make more precise recommendations about a procedure for vegetational recording in world-wide biological surveys. It does seem clear, however, that it would be better for the time being to collect a wide range of basic data and informed views through the use of a suitably designed *pro forma* rather than to select any one or more of the existing systems of vegetational classification. The form and detailed wording of such a *pro forma* will no doubt have to undergo modification with increasing awareness of what particular information about vegetation will be of most value for achieving the purposes of future surveys comparable with the check-sheet survey.

2.3 Soil types

2.3.1 Historical background

Problems closely analogous to those encountered in planning the treatment of vegetation-types in the check-sheet survey arose also in connection with soils. No single internationally accepted system of soil classification was available, and the organizers had to decide what kinds of information should be sought about the soils in surveyed areas. They needed to be sure that this information would be of real value for understanding the ecology of the areas and for assessing their land-use capabilities, including their suitability as nature reserves. They had also to keep in mind the improbability that the field-surveyors would be expert soil-scientists, and the likelihood that they might have had little or no special training in the description and classification of soils. Just as for the treatment of vegetation-types, therefore, a recording system was adopted which was judged sufficiently simple for use by non-experts but nevertheless capable of yielding information valuable for the purposes of the survey. This was one proposed by Dr D. F. Ball and is set out in the *Guide to the Check Sheet for IBP Areas* (Peterken, 1967). It is based on the system of Aubert & Duchaufour (Duchaufour, 1965). Soils are grouped into six broad categories: saline, ferritic, or organic soils, well-drained or poorly-drained, non-saline and non-ferritic soils with good profile development, and soils with weak profile development. Most of these are further subdivided to give thirteen groups in all, and a key is provided for their easy identification. In the *Guide* the key is preceded by descriptions of each of the four soil horizons to which reference is made and by a table showing the thirteen categories, the succession of horizons or profile types characteristic of each and the symbols by which they are designated in the key (see Appendix 3).

The field instructions accompanying the check-sheet, under Section 7(2) Soil, asked for information to be supplied in the two columns of p. 4 of the sheet. In the first, headed 'Soil type', the surveyor was to enter the code number of the Ball classification of the soil type occurring under each of the plant communities recognized, with the additional instruction that 'Where more than one soil type occurs under one Community, either the definition of the Community should be revised, or an explanatory note should be added' in the second column, under the heading 'Other notes'. In this column, too, 'Sub-types present should be mentioned, together with short descriptions of significant features, e.g., colour, humus content, depth'.

That the Ball classification did in fact present no serious difficulties for a majority of the recorders is suggested by the finding that about two-thirds of the sample of 200 completed check-sheets analysed in detail included entries for Section 7(2), and half of these supplied both the code numbers and additional notes on the soils of at least some of the plant communities recognized.

2.3.2 Problems of soil classification: R. F. Isbell

Cline (1949) pointed out that the principle of any classification is to organize our knowledge so that the properties of the classified objects may most easily be remembered, and their relationships understood, *for a specific objective*. In the present circumstances we are concerned with classifying soils in order to gain insight into the role soils may play in influencing the natural vegetation of an area. Apart from the obvious and sometimes overriding roles of the climatic environment and site drainage characteristics, there is the problem of knowing which soil properties are involved in determining the particular plant community occurring on a given soil. Relatively few soil properties seem likely to account for most edaphically-determined vegetational variation, but we are not yet in a position to erect a special-purpose soil classification for predicting vegetation-type even when allowance is made for the other relevant environmental factors such as climate. The picture is complicated further by the fact that vegetation itself is one of the important soil-forming factors and thus it becomes very difficult to separate cause and effect in many soil properties.

Mulcahy & Humphries (1967) emphasize the high predictive value of a general-purpose or 'natural' classification based on overall resemblances, allocation to one of its classes implying a great deal of information about the individual so classed. Doubts have been expressed about the existence, even the possible existence, of a general-purpose or universally applicable classification of soils (Gibbons, 1961, 1968; Avery, 1968; Butler, 1964). Nevertheless it seems that, in the present state of our knowledge, we are restricted to the use of some general-purpose classification if we wish to relate soils to particular plant communities; the problems of erecting an effective special-purpose classification appear for the time being insurmountable. A further point had to be taken into account by the organizers of the check-sheet survey, that the soils encountered would for the most part be examined by non-specialists. This gives rise to two questions: which of the many more or less general-purpose classifications available are likely to be most appropriate or most likely to predict meaningful relationships between soil properties and natural plant communities, and is it better to try to establish some simple 'open-ended' method of collection of soil data which may then be examined objectively by mathematical analysis?

Brief review of some existing soil classifications

It is appropriate to consider first the United States Department of Agriculture (USDA) system of soil classification which, until recently, has been commonly termed the 7th Approximation (Soil Survey Staff, 1960, 1975). There can be no doubt that it is the most comprehensive of the classifications in current use throughout the world, though it may be doubted whether it is strictly a general-purpose system. In the present context, however, it suffers a serious limitation which is admitted in the preface (1960): 'this is not a book for beginning students of soil classification'. It assumes knowledge of the basic and more or less generally accepted terminology for describing soils. The scheme, like many others, depends very largely on the identification of diagnostic horizons, many of them specified in considerable detail and a number of them requiring analytical data for their identification. It seems, in short, most unlikely that users without formal and practical experience in field pedology will be capable of using this system, even to the extent of classifying soils into the highest category of the classification (orders).

The FAO classification (1974) is in effect a legend for the World Soil Map. It is based largely on the 7th Approximation, many of its higher categories being almost identical and defined largely in terms of diagnostic soil horizons. It is somewhat simpler than the 7th Approximation, albeit less comprehensive, but is still not easy to use by people without considerable expertise in soil science. It has the further disadvantage that no provision is made for soils occupying less than 20% of an area large enough to be shown on a map at a scale of 1:5 000 000, so that some soils simply cannot be classified by this scheme.

Of soil classifications currently in use in Western Europe that proposed and used by the French (CPCS, 1967) has most appeal because it gives considerable attention to soils of tropical regions as well as to the better known temperate soils. It is based, essentially, on differences in soil morphology, but these are in general less specifically defined in the higher categories than in the USDA and FAO systems. Although this makes it somewhat easier for non-specialists to use, it also leads to a lesser degree of precision in many of the higher categories. It has the usual hierarchical structure with classes, subclasses, groups and subgroups of soils. For IBP it has the disadvantages that there is not yet a formal key, that descriptive phrases rather than specific names are used for most categories, and that analytical data are required for distinguishing lower-order categories.

The Russian soil classification (Tiurin, 1965) may be considered briefly because its basis differs in some respects from those discussed above. The classification is still characterized by a major emphasis on climatic and biotic factors for separating the higher categories, although more weight is given than formerly to features of the soils themselves. The scheme was developed primarily for Russian conditions and is not suited for world-wide adoption, even though it has been modified so as to be usable as a legend for the soil map of Eurasia (Tiurin, 1965).

Kovda, Lobova & Rozanov (1967) have proposed a preliminary historical–genetic classification for a world soil map on a scale of 1:5 000 000. It is claimed that soils are classified by properties and characteristics reflecting their evolution in time and not current environmental conditions, but the primary subdivision is still based essentially on broad climatic features. The authors freely admit, moreover, that knowledge of the evolution of soils is far from complete and that frequent resort to hypothesis is unavoidable.

A rather different approach to soil classification has been made in Australia in recent years with the development of a key for the recognition of Australian soils (Northcote, 1971). This was based, originally, on a study of about 500 soil profiles, mostly from the southern (temperate) part of the continent. The basis of the key is the concept of *profile form*, the term used to express the overall visual impact of the physical properties of the soil, i.e., those that can be identified readily and determined in the field. The only exception is the use of soil acidity trends, but these can also be determined readily in the field with a simple pH kit. The profile is thus regarded as a physical system, and those physical properties which are correlated with other features or properties – physical, chemical or biological – are used for distinguishing between groups at each step in the key, though the evidence for such correlations is nowhere stated or discussed. Northcote & Skene (1972) have shown distinct relationships between their key categories on the one hand and pH, salinity and sodicity on the other. Nevertheless it is largely through user-experience and familiarity with other and non-recorded properties such as nature of the parent material, depth, base-status, clay mineralogy, organic matter levels, etc., that some idea may be gained of overall soil fertility status and the relationship of specific soils to specific vegetation-types. This is clearly an area for future research.

The key has the advantage of simplicity of use, though some points, in particular the concept of pedological organization and the diagnosis of soil fabric, have caused difficulty to inexperienced users. The main disadvantage for use on a world-wide basis lies in its being based on a collection of Australian profiles which were not necessarily fully representative because of the state of local soil knowledge at the time. It would, therefore, be unreasonable to expect the differentiating criteria used at a particular level to be the most appropriate for a country with a greatly different array of soils: some unlikely bedfellows might result because of the choice of criteria adopted. But the principles underlying the key could be adapted for use in any other country.

So far we have considered various soil classifications mainly from the viewpoint of their relevance on a world scale and their successful use by a non-specialist. There seems little doubt that all those considered would be disqualified on one or other of these grounds. There is, however, a further vital consideration for any other classification we may wish to assess: is it likely to convey relevant information regarding plant communities on a world scale?

Some important soil properties of relevance to plant growth and distribution

The number of soil properties which may affect plant growth and distribution is not known but is likely to be large. They are, moreover, unlikely to act independently, so that even within a single uniform environment no general relationship can be expected between any one soil characteristic and vegetation-type or the presence of a particular plant species. Nor can it be expected that a particular soil attribute will be highly correlated with features of the natural vegetation in each of a range of environments. It is nevertheless relevant to the possibility of establishing some general relationships between soils and plant communities that some soil properties are highly correlated with others, and that soil properties are themselves largely products of environmental factors operating in the overall pedogenetic processes.

Before considering whether any existing or proposed soil classification may give useful information about the natural vegetation of a site and therefore about site-potential in more general terms, it will be valuable to list soil features which may influence the growth of plants.

Morphological properties

Of obvious importance are soil depth, degree of stoniness, and presence of unfavourable substrates such as hardpans. Surface conditions such as texture and structure may markedly influence water acceptance and the germination of seeds and emergence of seedlings. The porosity and permeability of subsoils may be very important for drainage and aeration and may often be inferred from such profile features as field texture, colour and degree of mottling.

Physical properties

A general assessment of physical properties of a soil may commonly be made in the field, laboratory measurement then being unnecessary. The most significant property is normally the soil's capacity to store water and supply it to the plant. This may often be, at least subjectively, assessed from a knowledge of the field texture, organic matter status and clay mineralogy. Related properties of importance are permeability, bulk density and aeration, again all open to subjective assessment in the field.

Chemical properties

Obvious factors here are the presence or absence of substances deleterious to plant growth, such as soluble salts and extremes of acidity or alkalinity; and also the levels of various elements known to be essential for plant growth. These can be quantified only by chemical analyses, and then not always satisfactorily in relation to plant performance. In a number of instances, however, the levels of at least some elements can be inferred from the nature of the soil itself.

Certain other soil properties can be of direct or indirect importance for plant growth. Thus high contents of 2:1 type clays may result in massive swelling and shrinkage in the soil with subsequent effects on plant roots. Conversely, high contents of kaolinitic clays and/or oxides of iron and aluminium are often features of soils of low fertility. Biological factors such as the presence of certain soil animals such as earthworms, and harmful soil organisms such as nematodes, may also have profound effects on plant behaviour.

It may be impossible for any general classification of soils, at least in the higher categories, to group together in an adequate way soils that are alike in all the ways listed above. It must be remembered, too, that soils having certain properties will affect plant growth differently with variations in the external environment. Thus a well-drained light-textured soil of low water-retaining capacity may be ideal for plant growth in a region with adequate and evenly distributed rainfall, whereas it would be much less favourable in an arid environment.

The Ball classification

It is pertinent at this stage to examine the classification due to Ball (1967) and to ask whether it is an adequate soil classification as such, i.e., does it tend to group together soils with similar properties and, more particularly, are the groupings likely to be relevant to the growth and occurrence of plant communities on a world scale? We must consider also if the scheme can be used satisfactorily by the non-specialist.

The first query concerns the adequacy of the definitions that accompany the key. The definition of the B horizons of soils requires the recorder to distinguish between the release of iron oxides without their transportation and the deposition of transported sesquioxides and/or clay. There is a large uncertainty as to whether clay or iron oxides are mainly transported or actually formed *in situ* (e.g., Brewer, 1968; Oertel, 1968), and the laboratory procedures for determining whether such constituents are transported or not are very difficult. A tentative alternative definition is as follows:

B or (B) Horizons lying between A horizons and the parent material C horizons, which show either, if designated B, a marked accumulation of clay and/or iron, and/or humus, or, if designated (B), little difference in these features between A and C but differ in colour, consistency, or structure.

Further definitions required to operate the key satisfactorily include the following:

(*a*) 'high concentration of alkaline salts'. Presumably high amounts of sodium bicarbonate and carbonate, but how high? Can the alkalinity be judged by pH, and do the high levels of alkalinity have to occur throughout the profile or only in some horizons?

(*b*) 'Saline soil with water-soluble salts.' Similar questions are raised as for (*a*) above.

(*c*) 'high (or normal) concentration of iron oxides'. This is particularly difficult to assess. Even if definite limits were given it is seldom possible to make a reliable estimate of iron oxide content without chemical analysis.

(*d*) 'Well drained (i.e., no evidence of strong impedance or waterlogging above 40 cm depth).' This is not adequate. Examples abound of poorly drained soils having an impeding layer deeper than 40 cm, and a soil with strong impedance at say 50 cm can hardly be called well drained.

A further matter for concern is that many of the alternatives in the key are not mutually exclusive:

(*a*) The impression given by the key is that only soils with a high concentration of alkaline salts will be saline, but this is not true. There are, moreover, many alkaline and saline soils with soluble salts in the lower horizons and high exchangeable sodium in the subsoil but not in the surface.

(*b*) Well-drained soils with a well-developed horizon sequence may not necessarily have 'moderate to strong biological activity', and the meaning of this phrase is not very clear.

Other problems arise from the heterogeneity of the ultimate groupings. Thus soils 'with high concentration of iron oxides' include such diverse soils as terra rossas (only mildly leached, with a high calcium status and generally base-saturated), and soils referred to as 'laterite', a term no longer applied to a soil but only to a soil-component: soils containing laterite are normally highly leached, with low calcium status and usually strongly base-unsaturated.

Finally, the Ball classification has a strong West European bias and is thus inadequate for classifying soils of many other regions. Some important and widespread soils are classified unsatisfactorily. The cracking clay soils (vertisols of the USDA and FAO schemes) may or may not be alkaline and/or saline and are therefore split at an early stage. Soils with high exchangeable sodium in the subsoil (but not necessarily in the surface) are also inadequately treated. Some with strongly sodic subsoils but without alkaline salts will be grouped in P_1 or P_2 irrespective of their subsoil salinity, which may often be considerable.

Any scheme which strives for simplicity and ease of operation must inevitably have shortcomings when trying to deal with the global array of soils. The main question is whether the Ball scheme could be modified to become more comprehensive and yet retain sufficient simplicity to enable a non-specialist to operate it successfully. I do not think this is possible. More definitions and specifications will make it more difficult for the non-specialist and will certainly involve at least some analytical data.

The scheme as it stands is unlikely to contribute much to an understanding of the relationships between soils and natural plant communities. The soil properties likely to have most relevance for plant growth cannot be taken into account at the level of generalization attempted, and the same is true of the higher levels of most other existing classifications, or any that one might attempt to devise.

A possible compromise

It is agreed that there is a need to collect information on soils that can be correlated, as far as possible, with vegetational information from the same set of sites, since IBP is concerned with 'the biological basis of productivity and human welfare'. There can be no doubt that biological productivity is to a large degree dependent on soil factors. The problem is really what relevant information can be collected readily by non-specialists in soil science.

Admitting from the start that we do not yet know all the soil factors that may be of most importance, there are some of obvious relevance to plant growth that might be collected readily. It seems most unlikely, on the other hand, that a simple and easily usable soil classification can be constructed, for reasons given earlier. It is therefore more profitable to consider what soil data we might consider collecting, if it can be done sufficiently simply.

(*a*) The parent material of the soil is of obvious importance in determining the kind of soil likely to be formed in a particular environment and may give a valuable guide to soil fertility status.

(*b*) Soil depth (i.e., thickness of combined A, B and C horizons) is obviously of importance and in most instances can be estimated within broad limits.

(*c*) Soil drainage status can usually be readily inferred in the field within broad limits and may be most useful in predicting vegetation-type because it may reflect a number of other important physical properties.

For a number of soil properties it is important to have information on both surface soil and subsoil, that is, normally, on the A horizons and the B horizons: either can independently exert considerable influence on plant performance. The most significant of these properties are:

(*a*) Soil texture, which, within fairly broad limits, has obvious implications for water-holding capacity, drainage, and sometimes nutrient status.

(*b*) Acidity, which is probably the most useful single guide to soil nutrient status when considered in relation to other information collected, and which may be estimated using a simple field pH kit.

(*c*) Soil nutrient status, an assessment of which is very difficult to obtain readily, though it may sometimes be inferred from other soil properties (see (*a*) and (*b*) above). Sometimes, too, the vegetation itself will give an indication when viewed in the light of previous experience. It seems worth-while to ask for at least a crude estimate, though without any great expectation of learning whether a given soil is likely to be high or low in any particular nutrient.

(*d*) Information on the likely presence of injurious amounts of salt will be available in some instances, and it is therefore worth-while to ask that it be recorded.

(*e*) Colour, which may often give a useful indirect indication of other

Table 2.8. *A possible* pro forma *for soil characterization*

Site no .... Date .... Observer(s) ....

Location .... (include latitude & longitude)

Region .... Country ....

Rainfall .... Elevation .... Site topography...................

Local national classification of soil ....

Soil parent material ....

Profile depth (D) .... Profile drainage status (DS)...................

Colour, Texture (T), Acidity (A), Salt (S), Nutrient status (N)

Surface soil (SS) ....

Subsoil (S) ....

Other prominent features ....

Explanatory notes

(*a*) For site topography record in simple terms the topographic nature of the soil site, e.g., steep hillslope, flat valley floor, etc.

(*b*) Profile depth refers to the combined thickness of the A, B & C horizons, i.e., depth before reaching relatively unaltered parent rock or parent sediment. Depths to be coded as follows: 1, shallow (0.5 m or less); 2, medium (0.5–1 m); 3, deep (> 1 m).

(*c*) Profile drainage status represents an integrated expression of run-off, soil permeability and internal soil drainage. It is coded as follows: 1, excessively drained; 2, well-drained; 3, poorly drained. In excessively drained soils water is rapidly removed from the soil and no mottling is present throughout the profile. In well-drained soils water is readily although sometimes slowly removed from the soil. Mottling may sometimes occur in the lower B or C horizons. In poorly drained soils water is very slowly removed from the soil which may remain wet for considerable periods. Mottling, often with gley colours, is usually common in the lower A, B or C horizons.

(*d*) For colour (moist soil) use well known terms but not such adjectives as 'chocolate', 'cinnamon', etc.; also indicate if mottled or not, e.g., mottled red and grey.

(*e*) Texture is assessed by manipulation of moist soil in the fingers and is assigned to one of three classes: 1, sandy (up to sandy loam); 2, loamy (loam-clay loam; sandy-clay loam); 3, clayey (heavier than clay loam). If gravel or stone is prominent use subscript G, e.g. 1_G.

(*f*) Acidity can be determined by a simple field pH kit and assigned to one of three classes: 1, acid (pH 6.0 or less); 2, neutral (pH 6.0–7.5); 3, alkaline (pH 7.5+).

(*g*) The likely presence of harmful amounts of salt can be coded as follows: 1, No; 2, Possible; 3, Yes.

(*h*) Inferred nutrient status: 1, low; 2, moderate; 3, high.

(*i*) Other prominent features may include a number of readily observable properties of both surface and subsoil, or of the profile as a whole, e.g., thick organic A horizon, strongly structured B horizon, presence of ironstone nodules throughout the profile.

properties that are less readily obtainable, e.g., a very dark A horizon often indicates at least a moderate amount of organic matter.

Such information as listed above can be readily arranged into a *pro forma* and simply coded (see Table 2.8). To a soil scientist this could be of considerable value. In addition the recorder should be asked to record the kind of soil in terms of the classification in general use in his country or region. This would enable a specialist to gain additional valuable information from the entry.

It is necessary to stress again the effect of environmental factors on some of the soil properties to which reference has been made above. The climate of an area will obviously be an important modifier in relation to plant growth, more particularly through the physical properties of the soil and its water relationships. The prevailing climate, considered in relation to soil parent material and soil morphology, may also give a valuable guide to likely nutrient status. Thus a well-drained soil formed from a siliceous rock in an area of high rainfall will inevitably be highly leached and of low nutrient status. In some circumstances, however, the present climate may not explain the characteristics of very old soils formed in a very different climatic regime. In such instances the local soil classification may help to clear up apparent inconsistencies between recorded soil properties and the present climate.

Additional information relevant to soil features will already have been collected on the IBP/CT check-sheets under Landform (Geomorphology). This will often be invaluable in assessing such soil properties as drainage status and water relationships, susceptibility to erosion, and whether a soil is likely to be highly leached and therefore relatively infertile.

Several studies of soil and ecology from north-eastern Australia may be used to illustrate various aspects of the problem of relating plant communities to soils. In each case the local soil classification is given (that of Northcote, 1971) together with the Ball classification and records of certain readily observable soil properties such as have been discussed above. Only the dominant soils are considered.

Table 2.9 is a diagrammatic east–west transect of 420 km across the lower Cape York Peninsula at about lat. 15° 45′ S. It is based on studies by Pedley & Isbell (1971). Rainfall ranges from *c.* 1700 mm in the east to *c.* 900 mm in the middle and rising to *c.* 1200 mm in the west. It is extremely seasonal and summer-dominant.

At this scale in this region where plant communities are generally well defined, it is obvious that the Northcote system shows important soil differences between most communities. There are some exceptions. Thus two different eucalypt communities (6 and 8) may occur on Gn 2.14 soils, perhaps a reflection of some soil fertility factor: the alluvial levee soils of community 8 have a slightly higher plant nutrient status.

2.3.2 Problems of soil classification

Considerable difficulty was experienced in using the Ball classification. It is clear from Table 2.9 that there is not only difficulty in assigning soils to a single category but also several different soils appear not to be separable. There are, further, some soils which it does not seem possible to allocate to a category. Most of these difficulties arise from deficiencies already pointed out above. The soils in communities 1, 5, 6 and 8, for example, contain from 3% to 8% Fe_2O_3 and may or may not have ironstone nodules, raising the question of whether they are to be regarded as ferritic, assuming that the iron content could be estimated in the field. It is similarly impossible to distinguish between F_4 (brown earth) and F_5 (podsol) with the present key. The drainage definitions also cause trouble. Communities 3, 7, 9 and 10 all have poorly drained soils which do not necessarily provide evidence for it within 40 cm of the surface. Lastly there is no place for the soils of salt pans because these are saline but not alkaline.

Use of the readily recorded information suggested on p. 102, including the parent material of the soil, provides one or more differentiating features in most cases, though these may merely reflect unrecorded properties such as water-holding capacity or nutrient status. But again there are some plant communities which are not distinguishable by their soils, e.g., 3 and 9, though the poorer drainage of the soil of 3 may be significant.

Table 2.10 is a diagrammatic transect, only about 12 km long, near Tully in north Queensland, at lat. 18° S. It is based on unpublished studies by L. J. Webb, J. G. Tracey, R. F. Isbell & G. G. Murtha. This region has a mean annual rainfall of 4300 mm, summer-dominent but with only a short dry period of a few months. The Northcote classification again affords in general a ready separation in soil characteristics between the various communities, though there are several instances of apparent failure. The soils of units 5 and 6 key out the same, as do those of units 7 and 8, although very different communities are found on the two members of each pair.

Here too difficulty is experienced with the Ball scheme. Categories F_4 and F_5 cannot be separated by the key as it stands, and in any case both categories include some diverse soils. The *pro forma* code provides distinctive features for each soil but the significance of the differentiating characteristic, e.g., colour, may not be readily apparent.

Table 2.11 is a diagrammatic transect from a much drier region in central Queensland (lat. *c.* 22 S) with *c.* 550 mm rainfall, summer-dominant but highly variable. It is based on studies by Galloway, Gunn & Pedley (1967) and Gunn (1967). In this instance the Northcote scheme works well for some communities but less well for others. The Ball scheme does not fare at all well. The first two units may or may not be considered to have ferritic soils and, if not, it is impossible to allocate them unequivocally to F_4 or F_5. The soils of communities 4, 5 and 6 may or may not be considered to be poorly drained

Table 2.9. *Diagrammatic transect across Cape York Peninsula, Queensland*

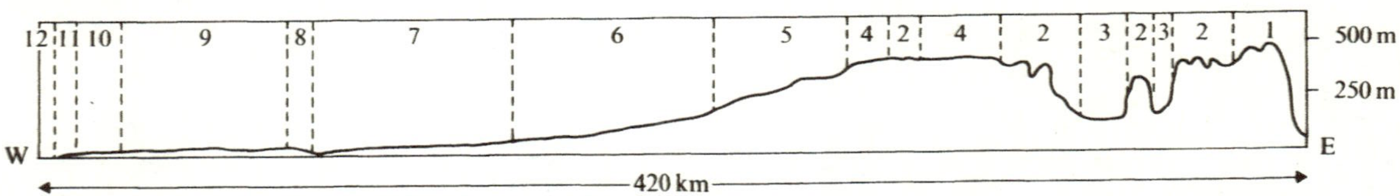

Community	Northcote key	Ball (1967)	Soil parent material	*Pro forma* code
1. Vine-forest	Um 4.41	F_4 or Fe	Slate, greywacké	D3 DS2 SS – Brown T2 A1 S1 N2 S – Red T2 A1 S1 N1
2. *Eucalyptus cullenii–E. dichromophloia* low woodland	Um 2.12	F_3?	Slate, greywacké	D1 DS2 SS – Pale brown T2 A2 S1 N1 S – Brown T2G A2 S1 N1
3. *E. leptophleba–E. polycarpa–E. alba–Melaleuca viridiflora* woodland	Dy 3.42–3.43	P_1 or P_2?	Alluvium	D2 DS3 SS – grey-brown T2 A1 S1 N1 S – mottled brown and red T3 A3 S2 N1
4. *E. dichromophloia–E. phoenicea* low open woodland	Uc 2.21	F_3	Sandstone	D1 DS2 SS – Grey T1G A1 S1 N1 S – Red-brown T1G A1 S1 N1
5. *E. cullenii–E. dichromophloia* woodland	Dr 2.11	F_4 or F_5, Fe?	Gneiss	D1 DS2 SS – Brown T2 A2 S1 N2 S – Red T3 A1 S1 N1
6. *E. tetradonta–E. sp. aff. polycarpa* open forest	Gn 2.14, 2.24	F_4, F_5 or Fe	Colluvium from sandstone, alluvium	D3 DS2 SS – Brown T1 A1 S1 N1 S – Red or yellow T2 A1 S1 N1

7. *M. viridiflora–Petalostigma banksii* low open woodland	Gn 2.94	F_5 or P_2	Alluvium	D3 DS3 SS – Grey T1 A1 S1 N1 S – Mottled grey and red T2 A1 S1 N1
8. *E. leptophleba–E. polycarpa–E. tessellaris* woodland	Gn 2.14	F_4, F_5, or Fe	Alluvium	D3 DS2 SS – Brown T1 A2 S1 N2 S – Red T2 A1 S1 N1
9. *E. microtheca–Bauhinia cunninghamii* low open woodland	Dy 2.33 Db 1.33	?	Alluvium	D3 DS3 SS – Grey T2 A2 S1 N1 S – Grey-brown T3 A3 S2 N1
10. *Panicum–Eriochloa–Sporobulus* grassland	Ug 5.28, 5.5	?	Alluvium	D3 DS3 SS – Dark grey T3 A2–A3 S2 N2 S – Dark grey T3 A3 S3 N2
11. Salt pans – some *Arthrocnemum* spp.	Uf 6.51	?	Alluvium	D3 DS3 SS – Grey T3 A2 S3 N1 S – Dark grey T3 A3 S3 N1
12. *E. papuana–E. polycarpa–Erythrina vespertilio–M. leucadendron* woodland	Ucl 11, 1.21	I_2	Dune sand and shells	D3 DS1 SS – Brown T1 A2 S1 N1 S – Brown T1 A2 S1 N1

Table 2.10. *Diagrammatic transect near Tully, North Queensland*

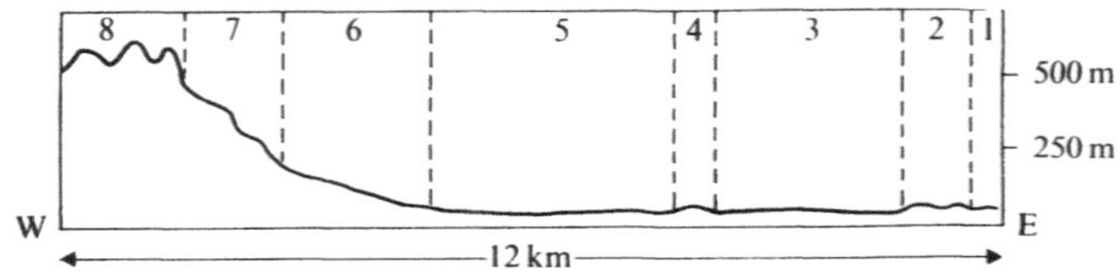

Community	Northcote key	Ball	Soil parent material	*Pro forma* code
1. Mesophyll vine forest ± *Eucalyptus tessellaris* and *Acacia aulacocarpa* var. *macrocarpa*	Ucl 22, 1.11	I_2	Dune sand and shells	D3 DS1 SS – dark brown T1 A1 S1 N1 S – yellow–brown T1 A1–2 S1 N1
2. *Melaleuca dealbata–Tristania suaveolens–E. tessellaris–E. intermedia* layered woodland (*c. Acacia* spp. and *Planchonia*)	Uc 5.11, Uc 4.21	F_3 or I_2	Dune sand	D3 DS 2–3 SS – dark grey T1 A1 S1 N1 S – yellow–brown T1 A1 S1 N1
3. *Melaleuca quinquenervia* open forest	Dg 4.41	P_2	Alluvium	D3 DS3 SS – black T2 A1 S1 N2 S – mottled grey and orange T3 A1 S1
4. Complex mesophyll vine forest	Gn 2.41–2.44	F_4 or F_5	Levee alluvium	D3 DS2 SS – dark brown T2 A1 S1 N3 S – brown T3 A1 S1 N2
5. *E. pellita–E. intermedia* woodland	Um 6.34	F_4 or F_5?	Alluvium	D3 DS2–3 SS – Dark grey-brown T2 A1 S1 N2 S – Yellow and red mottle T2 A1 S1 N1
6. Mesophyll fan palm vine forest	Um 6.34 Gn 3.74	F_4 or F_5?	Alluvium–colluvium	D3 DS2–3 SS – Dark grey-brown T2 A1 S1 N3 S – yellow–brown T2–3 A1 S1 N2
7. Mesophyll vine forest ± *Agathis* spp.	Gn 3.14, 3.74	F_4, F_5 or Fe	Granite	D3 DS2 SS – Dark brown T2 A1 S1 N2 S – Red or yellow T3 A1 S1 N1
8. *E. pellita–Casuarina torrulosa* open forest	Gn 3.74	F_5?	Granite	D2 DS2 SS – Dark brown T2G A1 S1 N1 S – Yellow T3G A1 S1 N1

Table 2.11. *Diagrammatic transect in Central Queensland*

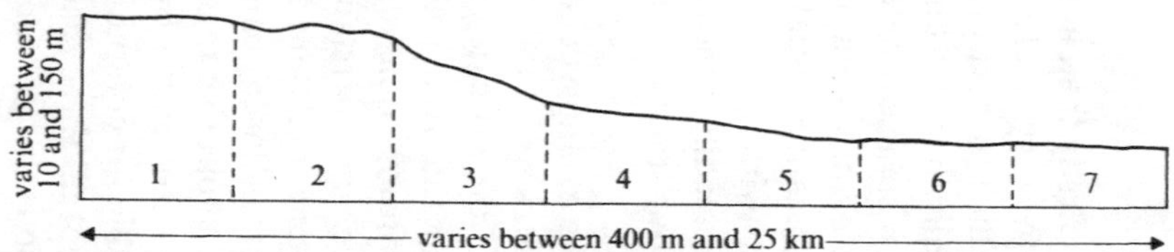

Community	Northcote key	Ball (1967)	*Pro forma* code
1. *Eucalyptus similis–E. papuana* woodland	Gn 2.12	Fe or F_4 or F_5	D3 DS2 SS – Red T2 A2 S1 N1 S – Red T3 A2 S1 N1
2. *E. melanophloia–E. crebra* open woodland	Gn 2.22	Fe or F_4 or F_5	D3 DS2 SS – Brown T1 A2 S1 N1 S – Yellow T2–T3 A2 S1 N1
3. *E. shirleyi* and/or *Acacia catenulata* low open forest	Um 1.2, Gn 2.12	I_1 or I_2?	D1 DS1–2 SS – Brown or red T2 A2 S1 N1 S – Red T2G A2 S1 N1
4. *E. normantonensis–E. thozetiana* woodland or open woodland	Dr 3.12, Db 1.43, Dy 2.22	F_5?, F_2?	D1 DS2–3 SS – Grey-brown T1 A2 S1 N1 S – Grey, brown or red T3 A3 S2 N1
5. *E. populuea–Eremophila mitchellii* open woodland or woodland	Dy 3.22, Db 1.13, Dy 3.43	P_1 or P_2?	D2–3 DS3 SS – Grey-brown T1 A2 S1 N1 S – Mottled grey and brown T3 A3 S2 N1
6. *E. cambageana–A. harpophylla* low open forest	Db 1.13, 1.33; Dd 1.13, 1.33	F_2?	D3 DS2–3 SS – Grey-brown T2 A2 S1 N2 S – Dark brown or dark grey T3 A3 S2 N2
7. *A. harpophylla* and/or *A. cambagei* low open forest	Ug 5.24	?	D3 DS2–3 SS – Grey-brown T3 A3 S2 N2 S – Grey T3 A3 S3 N2

by the definition, and as the soils may or may not be calcareous at depth they may fall into categories F_2, F_4, P_1 or P_2. The usual difficulty in allocating the deep cracking clay of 7 is again experienced.

In this transect the parent material is a variably truncated, deeply weathered, mainly argillaceous sediment. Many of the soils are extremely old and hence relationships between soils, parent mateials and present climate are not clear-cut. Even when parent material is neglected, it is, however, still possible to find some useful differentiating criteria by recording some simple morphological properties of the soils.

Conclusion

The evidence from these three examples suggests firstly that, at least under Australian conditions, the Northcote scheme is very useful for differentiating the soils of different plant communities. This is not surprising since the scheme was developed for use with Australian soils. It is also known from personal experience, however, that there are many instances where the same Northcote category will support widely different plant communities even within an area of uniform climate. This is not necessarily a weakness peculiar to the Northcote classification: it may occur with any general classification of soils. Even in such cases, however, it is readily possible to subdivide Northcote's key categories and thus to bring out some important aspects of soil–plant relationships.

The Ball scheme did not in general fare very well in the examples chosen, mainly because it cannot cater adequately for a world-wide range of soil types, though there are also some deficiencies in the definition of the key categories. It is likely to be adequate for a particular range of temperate-zone soils.

It is not claimed that the suggested procedure of recording some readily obtained but important soil properties will greatly advance our understanding of plant–soil relationships. At best it may enable some generalizations to be made for particular combinations of environmental factors including climate. But it does have the advantage of being somewhat less subjective than the Ball scheme. It can be applied to all soils, it should be used readily by the non-specialist, and the data are collected in a form suitable for computation and analysis by numerical methods. A major weakness of all three schemes is the difficulty of assessing the plant nutrient status of soils. In the transects described above an assessment is possible, but often through calling on past experience. Whether a non-specialist recorder could make a useful judgement is problematical. He might often be unable to do so on the basis of the recorded information, but in many instances it may be possible for him to seek a better qualified opinion from a local environmental specialist.

2.3.3 Further comments on soil recording for biological surveys

In the preceding section Dr Isbell has considered the many unresolved problems confronting those who seek a recording system for soils which will be suitable for use in undertakings such as the check-sheet survey. He draws attention to some shortcomings of the Ball classification, in particular its inadequacy for dealing with certain soil-types not found in the temperate zone and, more generally, he regards it as unlikely to contribute in any significant way to our understanding of plant/soil relationships.

The organizers were well aware that much more information about the soils of survey-areas would be desirable, but decided that it would be unrealistic to ask all recorders to supply data requiring expert knowledge of soils and especially quantitative data dependent on laboratory determinations. On the other hand they did not wish to dissuade those able to provide such information from doing so, as is made clear on pp. 23–6 of the *Guide to the Check Sheet for IBP Areas*. Here rather fuller guidance is given than is to be found in the field instructions accompanying the check-sheet. Recorders are invited to 'describe the soil on which each plant association is found in terms with which they are familiar', using the right-hand column of p. 4 of the check-sheet for this purpose. The descriptions 'should normally include some or all of the following factors, namely: parent material, texture, colour, base status, water regime (whether waterlogged or not), humus content, depth and horizon sequence'. It is then pointed out that much of this information can be conveyed very concisely by reference to the appropriate category in any widely known soil classification, an example cited being the term *gley-podsol*. This is defined by Kubiena in his *Soils of Europe* (1953) as 'a strongly acid, nutrient-deficient, ABC soil which generally shows raw humus or dystrophic moder formation as well as...a strongly pronounced bleached layer in the topsoil and one or more enriched layers in the subsoil', exhibiting also 'partial waterlogging and gleying'.

This is excellent as far as it goes, but the *Guide* recognized that inevitably problems would result from the use of many different systems of classification and the consequent variations in scope and terminology in the information supplied. It was, therefore, necessary to ask in addition that all recorders assign soils to their place in some single classification which, apart from its value in other respects, would serve as a basis for sorting and storing the survey data concerning soils. The system selected for this indispensable function, for reasons already explained (p. 95), was that proposed for the purpose, and at short notice, by D. F. Ball. In fairness to Dr Ball it must be recognized that the ultimate categories in his 'classification' are merely different combinations of answers to questions judged suitable for non-specialist recorders and framed to elicit information highly significant for site-assessment. The level of detail is such that virtually nothing is lost in the process of classification. Appendix 6 reports on the favourable response from recorders and explains how information on soils was handled at the data-centre.

It will be clear that the difficulties facing the organizers of the check-sheet survey would have been far less troublesome had there been available a soil classification at once applicable on the world-wide scale, and by non-specialist recorders, and also useful for the purposes of the survey. Dr Isbell has explained that there is still no such generally acceptable system, and the problem for those organizing future surveys remains largely unchanged. The suggestion made by Dr Isbell, that surveyors should be asked merely to record a standard set of features for all soils encountered, without attempting to assign each soil to its place in any particular classification, seems therefore to merit serious consideration. This is a proposal analogous to that made for recording vegetation-types in biological surveys (pp. 92–3), and it has comparable advantages in leaving it open to the centres handling the collected data to use the information in the most appropriate ways.

The basic problem is how to select the categories of information to be recorded, having in mind the dual requirement that the resulting data shall provide an adequate basis for meeting the aims of the survey and yet make no unreasonable demands on recorders who, for some time to come, will commonly have had no specialized training in soil science. Some attempt ought also to be made to incorporate sufficient flexibility to allow for possible changes arising from future research findings. Dr Isbell has devised a *pro forma* (p. 103) to meet these objectives, but it is appreciably more demanding than Ball's key and gives no guidance on the basic problem of the location of sampling points. A much more ambitious scheme is set out by Howard, Lindley & Hornung (1974). There is no doubt that a very great deal of further work will be necessary before a wholly satisfactory solution can be found. It is arguable that for the immediate future attempts should be made to draw up two variants of the same basic *pro forma*, the one making smaller demands on the time and expertise of recorders than the other, but the two being sufficiently related in design to enable a common initial level of analysis to be carried out, the analysis proceeding further where the more detailed and demanding variant has been selected.

Finally it may be supposed that the aims of future biological surveys will continue to be to collect information about survey-sites primarily as a basis for accurate assessments of their land-use potentialities, including their value for nature conservation as well as for commercial forestry and for various forms of agriculture. At the present time, information about certain features of the soil will often assist directly in making these assessments and will also serve to increase understanding of the indicator-value of all types of natural and man-modified vegetation. The experience of countries in which a highly developed floristic classification of vegetation (like that of Braun-Blanquet or related systems) has been in general use for some time is that very satisfactory assessments of land-use potential can be made without recourse to any detailed information about the soil. It is claimed, indeed, that the characteristics of the soil can to a very large extent be inferred from the vegetation.

We may conclude that in order to further the objectives of biological surveys there is urgent need in the first place for the more complete identification of those properties of soils that can be readily recorded in the field and will contribute substantially to the comprehensiveness and accuracy of site-assessments in relation to potential land-use. These should form the basis of as simple as possible a key, or *pro forma*, along lines suggested earlier by Dr Ball and more recently by Dr Isbell. In the second place the correlations between recorded features of soil and vegetation should be studied at data-collecting centres for biological surveys, in the expectation that very little direct information about soils will prove essential for the purposes of the surveys, as soon as adequate classifications of vegetation have been in universal use for long enough for their indicator-value to be understood clearly. Meanwhile, however, the use of a more detailed variant of any adopted key or *pro forma* would make it possible to broaden the scope of correlations between features of soil and vegetation and so test, more thoroughly than would otherwise be possible, the need for retaining direct recording of soil data. At the same time some of this additional information might contribute to soil science in a valuable way.

2.4 Other recording problems

Recording problems arose for the organizers of the check-sheet survey in connection with matters other than vegetation and soil. It was important to collect and store, for each survey-site, information on a number of features of an amount and kind that would both be adequate for the aims of the survey and allow useful comparisons with other sites. Some of the requisite data, such as those on precise location, altitude, aspect and slope, were readily obtainable either by the surveyors on the site or from sources such as maps; and here there were no real problems about the way in which the information should be recorded. Information on the climate of a survey-site was a very different matter. Only where full and long-term meteorological observations had been taken within or very close to the site could wholly adequate information be recorded for aspects of the climate of most significance for the purposes of the survey. The required on-the-spot information was in fact very rarely available and recourse had to be made to what could be obtained from the nearest meteorological stations, often at a considerable distance away and not always recording a sufficient range of climatic variables. It might then be necessary to rely on interpolation from national meteorological records and maps, or from maps showing the world distribution of climatic types, taking account of the general situation and topography of the site and its surroundings. It is interesting, in view of these difficulties, to note that a Technical

Meeting in the data-centre at Monks Wood recommended that climatic information should not be required for the check-sheet but might be sought later on a separate sheet.

Two points are noteworthy about climatic recording in the survey and, indeed, in future biological surveys. In the first place the responsibility for supplying the required information cannot lie to any important extent with the site-surveyor. It must be the responsibility of the local organizing body to obtain and record such relevant information as is available, whatever its source, and in doing so to ensure consistency of treatment throughout. The same allocation of responsibility should be made wherever the information to be recorded must come, not from direct observation on the site, but from any external source or sources.

Secondly, the *Guide* suggests that inferences as to the climate of a survey-site made from 'a published classification of climate and maps showing the world distribution of each climatic type' can be inserted by the data-processing centre. This is in line with the view, expressed earlier in connection with records of vegetation and soil, that it is inadvisable merely to invite the allocation of such complex features of ecosystems to their places in recognized classifications. It is usually preferable to ask for certain specific facts, either exclusively or in addition to the assignment of a class. The facts can then be translated at the data-centre into the terms of any classifications that may at the time be thought useful and still remain stored for future use.

An important major problem of recording through the medium of answers to questions on a check-sheet is one to which reference is made both in Chapter 1 (p. 14) and in the detailed analysis of answers in completed check-sheets in Appendix 6 – that questions inviting the choice of one out of two or a few stated possibilities may impose discrete classes where there is in fact continuous variation. Questions have been framed in this way for very good reasons (p. 11), but they have been found to cause difficulties for recorders asked, for instance, to class water as fresh, brackish or salt; landscapes as flat, undulating or hilly; and species as rare, threatened or relict. An important step towards a solution would be for local co-ordinating committees to ensure, through their training programmes, that all recorders use the terms for water of different grades of salinity, landscapes of varying degrees of relief and the like, consistently and similarly, and to explain their practice to the international data-processing centre. The terms descriptive of the status of species within a country fall into a different category. Here, editing by the local committee could correct errors of understanding or judgement on the part of the recorders, though this raises doubts as to whether this kind of question should in fact be amongst those left to be answered by recorders in the field. There must, indeed, always be some of the information sought about a survey-site which is best provided from external sources of reference, or by local experts, and the local committee should assume the responsibility for providing it.

3. Data-processing and the storage and retrieval of information

G. L. RADFORD

3.1 Introduction

The IBP/CT check-sheet was developed to further the declared objective of examining the range of ecosystems over the world and assessing the extent to which scientifically adequate samples of all the main types and their significant variants were already protected. It was designed to enable this information to be collected readily in the field and, as far as possible, to be processed economically and accurately. The earlier versions were designed for processing on a punched-card installation using a coding system proposed at the First General Assembly of IBP in 1964. Subsequent revisions of both check-sheet and coding system were given trial runs which led to the adoption of the Mark VII check-sheet in the standard *Level 2* form as the final version for use in the survey. In the years 1967–69 the main emphasis in the survey was on the distribution of the check-sheets, the appointment of National Organizers in as many countries as possible and the collection of completed sheets in sufficient numbers to enable trials of processing methods to begin. The Nature Conservancy offered the facilities of the Biological Records Centre at Monks Wood Experimental Station, near Huntingdon, UK, for handling the collected data, with advice from the Smithsonian Institution in Washington. An outpost of IBP/CT was therefore created at Monks Wood to receive completed sheets and to develop the necessary techniques.

The information in completed check-sheets was to be stored so as to furnish answers to questions on the status, content and interest of individual IBP areas and to serve as a permanent and continuously updated source of information about them which would constitute the start towards a more sophisticated system capable of handling a wider variety of data. It would also form a basis for the analysis of survey results, including investigations of the outcome of decisions taken in drawing up the final version of the questionnaire.

Three main types of enquiry about IBP areas were envisaged: What is known about a particular site or series of sites? In what sites does a particular feature or group of features occur? What emerges when information on a particular feature is summarized for all sites falling into a specific category (e.g., information about soil, vegetation or types of human impact in all sheets from a given country, a given altitude or a given climatic region)? In order to deal with enquiries of these kinds the data-base had to be capable of the following services:

(i) To provide basic site-documentation for all areas already protected as nature reserves or in some comparable way, and for additional areas proposed for similar protection, and in particular for use in the preparation by the IUCN of future editions of the *World Directory of National Parks and other Protected Areas.* It could also incorporate data collected in other surveys concerned with nature conservation or dealing with particular habitats and biomes. In this way all relevant information on sites of interest to conservationists would be made available at a central point to executive organizations and in a form permitting easy retrieval and facilitating comparisons. This was essentially a long-term objective.

(ii) To provide information of immediate relevance to conservation-action by government departments and international agencies like IUCN. This would be based on analyses of data at both national and international levels and would draw attention to deficiencies in the coverage of protected ecosystems in well-surveyed regions and assist in showing in what parts of the world consideration should be given to the need for further surveys.

(iii) To provide a data-store of increasing long-term scientific value with increase in its coverage of those countries which at present have little-developed conservation programmes, and of ecosystems, like tropical forests, about which little is known so far.

(iv) To provide data on the distribution and protection-status of threatened species, in particular those listed in the series of IUCN *Red Data Books.*

(v) As a short-term objective, to provide material for summaries of the work of the CT Section of IBP and for a critical assessment of its achievements.

In order to meet the above objectives, and therefore to form the basis for planning the data-handling procedures, it was regarded as fundamental that the system should:

(*a*) include all the information from each check-sheet;

(*b*) provide for the reproduction of check-sheets, in whole or in part, in answer to requests from a wide range of potential users;

(*c*) allow, as far as possible, for future needs to expand and revise the system;

(*d*) be, as nearly as possible, independent of any particular type of machine, so as to minimize problems of future transfer and compatibility;

(*e*) allow data-preparation to be kept simple and sufficiently rapid;

(*f*) check the data for logical consistency so as to reduce the need for preliminary editing and verification;

(*g*) be so adequately documented as to be usable by other organizations and individuals with approved access, and

(*h*) be managed with an efficiency that ensures minimal expense.

Periodic revision of the information would be sufficient, so that the system need not cater for continuous updating.

A sufficient number of completed check-sheets had been received by the autumn of 1969 to allow trials of the proposed data-processing system using the existing machinery of the Biological Records Centre. A sample of 200 edited sheets was taken and the information punched on standard IBM 80-column tabulator cards and also on paper tape for filing as an interim data-bank. The trials showed that the potential of the system was too limited and it was decided to adopt instead a computer-based system using magnetic tape for storage.

The advantage of an easily accessible computer, with the possibility of advice and assistance from consultant staff and the availability of a consultant systems-analyst and programmer familiar with the system, weighed heavily in favour of using the ATLAS 2 computer of the Computer-Aided Design Centre at Cambridge, UK. The case was strengthened by the registration of the Biological Records Centre as a user and the installation of an on-line console at Monks Wood, and in due course IBP/CT was registered as an independent user. There were nevertheless some disadvantages in using ATLAS 2, especially from the standpoint of compatibility and system design. The problem of compatibility was reduced by adopting FORTRAN as the programming language, even though this involved some sacrifice of efficiency in the data-bank.

3.2 Preparation of completed check-sheets for processing

A number of clearly defined stages are involved in the preparation of check-sheet information for storage in the data-base. They all involve processing the data in various ways but throughout the project the term 'check-sheet processing' has referred to the final computerized stage and this usage will be continued here. The whole series of preparatory stages is represented in the flow-chart shown in Fig. 3.1.

3.2.1 The serial prefix and the 'geo-code'

On receipt of a completed check-sheet it is first assigned a unique identifying prefix in two parts, the first based on the geo-code and the second an accession number. This prefix is used for both office and machine filing, each check-sheet and its supporting maps and literature bearing it.

The geo-code was devised by S. W. Gould, of the International Plant Index, Connecticut Experiment Station, USA, for indicating specific areas of land or ocean in a suitably concise way for handling by machine. It was initially designed to store coded information, for a museum record, on the place of origin of plant specimens and the geographical distribution of the taxa to

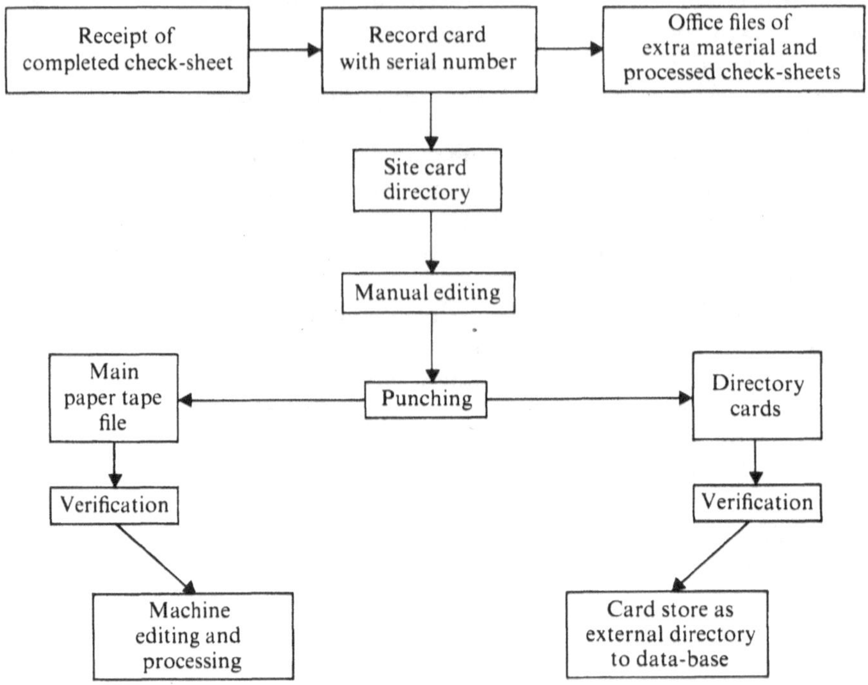

Fig. 3.1. Flow chart showing stages in the preparation of check-sheets for processing.

which they are assigned. It is based on a series of 4 × 4 rectangular grids, the first covering the entire surface of the globe but with the representation of land-areas favoured at the expense of ocean areas: islands on the continental shelf are treated as parts of the adjacent mainland. Fig. 3.2 shows how the letters K–Z label the sixteen squares of the primary or first-level grid and how the single-letter coding of the main land areas is made to correspond fairly closely with their location and relative size. The Pacific Ocean, on the other hand, is assigned only two grid-units, no more than the much smaller land-areas of Africa and North America.

The letters A–I of a 3 × 3 'overlay grid' (Fig. 3.3) may be used at this or at any subsequent stage to indicate either the whole set of sixteen grid-squares (E) or groups of four or eight contiguous grid-squares. The letter A indicates the four squares in the north-west corner, (K, L, O and P), and C, G and I indicate the remaining corner sets of four; while B and H refer to the eight squares of the northern and southern halves respectively, and D and F those forming the western and eastern halves.

Each unit of the first-level grid is next subdivided into sixteen second-level units which are assigned the same pattern of letters as before. Fig. 3.4 shows

K Canada Alaska	L Arctic Seas Atlantic Ocean (North)	M Europe	N North Asia
O United States	P Central America	Q Asia Minor	R South Asia
S Pacific Ocean (North)	T South America	U North Africa	V East Indies Australia New Zealand
W Pacific Ocean (South)	X Atlantic Ocean (South) Antarctica Antarctic Seas	Y South Africa	Z Indian Ocean

Fig. 3.2. The primary grid of the geo-code.

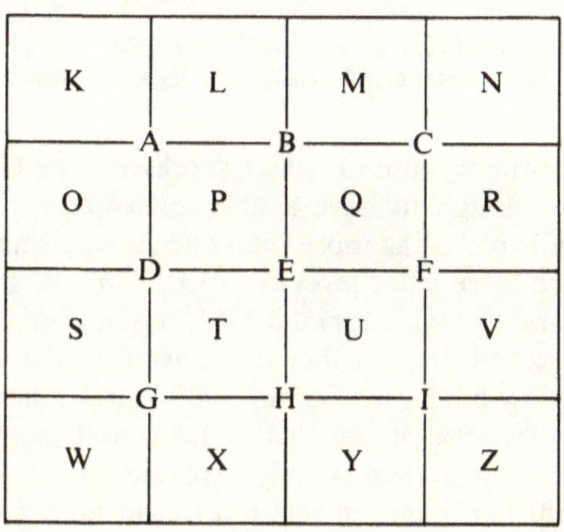

Fig. 3.3. The primary and overlay grids of the geo-code.

OK Washington A Oregon C California G Nevada I	OL Minnesota B Iowa H	OM Wisconsin A Michigan C Ohio G	ON Vermont Q New Jersey G New York A Maine N New Hampshire M Massachusetts R Connecticut U Rhode Island V
OO Idaho A Montana C Utah G Wyoming I	OP North Dakota A South Dakota C Nebraska H	OQ Indiana F Illinois D	OR District of Columbia Q Delaware R Virginia H Maryland C Pennsylvania A
OS Colorado C Arizona G New Mexico I	OT Kansas D Missouri F	OU Kentucky H West Virginia B	OV North Carolina B South Carolina H
OW Texas E	OX Arkansas C Louisiana I Oklahoma G	OY Tennessee B Alabama I Mississippi G	OZ Georgia B Florida I

Fig. 3.4. Subdivision of primary grid unit O, the USA, giving three-letter coding of all the states except Alaska (KKE) and Hawaii (SQH).

this subdivision for primary unit O, which represents the USA. At this level only Texas is large enough to have a unique two-letter coding, all other second-level grid-units including more than one state. Using the overlay grid Texas is coded at the three-letter level as OWE, while Wisconsin, Michigan and Ohio, which share the two-letter code OM, can be differentiated as OMA, OMC and OMG respectively. Further subdivision to the fourth level gives grid-units representing land areas of *c.* 3000 km², rather larger than an average county in the USA or the British Isles, and the process could be extended if required. The system is fully explained in two papers by S. W. Gould himself (1968, 1972) and, in relation to the survey, by F. H. Perring in an appendix to the Guide (Peterken, 1967).

The geo-code system has proved extremely useful in operating the data-base for the survey, both for office filing and also for computer-based storage and retrieval. National codes were used for filing except for those larger countries, like Australia, Canada and USA, from which high numbers of completed check-sheets were expected and for which coding was therefore at provincial or state level. All documents, including check-sheets, specific to IBP areas carry the appropriate serial prefix which combines the national or state coding with an accession number. Thus the first check-sheet received from Poland is filed under MUBE001, where the appearance as fourth letter of one of the letters A–I, in this case E, indicates the more detailed codes to be searched for when all information relating to Poland at the regional level is required.

Some IBP areas were subdivided, as recommended in the *Guide*, and this is indicated by symbols after the accession number. The letter M indicates a summarizing or 'master' check-sheet relating to the whole of the IBP area, e.g. OLBE 15M, while subsidiary or 'slave' sheets carry the letter S followed by the particular number of the subdivision, e.g., YQBE02S1, KUEE62S4. The length of the serial prefix is kept to eight characters, the limit for file codes on the ATLAS computer. The serials of all 'slaves' were added to the serial prefix for each 'master' check-sheet.

The geo-code was designed primarily for use on punch-card machinery and the development of procedures for sorting and reading cards has led to a high degree of versatility. The IBP/CT data-base was, however, established on a computer and the following comments relate mainly to that situation. In the data-base itself the serial prefix is used to designate the document-file in all operations, independently of its additional use in the check-sheet as one of the data-items. Once the country has been identified successfully by its national coding, the required state or county can be sought by unpacking the appropriate data-item. This still involves only four letters of the geo-code, whereas abbreviations of proper names to four letters would give a high probability of mis-sorting.

Although the geo-code system has proved useful in a variety of ways, there are nevertheless certain problems which have emerged during its use for the survey. The first reference volume became available in 1968 and included detailed codes for North, Central and South America. It had an index giving full regional and national codes for the rest of the world, including the oceans. Coding for the survey had to be started before the delayed appearance of the second reference volume (1972) in which there were revisions of land codes for the Old World and of oceanic codes. This resulted in the use in the IBP/CT data-base of some codes which differ from those in the second volume. Some change of practice in the use of the letter J has also been confusing. This letter, which occurs neither in the base grid nor in the overlay grid, was proposed by F. H. Perring for use in the fourth position to indicate an area of ocean exclusive of the islands in it, the islands themselves being given the normal land coding. In the index to the first volume, however, certain land areas in Borneo were assigned codes ending in J, though they were subsequently given conventional land codes.

3.2.2 Recording and editing the completed sheets

After a check-sheet had received its serial prefix an office record-card was completed for it, this showing the prefix and also the name and international conservation-class of the site. Any material appended to the check-sheet was also noted on the card. If this material was bulky or liable to damage it was marked with the serial prefix and removed to the office files, where it was stored with the check-sheets that had already been processed. For each site and its subdivisions, too, an 80–column tabulator-card was punched with the name of the site and its serial prefix. These cards constituted an up-to-date file for producing site-lists for reference and circulation.

The check-sheets were then edited for processing, special attention being given to their clarity, completeness and consistency. Most were typewritten but some were in longhand script and needed tidying up so that punch-operators unfamiliar with the technical language could read them without difficulty, and in some the general standard of legibility was very low. Spelling errors, especially in species names, were corrected as part of this early stage of editing.

For some contributions arrangements had to be made for translation. Sheets had been circulated in English, French and Spanish and translations were prepared in German, Polish, Russian, Serbo-Croat and Slovenian. Completed returns were received in Czech (175), Polish (167), French (160), Spanish (25), Russian (17), German (10) and Portuguese (3), the remainder being in English.

Each answer, if left incomplete by the surveyor, was made as complete as possible. Missing altitudes were entered from maps, material was transferred from appended notes, percentages were calculated for all rows and columns of the two-way table of types of land-relief, and other items were added or expanded wherever possible. Particular attention was given to the addition in some sections of coded or summarized information that would facilitate sorting and retrieval ('descriptors'). For example, in the section on the location of the site geo-codes were added for the country, state and county, or their equivalents; and in the soil section, summarizing information on the type, texture, colour, drainage, reaction and depth of the soil were extracted from the 'other notes' entry. For species named as of outstanding interest in the site (section 14, nos. 3 and 5) their taxonomic group and recorded status were added to the names. From entries to section 15, on 'exceptional interest' of the site, noteworthy items were extracted and summarized under eight different headings. To each specific type of human impact listed in the entry for section 16, too, was added wherever possible a brief annotation based on any additional notes appended to the check-sheet, and a summarizing statement for the whole section might also be added.

The third aim of the editing, consistency, involved the most difficult problems, but straightforward routine checks were performed at a later stage by use of the computer. The first check was on the Fosberg vegetation coding assigned to each plant community. In some instances codes were obviously inappropriate and had either to be replaced or omitted. The main reasons for this were that surveyors either failed to understand the plan underlying the Fosberg system or tended to cite the most nearly suitable code instead of indicating that a plant community did not fit into the scheme as circulated in the *Guide*. The next check was to ensure that equal numbers of entries were made in sections 7 and 8 under the three headings 'vegetation-types', 'soil-types' and the 'occurrence of similar communities elsewhere' in the country or state. If, for example, data on soil were available for only some of the plant communities, the number was adjusted by means of blank entries. The section on fresh water, too, had to be examined closely to ensure consistent use of the boxes.

One aspect of consistency gave particular cause for concern. Only one question in the check-sheet asked for an explicit indication of the *absence* of a feature. Answers to other questions raised the problem of distinguishing between the two classes of negative information: 'absent' and 'not known'. The questions to which this applied most seriously were those dealing with natural features that might or might not be present in a site: coastline, fresh, brackish or salt water, and adjacent water-bodies. These may be checked by reference to available maps, but the outcome is rarely clear-cut. There is further complication with questions inviting subjective judgements, such as on 'outstanding floral and faunal features', 'exceptional interest' and 'significant human impact'. The question concerned, or part of it, may be left unanswered because there is no available information or because the relevant feature has been stated, or judged, not to be present. In the latter case a whole series of factors may contribute to the subjectivity of the answer, including the background knowledge and personal interest of the surveyor, the lengths to which he goes to seek information in the field or from records, the completeness of the records in the literature consulted, and the surveyor's interpretation of the facts to hand in relation to the questions asked, particularly with respect to the 'degree of importance' of features.

The analysis of negative entries for whole sections showed that almost as many were left blank as were answered 'not applicable' or 'feature absent'. The policy for dealing with these was not wholly consistent and could have been improved when rules for processing were being established. The simplest measure would have been to introduce a set of indicators for positive, negative or blank entries to the sections or parts of sections concerned. This would have added significantly to the complexity of data-handling at all stages, and it seems questionable whether it would have real value. On the basis of the first 200 sheets received it was deemed unjustifiable to separate negative and blank entries, and the convention adopted was to discard answers without a positive entry except for those to section 11, dealing with fresh water: if this was indicated as absent the whole section was entered as a set of negative binary items. The fact that returns with negative and blank entries mounted respectively to 7% and 9% of the total seems, in retrospect, hardly to justify making an exception for section 11. The root of the problem, it must be conceded, lay in the design of that part of the check-sheet.

3.3 Problems of erroneous and incomplete information

Completed check-sheets were passed to the Data Centre either directly or through a national co-ordinating centre at which detectable errors were corrected. The thoroughness of such checks varied a great deal and on arrival at the Data Centre the sheets still covered a wide range in accuracy and completeness. While some errors of fact or omission were corrected in subsequent editing, some remained undetected and others might be introduced at any stage of preparation for processing. Their origin, nature and seriousness vary considerably, and the discussion below includes estimates of the likelihood of their detection and their effect on analyses if uncorrected.

3.3.1 Errors occurring prior to editing

Errors originating at the stage of data-collection and subsequent organization into a completed check-sheet may be divided into those of omission and those of fact. As has been pointed out in section 3.2.2, the check-sheet called for no consistent distinction between entries left blank because a feature was definitely absent and those omitted for any other reason, such as unavailability of information. Account can only be taken of positive records in such instances, and there is no reliable basis for assessing the extent to which there have been blank entries due to omissions or the seriousness of their nature. Their existence must be accepted philosophically.

Errors of fact arising during data-collection derive from the misinterpretation of questions and the misrepresentation of situations, the former exemplified by mistakes in Fosberg coding and in entries to the section on fresh water, referred to in section 3.2.2 above. The chance of detection of these errors is high, but the full correct entry can by no means always be reliably reconstructed. Misrepresentation may arise from misinterpretation, omission or careless slips, but a further source is the inappropriate choice of alternatives made in good faith by surveyors with inadequate background knowledge. This is rarely detectable and in the absence of organized controls in the survey no information is available as to its likelihood of occurrence. The seriousness of its influence on the interpretation of data must be considered high for detailed analytical studies.

Careless errors of spelling and the incorrect placing of ticks during data transfer or copying may be trivial or serious. In general, mistakes that pass undetected are trivial only when the sense remains unaltered. The spelling of generic and specific names is critical at the beginning of the word but less important elsewhere. While efforts are made to check at least generic names of unfamiliar species at the Data Centre, only those that appear irregular in some way are fully checked. Species names from all sections of the check-sheet are punched on tabulator-cards as part of the operation at the Data Centre and spelling errors likely to cause mis-sorting can be quickly detected in a final check by eye. A second record of a species that could not be found in the normal reference-works may reveal an error in spelling that immediately clarifies the situation.

3.3.2 Editorial errors

Errors may arise during editing for two main reasons, misinterpretation of surveyors' entries, and carelessness. Misinterpretations may be due to illegibility, to confused entries resulting in incorrect decisions by the editor, and to misunderstanding of surveyors' notes when assigning descriptors.

Legibility varied from the low standard of notes made in the field to neatly copied or typewritten information entered in the comfort of an office. Poor photocopying was often to blame for subsequent errors, even where the original was in typescript, and a number of check-sheets had whole pages completely illegible because of this. Corrections and alterations, too, especially arising from confusion or uncertainty in the field, may lead to misinterpretation and consequent errors during editing. In the sections of the check-sheet on soils and on features of exceptional interest, where coded or summarizing descriptors are added to facilitate sorting on the principal items in the entry, the context or emphasis of an entry may be misunderstood and a descriptor omitted or wrongly chosen: it would normally be omitted if there were clear grounds for doubt. Such errors can rarely be detected later.

Careless errors during editing include those of omission, where an obvious mistake on the part of a surveyor passes undetected, as well as of commission, as when wrong descriptors are added. These errors are normally detected during manual or machine verification and consistency checking.

3.3.3 Errors introduced by data-clerks

The stages of punching and verification by data-clerks are further potential sources of error. The clerks are often unfamiliar with the wide range of

Table 3.1. *The likelihood of occurrence and detection of check-sheet errors and the seriousness of their escaping detection*

Source	Nature	Likelihood of occurrence	Likelihood of detection	Seriousness
Surveyor, co-ordinator or copy typist	Omission	Uncertain	Low	Uncertain
	Errors of fact			
	misinterpretation	Low	Rather high	High
	misrepresentation	Uncertain	Low	High
	careless errors	Rather high	Variable	Variable
Editor	Misinterpretation and misrepresentation	Rather low	Moderate	High
	Careless errors	Rather low	Variable	Variable
Data-clerk	Misreading edited or unedited entries	Moderate	High	Variable
	Careless errors	Rather low	High	Variable

technical and scientific terms and in particular with the species names used in check-sheet entries and errors due to misinterpretation are frequent, especially when legibility is poor. Verification by a different clerk may not help to any degree and for this reason the checking of card-files for incorrect species-names is of special importance.

Table 3.1 summarizes this consideration of errors in check-sheets and gives some indication of their relative frequency and of the likelihood and seriousness of their escaping detection.

Approximate scales used for the indication of likelihood of occurrence of an error and its detection are the following:

Likelihood of occurrence		*Likelihood of detection*
1 in 5 check-sheets	High	1 in 2 errors
1 in 10 check-sheets	Rather high	1 in 5 errors
1 in 50 check-sheets	Moderate	1 in 10 errors
1 in 100 check-sheets	Rather low	1 in 20 errors
1 in 500 check-sheets	Low	1 in 50 errors

These scales must be considered in relation to the number of separate entries in each fully completed check-sheet. There are nineteen sections in the questionnaire, most of them divided into subsections and subdivided at up to five further levels. Each level may invite a free-style entry for which a descriptor must be supplied, and it is estimated that an average check-sheet comprises 4400 characters as input for computer-storage.

The final document for filing in the data-base has thus passed through a series of stages at each of which there is opportunity both to correct and to introduce errors. The general standard of error-detection is such that major efforts persist infrequently in relation to the total number of data-items in the system. The standard of verification is nevertheless not as high as could be achieved with more resources. Because the errors often involve scientific terminology the task of verification must fall heavily upon the data-editor, and he becomes in consequence the factor determining the rate of progress. The same is true of checking cards for residual spelling errors.

Once the check-sheets have been processed the data files may still contain errors although all the internal inconsistencies will have been removed. The only inconsistencies remaining will be those arising between check-sheets: a plant community may, for example, be entered as having no other known occurrences in one check-sheet, but as occurring occasionally elsewhere in another. Such discrepancies become apparent during comparative studies and, in the interest of unifying a national contribution, changes in processing may need to be made. The residual errors are classed as minor or major according to whether or not they would lead to incorrect sorting during information-retrieval. Major errors are detected by their effect on analyses. A list of minor errors is kept and major ones are corrected immediately. Because correction involves reprocessing, except in the case of character substitutions, the appropriate changes to remove minor errors are made only if the opportunity arises during updating.

From the check-sheets that were returned it soon became apparent that the degree of completion would vary from the barest minimum to a full set of answers with a great deal of supporting information. Therefore it was necessary to set some minimum level of completeness for incorporation into the data-base. Eventually it proved necessary to declare all but the first five sections of the check-sheet as optional for incorporation, so as to cater for cases in which there were sheets for subdivisions of the site but no master sheet summarizing information for the whole area. For these 'dummy master' check-sheets were written, these giving the total area, the extremes of altitude and the latitude and longitude as nearly as possible for the centre of the area. These dummy master sheets thus have entries only for the first five sections of the questionnaire. Apart from provision for whole sections and subsections to be absent, certain individual data-items are allowed to be incomplete or absent. The principal examples of these are Fosberg codes, soil codes, percentages for different types of coastal substratum, and tidal range. Processing arrangements for dealing with the Fosberg coding were the most sensitive and all examples of the various types of entries returned and the ways in which they were processed for storage are given in Appendix 6.

This degree of flexibility in the requirements for completeness places no constraints on the handling of the data, but the inclusion of check-sheets with information suspected or known to be either erroneous or incomplete demands a critical examination of the value of the data and of their reliability for different purposes. For this reason an appraisal was undertaken which might provide a basis for a reliability-rating. At the same time the data were examined with two further aims in mind: first, to review entries in those sections of the sheet, and particularly those concerning vegetation, over which there had been some change in policy for handling data or for which conventional modifications had to be made during editing. The second aim was to identify error or misinterpretations to deal with which involved the application of procedures based on accumulated experience with survey data. The outcome of this appraisal and the basis adopted for a reliability-index are considered in the following section.

3.4 Reliability rating

The special problems associated with the collection of the data, and with the detection and handling of omissions and errors in recording, point clearly to the need for some means of rating the information in check-sheets as more or less suitable for its intended uses. It was virtually impossible to confirm the accuracy or completeness of any sheet in detail, and no widespread trials were conducted from the IBP/CT centre to compare the results of independent surveys of the same sites by different recorders. This was for two main reasons. In the first place variability in methods of site-selection and conventions of recording at national level would have masked the effects of differences between the surveyors. Secondly, even with a close measure of agreement on the conventions to be observed between two surveyors of roughly equal capabilities, the two completed check-sheets would be likely to differ because of their different personal interests and experience. Without prior consultation on conventions the discrepancies would doubtless be still greater, the situation for each being comparable with that confronting a surveyor unfamiliar with a site and with no available information about it.

Variability in the nature and value of the data in different check-sheets derives principally from the following sources.

(1) Differences between and within national contributions arising from different methods of co-ordinating and standardizing site-selection and recording.

(2) Different sources of material, whether exclusively or mainly from field observations, exclusively or mainly from existing records, or about equally from both. The Canadian contribution, for example, involved the collection of field data in almost every instance but was supported by information from existing records and maps where available. The British entries were compiled from records made during a national conservation review having similar objectives to those of the IBP/CT Survey and were checked by individuals familiar with the sites, but no additional field work was undertaken specifically for completing check-sheets.

(3) Different degrees of completion of sheets. Apart from the omission of answers to particular questions by surveyors facing difficulties of terrain or

of access to existing records, incomplete returns were received in significant numbers for other reasons. The Japanese and Czechoslovakian contributions adopted a simpler approach to recording vegetation-types that that of the questionnaire; and the greatly reduced version of the check-sheet adopted by the American Institute of Biological Sciences, and used for recording most of the data relating to American State Parks, provided less information from each site than was sought in the normal check-sheet. Finally, the contribution from Australia was based initially on only the first three pages of the check-sheet and the subsequent pages for each site arrived too late for inclusion in the data on which synthetic studies were based.

(4) Different systems of recording. The French recorded the structural features of plant communities in a fundamentally different way from that forming the basis of the Fosberg classification, and in the version of the check-sheet used by the American Institute of Biological Sciences no structural information was supplied, the Fosberg coding being omitted without replacement.

(5) Different backgrounds of surveyors. The accuracy and completeness of recording for the survey varied for reasons ranging from differences in the training and specialized scientific knowledge of surveyors, affecting their capacity to overcome inherent conceptual and technical difficulties, to the incompetence or idleness of a minority.

(6) Loss of data. During the many stages between field recording and storage of the data at the Data Centre inevitably there were some losses of information, the principal causes being incomplete legibility and ambiguities or obscurities in entries.

A system of rating for reliability involved a consideration of each of these sources of difference between returned check-sheets in the amount and quality of the information provided and also of the criteria upon which an appraisal might be based. Such criteria must include the numbers of questions answered in whole or in part, the details of certain answers, and the style of key entries such as those on landscape and vegetation where the quality may be judged from maps and cross-references as being reasonable or otherwise. It was decided that a check-sheet could be regarded as contributing usefully to the analysis and assessment of the outcome of the survey if the completed answers included at least those to the first five sections of the questionnaire and also to the seventh. Because of the range in the content of entries to the seventh section, on vegetation, two levels were recognized: Level 2, for those with complete entries, and Level 1 for those in which Fosberg codes were assigned to all or most 'formation classes' but not to individual 'formations'. With a full set of completed answers a check-sheet would be rated in the highest category and, at the other extreme, with fewer than the first five together with the seventh it would fall into the lowest. In this rating for reliability attention was thus given to the number of questions answered and to details of certain

Table 3.2. *Reliability ratings for check-sheets*

	Purposes for which suitable			
Reliability grade	Detailed synthetic studies	Site classification and detailed vegetational studies	Studies of coverage and distribution	No synthetic studies
1 All sections complete, with full Fosberg codes	+	+	+	
2 Sections 1–5 and 7 complete, with full Fosberg codes		+	+	
3 Sections 1–5 and 7 complete, with part Fosberg codes			+	
4 Sections 1–5 or fewer complete				+

answers but not to the less objective criterion of the style of answers to key questions. This was, however, taken into account in certain instances of doubtful reliability of the data in incomplete returns.

Assessment on these lines led to the recognition of the following grades based on the purposes for which it was judged that the information in a check-sheet might reasonably be used.

Grade 1. Suitable for inclusion in detailed synthetic and comparative studies requiring complete and reliable answers to all questions.

Grade 2. Unsuitable for general synthetic and comparative studies but suitable for inclusion in site-classification and other detailed studies based primarily on vegetational data.

Grade 3. Unsuitable for general synthetic studies or for site-classification but suitable for inclusion in studies of the coverage and distribution of broad types of vegetation.

Grade 4. Not suitable for any of the above purposes but nevertheless having some value in site-documentation.

These grades and the restrictions associated with them are summarized in Table 3.2.

3.5 Data-storage

The data-storage system was designed to handle 10 000 or more check-sheets each requiring about 1000 words of computer storage, so that in all 10^7 words would need to be stored. This pointed to either magnetic disc or tape for permanent storage, all other forms of store being ruled out because of insufficient capacity, being too expensive, or both. Magnetic disc has the advantage that it is readily accessible at any point, but it is more expensive than magnetic tape. The contents of a magnetic tape can be read conveniently only in the order in which they were stored, but this was judged to be no serious disadvantage for answering the kinds of questions envisaged and it was accordingly decided to use tape.

The check-sheets having been edited for legibility, obvious errors and economy of wording, and brief descriptive phrases or 'descriptors' having been added at certain points to facilitate searching and sorting, then each is punched initially on paper tape in accordance with a schedule known as the 'blueprint', explained below.

The items of data contained in a completed check-sheet fall into the following types:

(1) *Positive integers*, e.g., the year in which the survey was carried out. Negative integers have occurred only in altitudes and so infrequently that it was not thought necessary to program the system to deal with them.

(2) *Strings*, arbitrary sequences of characters such as the names of reserves. These can include any combinations of letters, digits and punctuation marks and can be of any length, sometimes extending over more than one line.

(3) *Fixed strings*, non-arbitrary sequences of characters such as units of area, which can only be 'hectares', 'square kilometres', 'acres' or 'square miles'.

(4) *Binary items*, in answers recorded simply as one or other of two alternatives, like 'yes' or 'no'. These are really a special kind of fixed string but are treated separately because they occur so frequently.

(5) *Compound items*, combinations of items belonging to more than one of the above four types which are conveniently treated as constituting an additional type of data-item. An example is the area occupied by a plant community, which is an integer of type 1 followed by a fixed string of type 3 (the unit of area).

The blueprint specifies how many sections there are in a completed check-sheet, how many subsections each section includes, how many parts there are in each subsection and so on through six such levels down to the individual items. For each individual item the type of data it contains is then indicated by citing the appropriate number from the above list. This is followed by a sequence number together with other numbers indicating whether the item is optional, whether it can be repeated and, if so, whether the repetitions are of fixed or variable number and how many there are. The sections, subsections and parts are all treated as compound items which, within subsections, can be repeated or omitted in their entirety. The blueprint also contains lists of all the fixed strings and shows which is the appropriate list for a particular data-item. The following examples illustrate the procedure.

3 Data-processing, storage and retrieval

(*a*) The answer to Section 2, subsection 1 of the check-sheet contains first the name or names of the reserve. It has sequence number 1 and is an arbitrary string of type 2. It can be repeated any number of times, according to how many names there are. This is expressed as −1. If the number of repeats was fixed, then that number would replace −1. Lastly, this item cannot be omitted, a fact indicated by 0 (rather than 1, which would imply an optional item). Hence the relevant part of the blueprint reads:

1 2 −1 0

(*b*) The answer to Section 4, subsection 3 gives the class of site, whether A, B, C, or D (p. 139). It is the third item in the section, it is a fixed string (type 3), it cannot be repeated so that the number of times it is expected is 1, it is not optional (0), and the letters A, B, C, D are given in list 5 of the set of lists of all the fixed strings. Thus the blueprint, at this point, reads:

3 3 1 0 5

The paper tape is punched according to a set of rules which ensure that the punched data will exactly match the requirements of the blueprint. If the blueprint is changed for any reason, the rules for punching must also be amended. Every section begins on a new line and is preceded by a dollar sign. Every subsection begins with a dot. Integers are punched as they stand, strings with an asterisk to mark the end, and fixed strings with at least two blanks at the end (because a fixed string may contain a single blank). Missing items are denoted by a minus sign which may be omitted if the item comes at the end of a subsection. The number of compound items, where it is variable, is punched with a preceding left bracket to distinguish it from other integers. When punching has been completed the tape is verified and any errors noted for subsequent correction on the computer. All the remaining stages of processing are carried out on the computer.

The paper tape is read into the computer as a binary file on magnetic disc, and it is at this point that any errors detected during verification can be corrected, using an EDIT routine. The processing programmes are then run, the first stage being to identify any residual errors of format, such as the omission of a compulsory item. On average one such error escapes notice on each check-sheet, but in the computer any failure to match the blueprint specification is registered as an error. The programme then skips to the next section or subsection to look for other errors, and unless more than one occurs in any subsection the programming allows all errors in format to be found in a single run. When all the programmes have been run and the errors corrected, the final version is ready for storage. A corrected copy of the paper tape is punched as a safety store for the data and for occasional use in answering queries directly, and a listing is also made for the office files. Five series of cards are also punched, each with a copy for safety, for every corrected check-sheet read into the computer. The first of these is for the site-card index (p. 122). The remaining four are for external 'directories' or indexes to particular parts of the check-sheets. One holds the names of all individual species listed in community descriptions and indicates the sites for which they have been recorded, and a second lists the scientific or vernacular designations of the communities described, e.g., Littorellion, tundra, *campo cerrado*. The remaining two give the names of rare, threatened, endemic and relict species of animals and plants respectively, with the sites for which they are listed. These directories make it possible to limit searches to the relevant individual check-sheets when there are enquiries concerned with particular species or community types (p. 115).

The data file on magnetic disc may now be transferred to magnetic tape for permanent storage and the disc file can be deleted. The writing to magnetic tape is effected by using a program which can add or delete files and keep track of the position, length, serial number and date of creation of each check-sheet. Each ATLAS tape has a capacity for just over 4000 blocks of stored data. Since the modal amount for individual sheets is four blocks, the total number of returned check-sheets (*c*. 2800) could be stored on three tapes. Safety copies are kept, so that the requirement becomes six tapes. In practice additional tapes are used so that contributions from each continent can be kept together. This reduces the number of files that need to be searched during the retrieval of information for certain kinds of enquiry.

The storage in memory is of connected lists of data of the following kind. The first word gives the number of sections, nineteen. The next nineteen words give the numbers of the words at which the successive sections start, so that since Section 1 starts at word 21, word 2 will contain the number 21. There being four subsections in Section 1, word 21 will contain a 4 and words 22–25 will refer to these subsections 1–4. This goes on until an actual item of data is reached. The strings are then stored as the characters of which they are composed, but all other types of data-item appear as integers. Fixed strings appear as the appropriate number: for class C sites, for example, their class is stored as the number 3, C being third in the series A, B, C, D. For a compound item the word contains merely the number stored to indicate where the list of parts begins. This type of arrangement is termed a 'list structure', with lists of variable size. The absence of an item, whether a section, subsection or part of a subsection, is denoted as −1. Every list in this structure includes its length, this being important because it allows comparisons of check-sheets stored according to the specifications of different blueprints. If, for instance, a subsection 5 were to be added to the existing four of Section 1, a program trying to find subsection 5 in a check-sheet dating from before the addition could tell that there were in fact only four subsections and that subsection 5 must therefore be missing. In a check-sheet of the new type, subsection 5 will either be present or its absence will be indicated by a −1, so the program operates successfully with either kind of check-sheet.

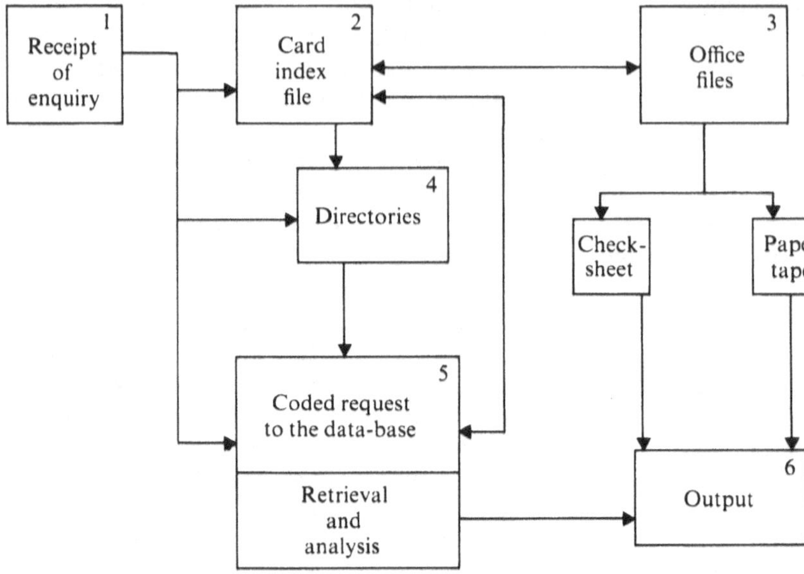

Fig. 3.5. Flow chart showing pathways for retrieval of information from the data-base.

3.6 The information retrieval system

The data-base is so organized that several different pathways can be used in the retrieval of information according to the nature of the enquiry. The pathways are shown in the flow-chart above.

On p. 115 reference was made to three main types of enquiry to which answers might be sought from the data-base:

(1) What is known about a particular site or series of sites?

(2) In what sites does a particular feature or group of features occur?

(3) What emerges when information on the occurrence of a particular feature is summarized for all sites falling into a particular category (e.g., information about soil, vegetation or types of human impact in all check-sheets from a given country, a given altitude or a given climatic region)?

The first two of these types of enquiry involve simple retrieval techniques, and for many questions of the first type an office indexing system and files would quickly furnish the information. Copies of whole check-sheets or particular parts are sometimes required. At present this is most economically effected by using a photocopier, though facilities for microfilming would greatly ease the operation if repeated copying were necessary. Reference to a simple punched card file gives the names of sites for which information is available from any country, and copies can be prepared for circulation. This file is updated as additional check-sheets are put into the data-base.

The second type of enquiry would be extremely laborious to answer without mechanical aids unless both the number of check-sheets and the number of features under consideration were low. The technique required is precisely that used in mechanical sorting, a combination of logical operators such as 'AND' and 'OR' in expressions of varying complexity. A group-select card sorter allows the combination of up to eight characters in any one search using binary punched cards and this is a facility that would rarely prove inadequate in answering enquiries of this type. It was for this reason that considerable efforts had been made earlier to design a practical card-based system until the sheer bulk of data and the requirements for flexibility ruled out this possibility.

The third type of enquiry involves both searching and computing. It was made a requirement of the system that in addition to the facility for searching for combinations of up to eight characters, standard arithmetical computation should be available for analysis of the data. It was borne in mind that the possibility of incorporating additional information, such as climatic data, into analyses would be desirable: any complementation of data already in the system greatly increases the potential for analysis and for practical application.

Examples of the handling of the three main types of enquiry are given below, the figures after the questions being those shown on the flow-chart.

(1) What is known of the Serengeti National Park? (1, 2, 3, 6)

Here the enquiry is concerned only with a single site. Reference can be made to the record card to see whether there is a check-sheet and, if so, whether there is any additional material in the office file (p. 122). All relevant material can be consulted and copied for dispatch if necessary.

(2) What are the plant communities supported by peat soils between the tropics? (1, 5, 6)

This type of question allows no reduction in the amount of searching through the use of the directories. Each check-sheet must be examined to find whether its site lies within the tropics and, if so, whether peaty soils occur in it. For each occurrence of peat the details of locality and vegetation, together with full soil notes, have then to be listed.

(3) What is the distribution of plant communities dominated by, or with a significant amount of, *Pinus banksiana* Lamb. in North America? (1, 2, 4, 5, 6)

Reference to the index file gives the geo-code of all countries of North America to be included in the search. For the purpose of this enquiry the countries will be Canada and the USA, and all check-sheets having either K or O as the first character of the serial prefix need to be included. All records of *P. banksiana* in the directory of species names are then extracted and a paper tape punched for those with the appropriate serial prefixes. The tape is then input to the computer and used to limit the searching routine to the relevant check-sheets. The punched tape also references the particular numbered community within each check-sheet, so the relevant communities can be consulted directly. The program extracts a listing giving the site name and serial prefix, its latitude and longitude, its range of altitudes, and details of the plant community and soil type.

(4) What is the distribution of logging, grazing and burning in Class A sites in the USA? (1, 2, 5, 6)

Reference to the card index will show that all USA sites have a serial prefix beginning with the letter O. This can be incorporated into the search programme, but no help can be obtained from the directories. All sites of Classes B, C or D are omitted at the next programming stage, and then a series of arrays is used to compile a summary of the impact entries for each Class A site from which details of the occurrence of logging, grazing and burning can be extracted.

In order to test the capability of the data and of the retrieval system to deal with distribution studies, an investigation was made of the distribution of four species of *Pinus* in North America on the basis of information collected during the survey. The coverage of North American plant communities by completed check-sheets is quite good, particularly so for those represented in National Parks, National Monuments and Research Natural Areas in the USA and in sites proposed as Ecological Reserves in Canada. The four species of *Pinus* chosen for the analysis have well-documented distributions in North America: *P. banksiana* Lamb. and *P. strobus* L. in the north-east and *P. edulis* Engelm. and *P. ponderosa* Douglas ex P. & C. Lawson in the south and west (Mirov, 1967; US Forest Service, 1972). The intention was to compare findings from the survey data against published distributions.

The search program was written to use a separate directory tape for each species, indicating the relevant check-sheets and community numbers. From each listing the locations were plotted on a small-scale map and compared with the published maps. Information on altitude and soils was examined in relation to the data for distribution to see how far it might be possible to infer the different ecological requirements of the four species on the basis of the available data, though it was realized that inadequacies of the sampling methods, and a lack of information on the status of each species in the communities where it occurs, might restrict the value of the results. Specialized studies of species distribution were well outside the original objectives of the survey.

In general the data extracted for these four species of *Pinus* showed quite clearly the main features of their distributional ranges and their environmental requirements and demonstrated the potential value of the data-base for such purposes. The comparison between the analysis of the distribution data based on the survey with published distributions was presented in display form at the Second World Conference on National Parks at Grand Teton in September 1972. The outcome was least satisfactory for *P. banksiana*, which extends from Nova Scotia and northern New England westwards to the Mackenzie River and Alberta. Check-sheet records are mainly from the southern part of its enormous range and show it as occurring almost exclusively as closed forest on podsolic soils. This is certainly true for most of its range, but it thins out further north and becomes more scrubby. *P. strobus* goes further south than *P. banksiana*, extending along the Appalachian Mountains to the northern tip of Georgia, but their two ranges overlap considerably in eastern Canada and the New England states and in an area round the Great Lakes, though *P. banksiana* reaches both further north and further west than *P. strobus*. The data from the survey show the two species as similar in forest structure and soil type and bring out the widespread homogeneity of structure and soil type over vast areas of coniferous forest in the boreal floristic province of North America. *P. ponderosa* was shown to occur on a wide range of soils over a broad altitudinal zone and in a corresponding variety of structurally different vegetation-types, suggesting a wide tolerance of environmental conditions. *P. edulis*, occupying only the south-eastern part of the range of *P. ponderosa*, is very largely in open forest and on poorer soils.

For all searching programs two basic points have first to be considered when a question is put to the information-retrieval system: does every check-sheet have to be searched, or can the scope of the search be limited by the use of the directories; and can the questions be answered simply by extracting relevant details from each check-sheet and printing them at once, or have further calculations to be made after searching has been completed?

There is a standard main program for searching magnetic tape, but at least one additional subroutine must be written for each specific question. If the search is limited by the use of the directories, then the main program reads only those check-sheets that are specified. The procedure is to go through the directory or directories by hand, looking for the relevant entries and extracting the serial numbers of the check-sheets involved, the numbers being punched as data for the guidance of the searching program. If computations are to follow the searching, then the searching routine must arrange to store retrieved data and a second additional routine must be supplied to effect the analysis and to print the results. It is sometimes convenient also to have a third subroutine to print headings and to set initial values of variables and constants.

The searching routine is normally the most complex of the three, and there are two programming aids available:

(1) In references to different items in the check-sheet it is technically most

convenient to use a sequence of integers, currently six in number. This is troublesome when writing the subroutine, making it more difficult to read and understand. Every item therefore has been given a name which may in some instances have a numerical subscript. Thus the second plant community could be designated COMMUNITY 2. Such names are converted to numerical form by a translating program ('macrogenerator') before the retrieval program is run.

(2) When writing the searching subroutine there are certain operations that are required frequently, such as testing whether an optional item is present or not. A collection or 'library' of such operations has been written and this has proved very useful. It includes subroutines which extract information from the check-sheets and routines for tabulating and for printing reports.

The effect of these procedures to to make the FORTRAN program written for information-retrieval retain some of the flavour of the original English of the query. Such programs have been written successfully by persons having only a basic knowledge of FORTRAN with some documentation for the macrogenerator and the library of subroutines.

4. The outcome of the check-sheet survey

R. J. DE BOER & G. L. RADFORD

4.1 Statistical review of completed returns

Over 13 000 check-sheets have been distributed to all parts of the world. The normal procedure was to organize the survey in a participating country through the National IBP Committee, or, where none existed, through a national organizer who agreed to act in that capacity. Early responses to requests for survey results were encouraging and the impression was gained from correspondence that a good coverage would be achieved and would include areas from which information was most needed. In the event, by 1973, check-sheets had been returned for 2930 'IBP Areas', these being defined in the *Guide* (Peterken, 1967) as sites falling within any of the following four categories:

Class A Sites includes in the United Nations List of National Parks and Equivalent Reserves.

Class B Sites mentioned in Chapter V of the UN List as having been considered for inclusion but rejected on one or more criteria.

Class C All other sites which are at present protected.

Class D All other sites of interest to conservationists and biologists which are at present unprotected.

It is evident that Class D areas are potentially very numerous and some principles had to be laid down to guide the selection of the most important of them for the purposes of the survey. The *Guide* recommended basing selection on a consideration of

the uses to which the information on such sites will be put, namely:

(1) to discover sites which contain plant associations of which no example (or an insufficient number of examples) is as yet protected. Some of these sites may as soon as practicable be recommended for protection, in order to fill gaps in the series of safeguarded vegetation formations. In particular countries, therefore, sites which contain a vegetation formation not known in any reserve in the country should be included;

(2) to enlarge the range of ecosystems safeguarded. Full information about each plant association on record will enhance IBP/CT's ability to ensure that adequate samples (including samples of each main variant) of each vegetation formation are protected;

(3) to place in the data-store information about sites to which some official conservation status may be accorded in the future.

4 *The outcome of the check-sheet survey*

The total number of check-sheets returned, 3010 – rather fewer than a quarter of those issued – was in fact larger than the number of IBP Areas surveyed. This was because some of the areas were subdivided, either on the ground of very large size or because they included a wide range of vegetation formations, land-forms, types of human impact or the like. Guide-lines for the subdivision of IBP Areas were given on pp. 12–13 of the *Guide*.

Of the sheets returned, those for 320 IBP areas provided insufficient information to be utilizable for comparative purposes, although perhaps of some value as a basis for documentation of the sites to which they refer. These were sheets whose 'reliability' (pp. 128–30) was assessed as no higher than Grade 4. This left 2481 sheets of reliability-grades 1–3 for 2610 IBP areas.

4.1.1 Participating countries

Check-sheets were returned from fifty-five 'countries', using the term in a broad sense and treating all the West Indies as a single unit and similarly all the island groups of the Indian Ocean and all those of the Pacific Ocean as two further units. In order to secure more precise information on the geographical distribution of the returns and to allow comparisons of coverage of different types of vegetation, soil, climate, etc., between areas of more restricted and more nearly equal size, the larger 'countries' were subdivided into 'administrative geographical units' (AGUs) – states, provinces, socialist republics, island groups and so forth. Table 4.1 shows the outcome of this subdivision.

It will be seen that completed check-sheets were returned from 155 AGUs. Those from 140 AGUs were utilizable, the average number of useful returns per AGU being about 18.5.

Table 4.1. *Subdivision of 'countries' into 'administrative geographical units' (AGUs) as a basis for mapping the distribution of completed check-sheets* (*see Fig. 4.1, pp. 142–3*)

'Country'	Number of AGUs	Number from which sheets were returned
Australia	7	5
Canada	15	13
Continental USA	49	43
Czechoslovakia	2	2
India	—	8
Indian Ocean	—	3
Pacific Ocean	—	13
Spain	2	2
United Kingdom	3	3
USSR	—	15
West Indies	—	4

Table 4.2. *World distribution of administrative geographical units (AGUs) by grades based on the numbers of surveyed IBP Areas per AGU (see below and the legend to the map on p. 142–3)*

Region	Number of partici-pating AGUs	Grade						
		1	2	3	4	5	6	7
N. America	56	1	4	4	7	10	25	5
C. and S. America	7	—	—	—	—	—	3	4
Europe	30	2	5	2	4	6	10	1
Africa	13	—	—	1	2	1	6	3
Asia	24	—	1	—	—	2	21	—
Australia	5	1	1	—	—	—	3	—
Indian Ocean	3	—	—	—	—	1	1	1
Pacific Ocean	13	—	—	—	—	—	13	—
Atlantic Ocean	4	—	—	—	—	—	3	1
Antarctica	—	—	—	—	—	—	—	—
Totals	155	4	11	7	13	20	85	15

4.1.2 Grading of AGUs from which returns were received

For mapping the distribution of returns, AGUs were grouped into the grades of an approximately geometrical scale on the basis of the numbers of IBP Areas for which useful returns were received. The legend to the map (Fig. 4.1) gives these grades which indicate the decreasing number of returns; the shading reflects this. Of the 155 participating AGUs eighty-five, or more than half, fall into grade 6, as shown in Table 4.2.

4.1.3 World distribution of participating AGUs

Table 4.2 shows that of the thirty AGUs from which more than ten completed check-sheets were received, and which therefore fall into one or other of the first four grades shown on the map, twenty-four are either in Europe or in North America. The remaining 114, from each of which there were ten or fewer returns, are more widely distributed over the world.

From Table 4.3 it can be seen that the total number of IBP Areas from which useful information was received was 2610. 2016 of these, or about three-quarters, were in Europe and North America. Substantial information came also from Australia with 310 utilizable returns and Asia with 151. The total number of returns from Africa included a valuable contribution of forty-six completed sheets from Tunisia.

It is clear from the map (Fig. 4.1) and from Table 4.3 that more than 80% of the IBP Areas from which useful information was received are in temperate

Fig. 4.1. IBP Areas from which satisfactorily completed check-sheets were returned.

For each administrative geographical unit (AGU) the number of IBP Areas from which satisfactorily completed check-sheets were returned is indicated by one of the shades shown in the key. The numbers of returns defining successive shading grades are in an approximately geometrical series.

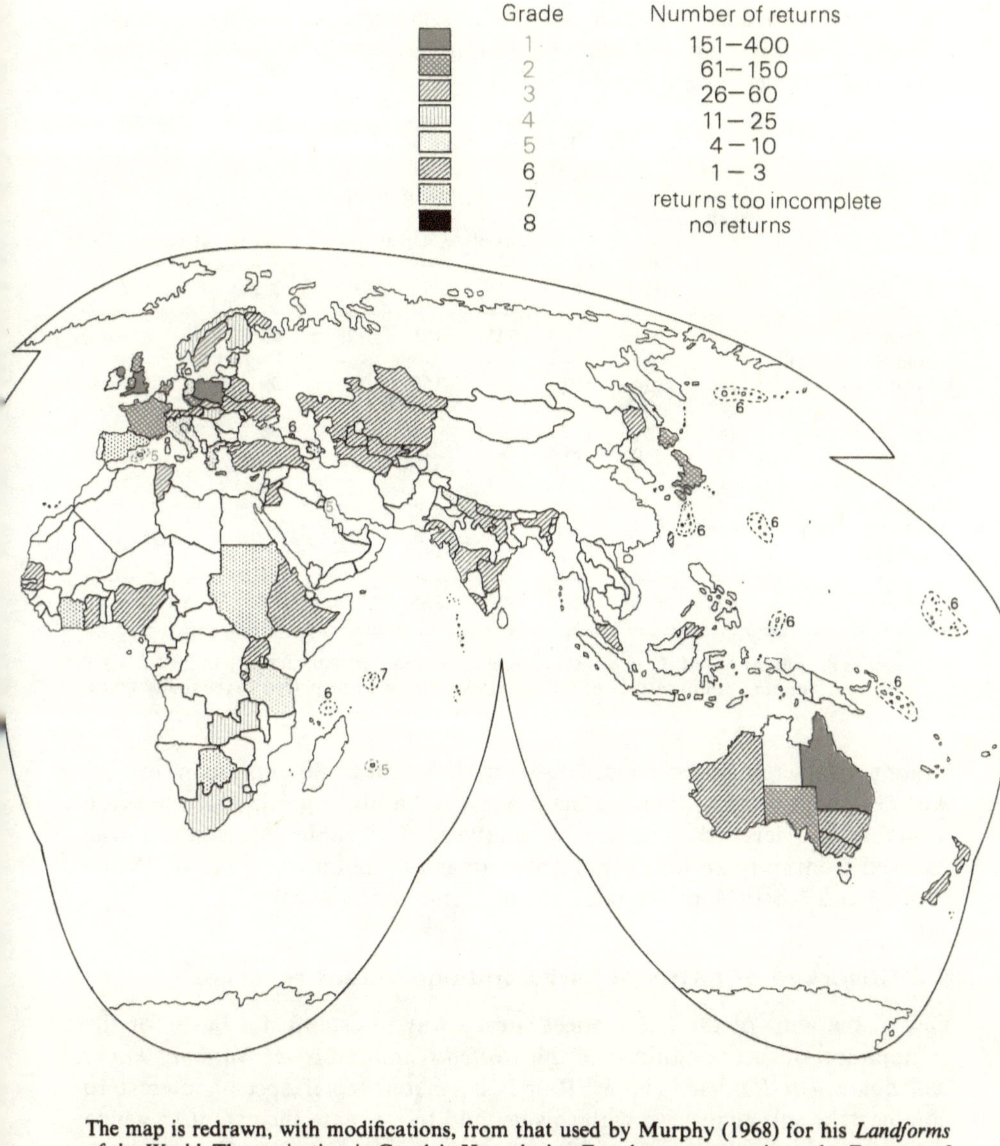

The map is redrawn, with modifications, from that used by Murphy (1968) for his *Landforms of the World.* The projection is Goode's Homolosine Equal-area at a scale at the Equator of 1:57 000 000. The map gives true distances on mid-meridians and parallels 0° to 40°. This projection is very satisfactory in meeting the requirements that area, shape, scale and compass-bearings should be preserved and that the maps should be easy to draw. Base maps were made available on the scale required for this volume by the University of Chicago Press Ltd, USA.

Table 4.3. *World distribution of IBP Areas from which returns were received and (columns headed 1–7) their distribution amongst the differently shaded grades of the map (Fig. 4.1) denoting the total numbers of useful returns from the AGUs in which they are situated*

	Numbers of IBP Areas from which returns were received								
	Total returns	Useful returns							
			Grades of AGUs in which Areas are situated						
Region		Total useful	1	2	3	4	5	6	7
N. America	1087	1002	199	393	181	125	63	41	—
C. and S. America	31	4	—	—	—	—	—	4	—
Europe	1146	1014	367	372	158	66	39	12	—
Africa	128	96[a]	—	—	46[a]	33	6	11	—
Asia	161	151	112	—	—	—	6	33	—
Australia	333	310	241	61	—	—	—	8	—
Indian Ocean	17	7	—	—	—	—	6	1	—
Pacific Ocean	22	22	—	—	—	—	—	22	—
Atlantic Ocean	5	4	—	—	—	—	—	4	—
Antarctica	—	—	—	—	—	—	—	—	—
Totals	2930	2610	919	826	385	224	120	136	—

[a] Questions in sections 1–5 of the check-sheet were not completed satisfactorily in the forty-six sheets from Tunisia, but in respect of the remainder they may be placed in reliability grade 2.

regions. Of the remainder about 200 are in the tropics, 130 of these in tropical Australia with sixty of them in the highest 'reliability' grade and therefore available for detailed comparative analyses. Utilizable information was received from only about twenty polar areas of Scandinavia, USSR (Murmansk) and North America (Canadian Arctic and Alaska).

4.2 Coverage of national parks and equivalent reserves

One of the aims of the check-sheet survey was to establish a basis for the compilation of future editions of the *United Nations List of National Parks and Equivalent Reserves* (NP&ERs). It is, therefore, of special interest to compare the survey returns with this list and to estimate the extent to which the returns cover all countries with one or more NP&ERs (the Class A areas of the survey) and also what fractions of the total number and total area of NP&ERs are represented in the returns.

Table 4.4 gives information extracted from the 1973 edition of the *UN List*, based on data available on 30 June 1972. The thirteen 'Main areas' of the left-hand side replace the Regions of Table 4.3. Some 71% of the earth's

Table 4.4. *World distribution and areas of National Parks and Equivalent Reserves (NP&ERs), from the 1973 United Nations* List

	Main areas	Land area ($\times 10^9$ ha)	% of world total	Numbers of NP&ERs	% of world total	Area of NP&ERs ($\times 10^6$ ha)	% of land area	% total area of NP&ERs
1	N. America	2.1	14	199	20	18.210	0.9	17.5
2	C. and S. America	2.3	15	98	10	9.817	0.4	9.4
3	Europe	0.7	5	212	22	3.717	0.5	3.6
4	Africa	3.2	21	157	16	53.784	1.7	51.6
5	N. Asia	1.6	11	29	3	2.299	0.1	2.2
6	S.W. Asia	1.2	8	41	4	1.121	0.1	1.1
7	E. Asia	1.1	7	17	2	1.666	0.2	1.6
8	S.E. Asia	0.5	3	52	5	2.713	0.5	2.6
9	Australia	0.8	5	121	12	7.667	1.0	6.3
10	Indian Ocean	0.1	1	12	1	0.618	0.6	0.9
11	Pacific Ocean	0.1	0	30	3	2.564	2.6	2.1
12	Atlantic Ocean	0.1	1	11	1	0.139	0.1	0.1
13	Antarctica	1.3	8	0	0	0.000	0.0	0.0
	Totals and means	15.1	99	979	99	104.315	0.7	99.0

surface is covered with water. The remaining 29% is land, ranging from tiny islands to vast continental masses and totalling about 15 billion hectares, distributed as shown in Table 4.4.

The 1973 *UN List* includes 979 reserves, all but ten of them exceeding the recommended minimum area of 1000 ha. Of this total no fewer than 411 (42%) are in Europe (212) and N. America (199), with the highest concentrations in Great Britain (18, of which 13 are in Scotland) and Sweden (12) amongst European countries, and in California (22) amongst the states and provinces of N. America.

The reserves in the *UN List* vary very widely in size. Some, like the Eldey Island Nature Reserve in Iceland (1.5 ha) and the Cochon Island Nature Reserve in the Falkland Islands (7.5 ha), are far below the suggested minimum area of 1000 ha, while others cover several million hectares. It is noteworthy that, out of twenty-two reserves exceeding one million hectares in area, fourteen are in Africa, the largest being the Central Kalahari Game Reserve of 5 280 000 ha. The same point emerges clearly from the figures in columns 5 and 7 of Table 4.4, which show the NP&ERs of Africa as covering nearly 54×10^6 ha, over half the total area of NP&ERs throughout the world. The table also reveals that Europe has not only the highest number of

Table 4.5. *World distribution of NP&ERs from which check-sheets have been returned and the areas covered by the returns. Based on the United Nations 1973 List of National Parks and Equivalent Reserves*

	Main areas	Numbers of NP&ERs surveyed: Reliability-grades 1–3	4	Total	Numbers surveyed as % of all NP&ERs	Areas of surveyed NP&ERs (10^6 ha)[a]: Reliability-grades 1–3	4	Total	Surveyed areas as % of areas of all NP&ERs
1	N. America	63	48	111	56	9.504	2.255	11.759	65
2	C. and S. America	3	0	3	3	0.020	0.000	0.020	0
3	Europe	55	7	62	29	2.270	0.029	2.299	62
4	Africa	19	1	20	13	10.371	1.036	11.407	21
5	N. Asia	0	0	0	0	0.000	0.000	0.000	0
6	S.W. Asia	6	3	9	22	0.224	0.180	0.404	36
7	E. Asia	14	0	14	82	1.303	0.000	1.303	78
8	S.E. Asia	2	0	2	4	0.069	0.000	0.069	3
9	Australia	49	0	49	40	2.007	0.000	2.007	26
10	Indian Ocean	0	0	0	0	0.000	0.000	0.000	0
11	Pacific Ocean	4	0	4	13	0.099	0.000	0.099	4
12	Atlantic Ocean	1	0	1	9	0.006	0.000	0.006	4
13	Antarctica	0	0	0	—	0.000	0.000	0.000	—
	Totals and means	216	59	275	28	25.873	3.500	29.373	28

[a] Where a National Park or Equivalent Reserve has been surveyed only in part, the area given is of the surveyed part, not of the whole reserve.

NP&ERs in any 'main area' but also, with 22% of the total number on only 5% of the world's land surface, by far the highest density. Next comes Australia, with 12% on 5% of the land surface, and N. America with 20% on 15% of the land surface. On the other hand Asia has only 14% of the world's NP&ERs on 29% of the total land area, but this estimate takes no account of mainland China, constituting over one-fifth of the land area of Asia but with no information being available about its NP&ERs. With respect to areas of NP&ERs, rather than numbers, Africa leads with 51.6% of the total area throughout the world, with N. America (9.4%), Australia (6.3%) and Europe (3.6%) following in that order.

Table 4.5 shows that 275 NP&ERs, or 28% of the 979 in the 1973 *UN List*, have been covered by check-sheet returns. As far as numbers are concerned the largest contribution from any 'main area' was that from N. America, with 111 surveyed, 56% of the total of 199 (Table 4.4), all but one from the USA. Amongst these, however, are forty-eight from which the returns were rated

Table 4.6. *World distribution of 'administrative geographical units' (AGUs) participating in the survey and of the numbers and percentages of the National Parks and Equivalent Reserves (NP&ERs) which have been surveyed*

	Main areas	Numbers of AGUs with NP&ERs: Total	Taking part in survey	%	Numbers of NP&ERs: Total (A)	In participating AGUs (B)	% of total	Surveyed	% of (A)	% of (B)
1	N. America	47	46	98	199	198	99	111	56	56
2	C. and S. America	15	7	47	98	56	57	3	3	5
3	Europe	36	23	64	212	153	72	62	29	41
4	Africa	31	11	35	157	84	54	20	13	24
5	N. Asia	6	5	83	29	28	97	0	0	0
6	S.W. Asia	15	7	47	41	18	44	9	22	50
7	E. Asia	3	2	67	17	16	94	14	82	88
8	S.E. Asia	9	3	33	52	4	8	2	4	50
9	Australia	7	5	71	121	111	92	49	40	44
10	Indian Ocean	2	0	0	12	0	0	0	0	0
11	Pacific Ocean	8	6	75	30	27	90	4	13	15
12	Atlantic Ocean	5	1	20	11	1	9	1	9	100
13	Antarctica	0	0	—	0	0	—	0	—	—
	Totals and means	184	116	63	979	696	71	275	28	40

as falling in reliability-grade 4 (p. 130), the remaining sixty-three being placed in grades 1–3 as providing data utilizable in world-wide comparisons. Comparable numbers of check-sheets completed satisfactorily came also from Europe and Australia. The total number from which grade 1–3 returns were received was 216, or 22% of all listed NP&ERs. From the standpoint of areal coverage, Table 4.5 shows that 29.4×10^6 ha, or 28% of the total area of listed NP&ERs, was covered by returns, and 25.9×10^6 ha by utilizable returns in reliability-grades 1–3. N. America and Africa show the largest areas surveyed, the former leading slightly on the basis of all returns but Africa standing first if only returns in reliability-grades 1–3 are considered. The areas covered incompletely comprise forty-eight reserves in N. America but only one in Africa. The highest coverage of NP&ERs was from E. Asia, with a numerical coverage of 82% and an areal coverage of 78%. This was due mainly to the Japanese contribution which included thirteen Class A areas out of a total of fourteen. Australia was second to E. Asia as far as numbers are concerned,

Table 4.7. *World distribution of National Parks and Equivalent Reserves (NP&ERs) in participating administrative geographical units (AGUs), by area*

	Main areas	Area of NP&ERs in participating AGUs (10^6 ha)	% of total area of NP&ERs
1	N. America	18.207	100
2	C. and S. America	6.635	68
3	Europe	3.051	82
4	Africa	29.243	54
5	N. Asia	2.283	99
6	S.W. Asia	0.528	47
7	E. Asia	1.622	97
8	S.E. Asia	0.510	19
9	Australia	7.067	92
10	Indian Ocean	0.000	0
11	Pacific Ocean	2.543	99
12	Atlantic Ocean	0.006	4
13	Antarctica	0.000	—
	Total and mean	71.695	69

with 40% of the listed NP&ERs represented in the returns. Europe and N. America follow E. Asia in areal coverage, with 61% and 52% respectively.

Tables 4.6 and 4.7 indicate that 116 'administrative geographical units' (AGUs, p. 140) took part in the survey, this being 63% of the 184 shown in the 1973 *UN List* as having at least one National Park or Equivalent Reserve. These 116 had in all 696 NP&ERs, or 71% of the world total of 979 (Table 4.4). Of the 696 NP&ERs 275 were surveyed, the 216 of them from which returns were placed in reliability-grades 1–3 being 31% of all those in participating AGUs but only 22% of the world total of 979. The area of all NP&ERs in participating AGUs, just short of 72×10^6 ha, was 69% of the total area, 104×10^6 ha, of NP&ERs throughout the world.

It emerges, then, that about 28% of both the total number and the total area of all NP&ERs were surveyed, but returns utilizable for comparative purposes came from only 22% and 25% respectively. These low proportions arise from the two facts that only 71% of all NP&ERs are in participating countries and only about 40% of those were in fact surveyed. In order to extend the coverage the most promising course would be to concentrate in the first instance on 'countries' which have participated in the survey but have left a high proportion of their NP&ERs unsurveyed. In some regions, however, the aim must be to induce additional countries to participate. This is notably true of Africa, where twenty-one countries with seventy-three NP&ERs have so far failed to take part. The position could be still further improved if some of the larger NP&ERs were surveyed more completely by covering additional sub-areas.

Table 4.8. *Regional distribution of NP&ERs for which check-sheets have been returned*

	Numbers of NP&ERs surveyed			
Region	All grades	% of total	Grades 1–3	% of total
Polar	15	5.5	15	5.5
Temperate	217	78.9	159	57.8
Tropical	43	15.6	42	15.3
Total	275	100.0	216	78.5

It is of interest to consider how far check-sheet returns are equally representative of different latitudinal zones. Taking first the broad division into polar, temperate and tropical regions, with boundaries at *c*. 23° 30′ and 66° 30′ on either side of the equator, Table 4.8 shows that, of the 275 NP&ERs that have been surveyed, fifteen or about 5.5% lie in polar regions and only forty-three (15.6%) in the tropics. The remaining 217 (78.9%) are in temperate regions. The polar NP&ERs comprise fourteen in Scandinavia and one in European USSR, so that all are in Europe. The tropical areas are more widely scattered, but nineteen of them are in Queensland, Australia. All but one of the incompletely surveyed areas, placed in reliability-grade 4, are in temperate regions.

Fig. 4.2 presents a more detailed analysis of latitudinal representation. Fig. 4.2(*a*) shows for each of eighteen zones of latitude, 10° wide and extending from North Pole to South Pole, the numbers of surveyed NP&ERs expressed as percentages of the total number surveyed, 275. Fig. 4.2(*b*) differs in showing the combined areas of the surveyed NP&ERs, zone by zone, as percentages of the total area of all 275 that have been surveyed. In both Fig. 4.2(*a*) and (*b*) the shaded areas show for each zone its land-surface as a percentage of the land-surface of the whole world. The broken lines include, but the continuous exclude, those NP&ERs for which the returns have been placed in reliability-grade 4 as unsuitable for use in world-wide comparisons.

A concentration of numbers of surveyed NP&ERs is apparent between latitudes 30° N and 50° N, where 48% of the total returns come from only 20% of the world's land-surface. This becomes appreciably less marked when only returns in reliability-grades 1–3 are included. Interestingly, too, when areas rather than numbers are considered the concentration shifts northwards, utilizable returns reaching a maximum in the northern hemisphere between 50° N and 60° N. This is because there are many relatively small NP&ERs between 30° N and 50° N, but fewer and larger ones between 50° N and 60° N.

In contrast with this concentration there is a relative under-representation in returns from the northern tropical and equatorial zones, from 30° N to 10° S, both in numbers and in areas. The minimum lies between 10° N and

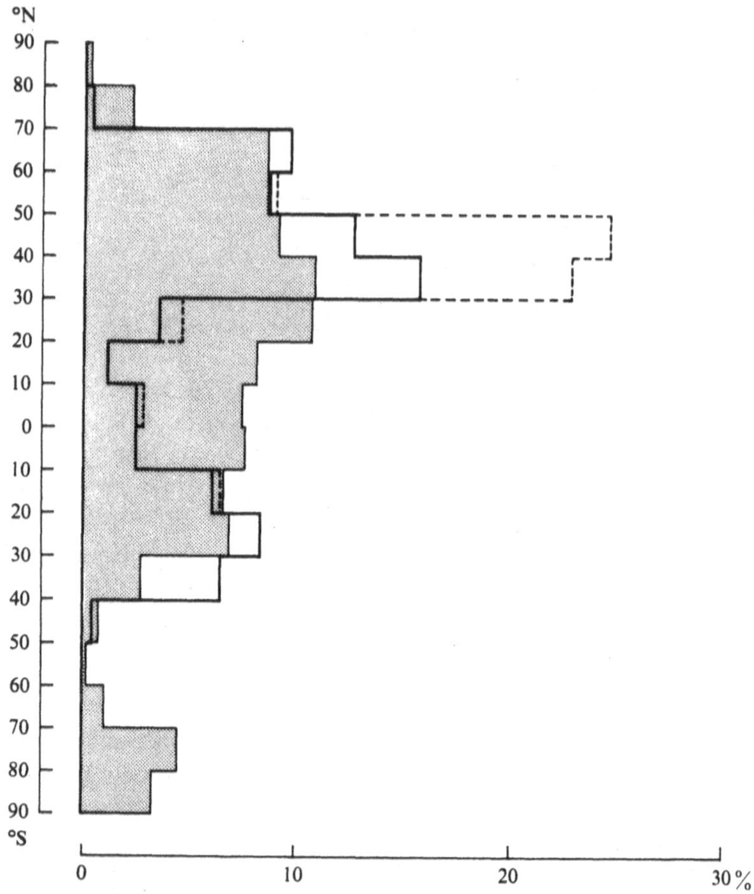

Fig. 4.2(*a*). Numbers of the NP&ERs in each latitudinal zone 10° wide as a percentage of the total number surveyed. For explanation, see p. 149.

20° N, where there are only three NP&ERs totalling less than 100 000 ha and each returning a single check-sheet. Between 10° S and 40° S the representation improves, there being a concentration in numbers and strikingly so in surface-area between 20° S and 30° S. This is due mainly to the contribution from Queensland (Australia), covering twenty Class A areas and about 7.7×10^6 ha.

Northern polar regions are poorly represented, surveyed areas being restricted to the European Arctic. Almost all land south of 60° S is part of the Antarctic continent, which not surprisingly lacks NP&ERs completely.

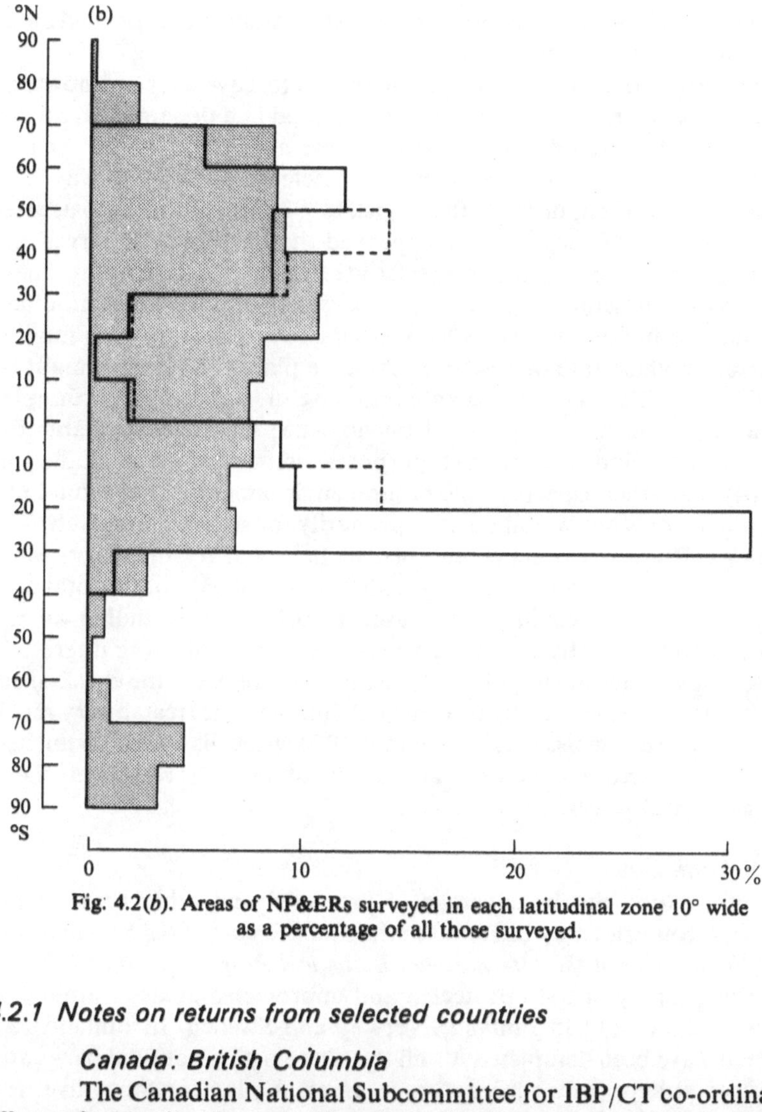

Fig. 4.2(*b*). Areas of NP&ERs surveyed in each latitudinal zone 10° wide as a percentage of all those surveyed.

4.2.1 Notes on returns from selected countries

Canada: British Columbia

The Canadian National Subcommittee for IBP/CT co-ordinated the efforts of ten regional panels. The check-sheet survey was supervised by these panels within guidelines for consistency in completion agreed at federal level. The panels included representatives from federal and provincial agencies and most included a large proportion of university scientists. Check-sheets were completed both by members of the panels and by other suitably qualified surveyors, mostly university staff and graduates. The contribution from British Columbia was similar to the others in this respect but is of particular interest because of its success in leading to the establishment of protected areas identified during the operation of the survey.

The intention of the federal subcommittee was to have sites of known or suspected biological interest in each region surveyed and documented and to organize an active search for others. From these a selection would then be made of areas to be set aside as reserves or protected in some other way. The aims were widened when, in 1971, the Legislature of British Columbia, after discussions extending over six years, approved the Ecological Reserves Act authorizing areas of Crown Land to be set aside for ecological purposes. These would be areas representative of natural ecosystems or ecosystems modified by man, but offering opportunities for studying recovery from such modification; areas in which rare or endangered native plants or animals might be conserved in their natural habitats or containing unique and rare examples of botanical, zoological or geological phenomena; and 'areas suitable for scientific research and educational purposes associated with studies in productivity and other aspects of the natural environment'. They would not be recreational areas but would be used primarily for scientific research. The first 'ecological reserves' were established in 1971 and by 1975 there were sixty-five in British Columbia, totalling *c.* 45 000 ha or 0.05% of the total area of the province. This 'would be increased to 0.19% by including all the potential reserves currently under high pressure from conflicting interests'.

Because of this determined policy of creating new reserves the check-sheet survey in Canada, and especially in British Columbia, was treated very much as a vegetational reconnaissance. Up to mid-1974 sheets had been completed for 135 areas in British Columbia and many of these have already been declared ecological reserves.

Scandinavia

Information on the coverage of protected and other areas in the Scandinavian countries may be found in the *Final Report of the Scandinavian National Committees of the International Biological Programme* (Oslo, 1975). A check-sheet survey of both protected and unprotected areas of biological interest was carried out in Finland, Norway and Sweden. In Finland IBP check-sheets have been completed for all the nine existing national parks and all the fifteen strictly controlled nature reserves. A plan for thirty-five new national parks and the enlargement of eight of the present parks was published in 1973, and there have also been proposals for eleven new nature reserves and the enlargement of four of the existing reserves. In Norway check-sheets were completed for fourteen national parks and reserves, and Sweden returned thirty-three sheets covering national parks and other areas of conservation value.

Japan

For Japan information relevant to coverage was published in *JIBP Synthesis Volume 8* (Numata, Yoshioka & Kato, 1975), in the summary and conclusion of which it is stated that 'There are twenty-six national parks and forty-six quasi-national parks in the country', ranging from subarctic through temperate to subtropical climates and from coastal to high mountain areas and occupying about 3×10^6 ha or 8% of the total area of the country. There are also 287 'prefectural parks' covering about 2×10^6 ha or 5% of the whole country, 'national monuments' for protecting natural vegetation and rare plants, often too small to ensure effective conservation, and 'reserve forests', tending again to be of inadequate size. 'Most sites of biologically important vegetation, flora and fauna have been recorded on check-sheets, and at the very least they should be protected from destruction' as a consequence of the IBP project.

France

The French IBP/CT Committee translated the check-sheet into French and sought also to adapt it to local conditions. Several hundreds were circulated, but some intending participants found difficulty in answering certain of the questions. Nevertheless about 170 sheets were completed. These came from areas all over the country, though the south-west, the northern Massif Central, the Pyrenees and the Alps were underrepresented. Some check-sheets were returned from the Ivory Coast, Senegal and French territories in the Southern Hemisphere and the Antarctic. This was regarded as a very satisfactory outcome of the French effort.

4.3 Coverage of vegetation-types

4.3.1 Introduction

It was explained earlier (pp. 15–16) that at the time the check-sheet was issued no universally applicable and acceptable system of vegetational classification was available which was sufficiently detailed for all the purposes of the survey. It therefore became necessary to seek an interim solution to the problem of recording vegetational information. It was decided that Fosberg's system (Appendix 2) should be adopted as a means of securing a standardizable initial stage of recording which would permit the storage of certain comparable data for all surveyed sites. It was recognized, however, that the information thus obtained would not be sufficient for the full purposes of the survey. It would be essential to go further to reach the necessary level of detail for ecological inventory and for decisions on plant communities meriting protection. Recorders were, therefore, asked not merely to allocate the vegetation-types of survey-sites to their Fosberg classes but also to list the individual plant communities present, using any system prevalent within their own country or region.

There were difficulties over the use of the Fosberg system in the survey, even though it was explained at length in the *Guide* and a key supplied for allocating each vegetation-type to its correct class and for giving it the appropriate coding. These difficulties encountered by recorders, and the consequent problems for those analysing the returns, are considered in detail in Appendix 6.

4.3.2 Coverage of Fosberg formations

The Fosberg classification, as published in the *Guide* (Peterken, 1967), comprised 194 formations some of which were further divided into subformations to give 225 categories in all. Of these fifteen formations and one subformations were considered by Fosberg himself as of doubtful existence. The use of the system in the survey and for certain other purposes led to its issue as a reprint (1971), with the addition of a further thirty-one formations and four subformations. It is this reprint that is reproduced as Appendix 2 of this volume. During the examination of completed check-sheets it became clear that still other additions would be necessary.

Table 4.9 shows the two 'formation-groups' (with the top layer of the vegetation evergreen or seasonally green respectively) of each 'formation-class' of the system. Each 'formation-group' comprises one or more 'formations', some of which are further divided into 'subformations'. Table 4.10 will also serve as a key to Figs. 4.3 and 4.4.

The extent to which the formations of the Fosberg system are represented in completed check-sheets is shown in summary in Table 4.10 and in greater detail in subsequent tables and figures.

Table 4.10 shows that 207 formations were recorded in check-sheets out of a grand total of 261, or 79.3%. Omission of formations of doubtful existence raises the coverage to 82.5%.

Figure 4.3 shows the extent to which the individual formations of the Fosberg system, as revised in January 1974, are represented in completed check-sheets, the symbols being those of Table 4.10.

The numbers of Fosberg formations recorded from each of the thirteen 'main areas' of the world (p. 144), and the extent to which individual formations are recorded from only a single area or from two or more, are shown in Table 4.11 and Fig. 4.4 respectively.

Table 4.12 supplements this distributional information by showing the numbers of 'formation-classes' and 'formations' reported in completed check-sheets from each of the eighteen latitudinal zones, 10° wide, extending from North Pole to South Pole. This information is based on 141 NP&ERs selected from the total of 275 that were surveyed because the check-sheets

Table 4.9. *List of 'formation-classes' and their 'formation-groups' in the Fosberg system. Those in brackets are of doubtful existence* (*p. 286*)

Formation-classes	Formation-groups 1		Formation-groups 2	
(*a*) *Closed vegetation* (*with crowns or peripheries of the plants of at least one layer mostly touching or overlapping*)				
1A	1A1	Closed evergreen forest	1A2	Closed deciduous forest
1B	1B1	Closed evergreen scrub	1B2	Closed deciduous scrub
1C	1C1	Closed evergreen dwarf scrub	1C2	Closed deciduous dwarf scrub
1D	1D1	Open evergreen forest with a closed lower layer	1D2	Open deciduous forest with a closed lower layer
1E	1E1	Closed evergreen scrub with scattered trees	1E2	Closed deciduous scrub with scattered trees
1F	1F1	Closed evergreen dwarf scrub with scattered trees	1F2	Closed deciduous dwarf scrub with scattered trees
1G	1G1	Open evergreen scrub with closed ground cover	1G2	Open deciduous scrub with closed ground cover
1H	1H1	Open evergreen dwarf scrub with closed ground cover	1H2	Open deciduous dwarf scrub with closed ground cover
1I	1I1	Evergreen tall savanna	1I2	Deciduous tall savanna
1J	1J1	Evergreen low savanna	1J2	Deciduous low savanna
1K	1K1	Evergreen shrub savanna	1K2	Deciduous shrub savanna
1L	1L1	Evergreen tall-grass	1L2	Seasonal tall-grass
1M	1M1	Evergreen short-grass	1M2	Seasonal short-grass
1N	1N1	Evergreen broad-leaved herb vegetation	1N2	Seasonal broad-leaved herb vegetation
1O	1O1	Closed bryophyte vegetation	1O2	Closed lichen vegetation
1P	1P1	Evergreen submerged meadow	1P2	Seasonal submerged meadow
1Q	1Q1	Evergreen floating meadow	1Q2	Seasonal floating meadow
1R	—	—	1R2	Ephemeral meadow
1S	—	—	1S2	Seasonal algae vegetation
(*b*) *Open vegetation* (*plants or tufts not touching but crowns separated by not more than their diameters: plants, not substratum, dominating landscale*)				
2A	2A1	Evergreen steppe forest	2A2	Deciduous steppe forest
2B	2B1	Evergreen steppe scrub	2B2	Deciduous steppe scrub
2C	2C1	Evergreen dwarf steppe scrub	(2C2	Deciduous dwarf steppe scrub)
2D	2D1	Evergreen steppe savanna	2D2	Deciduous steppe savanna
2E	2E1	Evergreen shrub steppe savanna	2E2	Deciduous shrub steppe savanna
2F	2F1	Evergreen dwarf shrub steppe savanna	2F2	Deciduous dwarf shrub steppe savanna
2G	2G1	Evergreen steppe	2G2	Seasonal steppe
2H	(2H1	Open bryophyte vegetation)	2H2	Open lichen vegetation
2I	(2I1	Evergreen open submerged meadow)	2I2	Seasonal open submerged meadow
2J	(2J1	Evergreen open floating meadow)	2J2	Seasonal open floating meadow
2K	—	—	2K2	Open seasonal algae vegetation
(*c*) *Sparse vegetation or desert* (*plants so scattered that substratum dominates landscape*)				
3A	3A1	Evergreen desert forest	3A2	Deciduous desert forest
3B	3B1	Evergreen desert scrub	3B2	Deciduous desert scrub
3C	3C1	Evergreen desert herb vegetation	3C2	Seasonal desert herb vegetation
3D	(3D1	Evergreen sparse submerged meadow)	3D2	Seasonal sparse submerged meadow

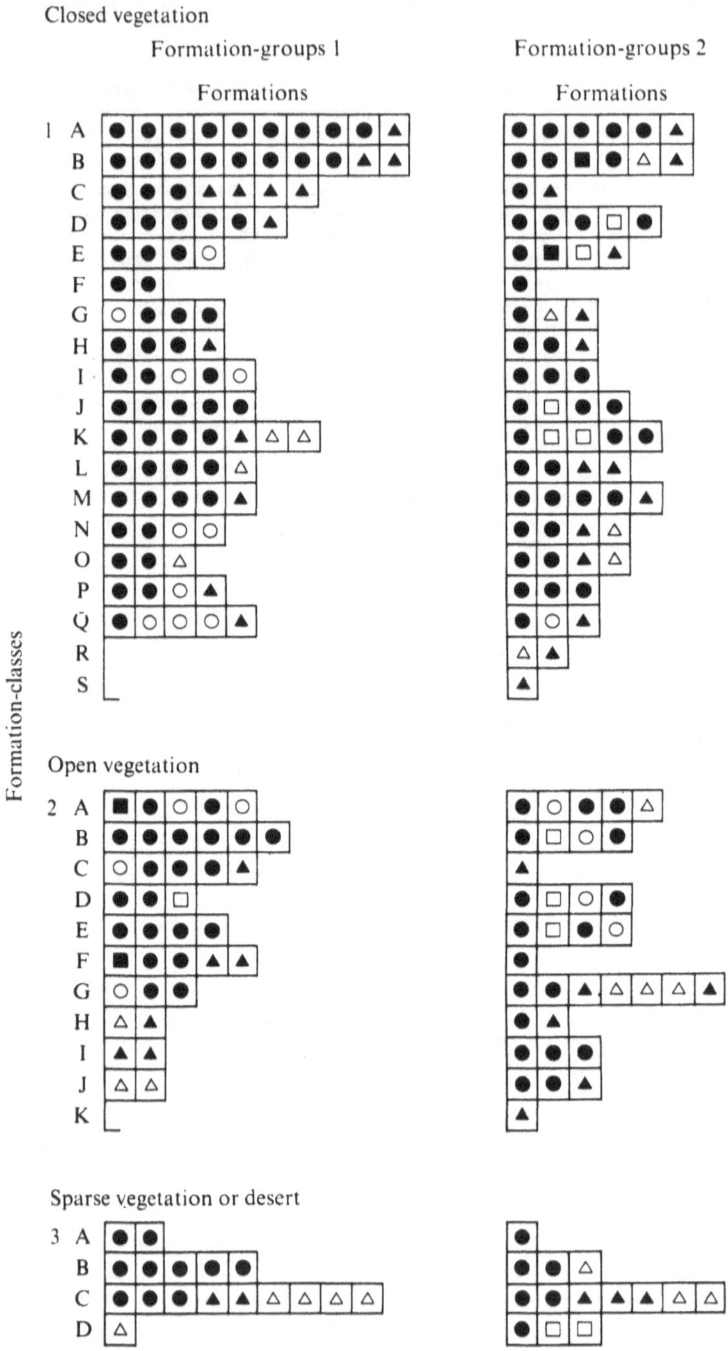

Fig. 4.3. Coverage of Fosberg formations in completed check-sheets. For symbols see Table 4.10.

Table 4.10. *Summary of coverage of Fosberg formations in completed check-sheets and key to symbols used in subsequent tables and figures*

	Recorded in check-sheets	Not recorded in check-sheets	Totals
Originally published with examples	● 159	○ 19	178
Originally published but of doubtful existence	■ 4	□ 11	15
Additional formations	▲ 44	△ 24	68
Totals	207	54	261

Table 4.11. *Numbers of different Fosberg formations recorded from each of the thirteen 'main areas' of the world*

	Number of recorded formations		
Main area	●	■	▲
N. America	135	2	13
C. and S. America	15	0	1
Europe	99	1	29
Africa	107	1	4
S.W. Asia	26	1	0
N. Asia	18	0	0
E. Asia	58	0	10
S.E. Asia	11	0	0
Australia	40	0	0
Atlantic Ocean	12	0	0
Indian Ocean	20	0	1
Pacific Ocean	48	0	4
Antarctic	0	0	0
Totals	159	4	44

received from them were the most suitable for distributional studies of this kind. In all, 178 Fosberg 'formations' belonging to thirty-three 'formation-classes' have been recorded from these 141 NP&ERs. Their distribution over the latitudinal zones shows little correlation with the numbers of check-sheets received from the zones. This may reflect real differences in vegetational diversity but in addition other factors may well be operating, such as differences in mean areas of the NP&ERs in different zones and also differences in the criteria adopted in the selection of areas for survey and in the training and relevant experience of the surveyors. The latitudinal distribution of individual 'formation-classes' shows that those reported from most of the zones are closed forest or scrub (1A and 1B) or open forest with

Table 4.12. *Distribution of recorded 'formation-classes' and 'formations' over the 10° zones of latitude*

		Numbers of		Numbers of Fosberg	
Zone	Latitude	NP&ERs	check-sheets	'formation-classes'	'formations'
1	90°–80° N	0	0	0	0
2	80°–70° N	1	1	8	13
3	70°–60° N	23	25	22	56
4	60°–50° N	19	27	27	74
5	50°–40° N	17	49	22	55
6	40°–30° N	22	36	24	69
7	30°–20° N	4	4	10	20
8	20°–10° N	1	1	1	4
9	10° N–0°	4	4	15	25
10	0°–10° S	4	10	15	40
11	10°–20° S	15	15	11	19
12	20°–30° S	16	16	8	11
13	30°–40° S	14	14	17	31
14	40°–50° S	1	1	5	7
15	50°–60° S	0	0	0	0
16	60°–70° S	0	0	0	0
17	70°–80° S	0	0	0	0
18	80°–90° S	0	0	0	0
Totals		141	203	33	178

closed lower layers (1D), shrub savanna (1K) and closed grassy or herbaceous vegetation (1M and 1N). Open steppe (2G) was the only formation-class of non-closed vegetation recorded for as many as half the zones. At the other extreme were some types of floating and submerged aquatic vegetation each recorded from a single zone. It is clear that the information so far available is too incomplete to warrant more detailed consideration of these and many other points of interest.

The most widely recorded Fosberg formations are listed in Table 4.13, those marked with an asterisk having recognized subformations. It is of interest that 1A1 1, 'Multistratal evergreen forest' ('rain forest') should be recorded from eight of the thirteen main areas, the countries involved being the following, those italicized lying within the range of tropical rain forest as defined by Richards (1952):

N. America – Olympic Peninsula (Washington)
C. and S. America – *Panama Canal Zone*
Africa – *Uganda*, *Nigeria*, Ethiopia, South Africa
S.E. Asia – *Sabah*

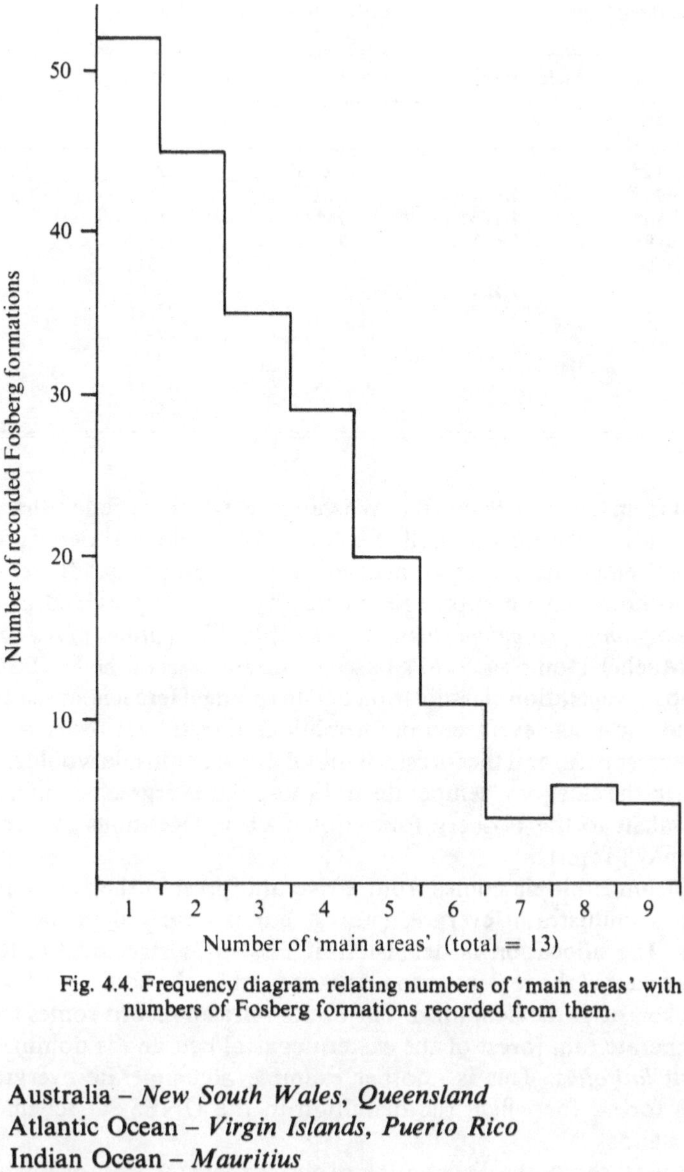

Fig. 4.4. Frequency diagram relating numbers of 'main areas' with numbers of Fosberg formations recorded from them.

Australia – *New South Wales, Queensland*
Atlantic Ocean – *Virgin Islands, Puerto Rico*
Indian Ocean – *Mauritius*
Pacific Ocean – *Solomon Islands*

Fosberg's definition is broad and, the classification being purely structural, it is not possible to exclude subtropical or temperate (rain) forest where it is tall, multistratal and predominantly orthophyllous. The least appropriate

Table 4.13. *Fosberg formations most widely recorded in IBP areas*

Number of 'main areas' from which the formation is recorded			
9	8	7	6
1A12*	1A11	1A15*	1A13
1A17*	1A14	1E12	1B11
1B14*	1A16*	1M22	1C13
1B18*	1B13*	2G21	1D14
1M21	1C12		1D21
	3B12		1L12
			1L21
			1M13
			1M23
			1O11
			3C21

example is that from Olympic Peninsula, Washington, which is predominantly sclerophyllous rather than orthophyllous. A suitable ecological description for it might be 'Temperate evergreen needle-leaved ombrophilous forest'. It is interesting to note that no appropriate category for it is provided in the *Tentative physiognomic-ecological classification of plant formations of the earth* (Ellenberg & Mueller-Dombois, 1967), based on discussions of the UNESCO working-group on vegetation classification and mapping. Here it is stated that 'temperate and sub-polar evergreen ombrophilous forests' are restricted to the southern hemisphere, and the forests of the Olympic Peninsula would have to be included in the category 'temperate and sub-polar evergreen coniferous forests', equivalent to the Fosberg formation 1A17a, 'Resinous evergreen narrow sclerophyll forest'.

The record from Ethiopia comes from a riverain forest in the Rift Valley and qualifies as multistratal evergreen forest but is clearly dependent on ground-water. The allocation is defensible if Fosberg's structural criteria alone are followed, but the real intention is made clear by the addition of 'rain forest' in brackets after the definition. The South African record comes from the warm temperate rain forest of the eastern coastal belt and is dominated by *Podocarpus latifolius*. This is another example of temperate evergreen ombrophilous forest, for which the definition in the UNESCO scheme is suitable as it stands.

Figure 4.5 (*a*)–(*l*) shows the distribution of check-sheet records for Fosberg formations by geographical regions ('main areas'). It is immediately clear that there is a heavy bias in favour of N. America, Europe, Africa and E. Asia, so that inferences about the world distribution of individual formations are likely to be misleading. It is of interest in this connection to compare the number of formations, forty, recorded from 331 IBP Areas in Australia with

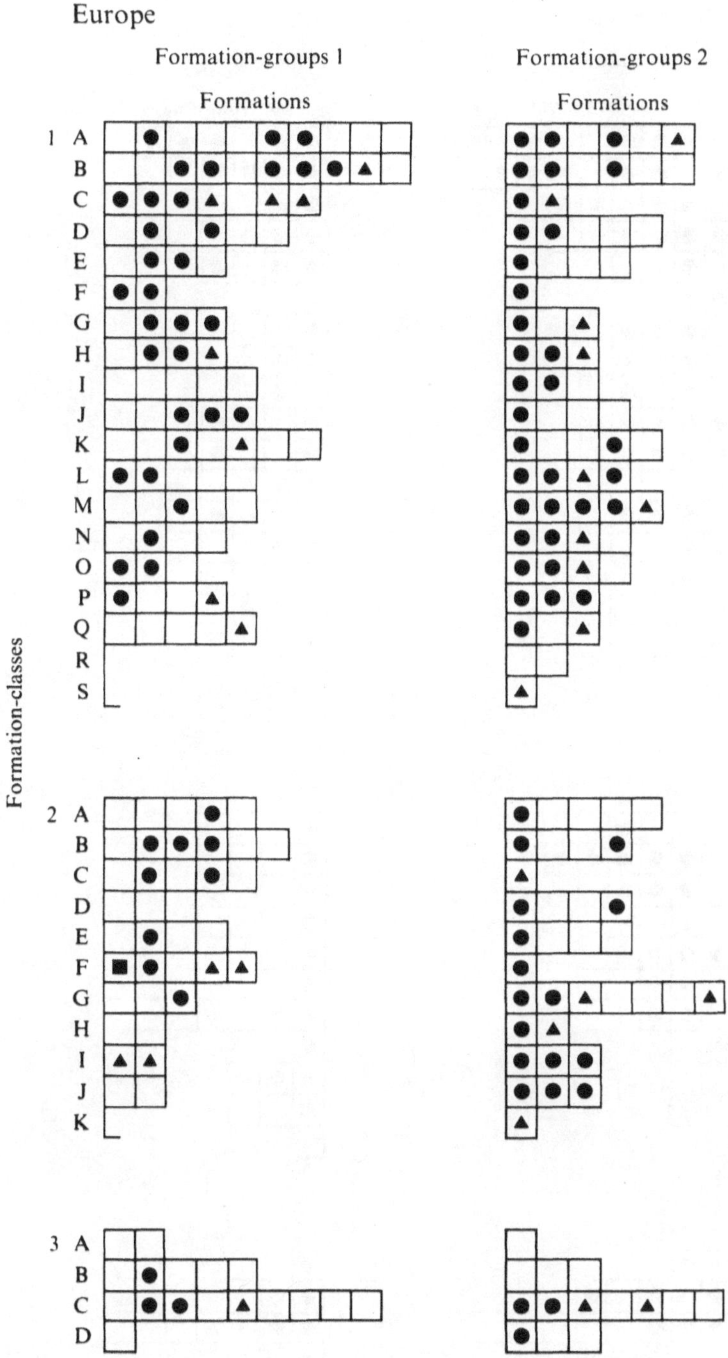

Fig. 4.5 *a–l*: distribution of check-sheet returns for Fosberg formations by geographical regions ('main areas').

Fig. 4.5 *a*

For letters and symbols see Table 4.9 (p. 155) and Table 4.10 (p. 157).

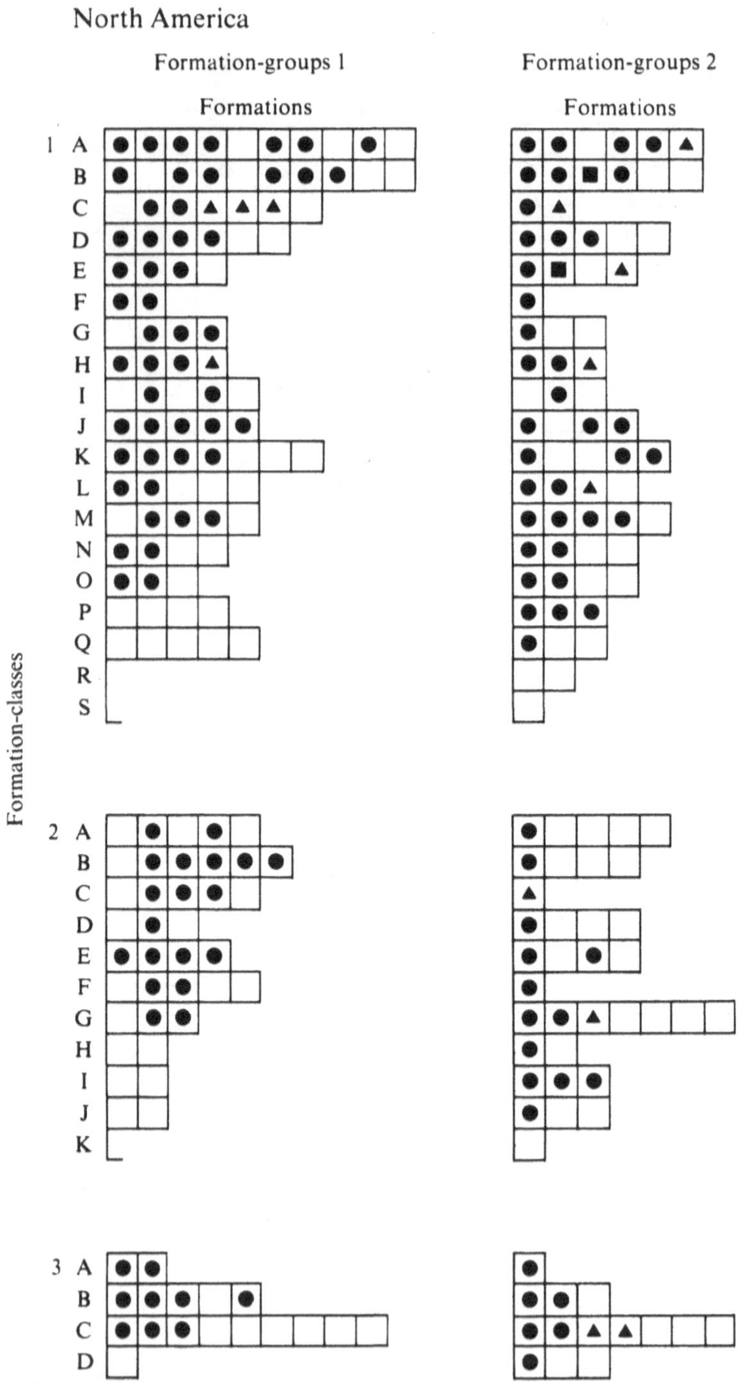

Fig. 4.5 *b*
For letters and symbols see Table 4.9 (p. 155) and Table 4.10 (p. 157).

Central and South America

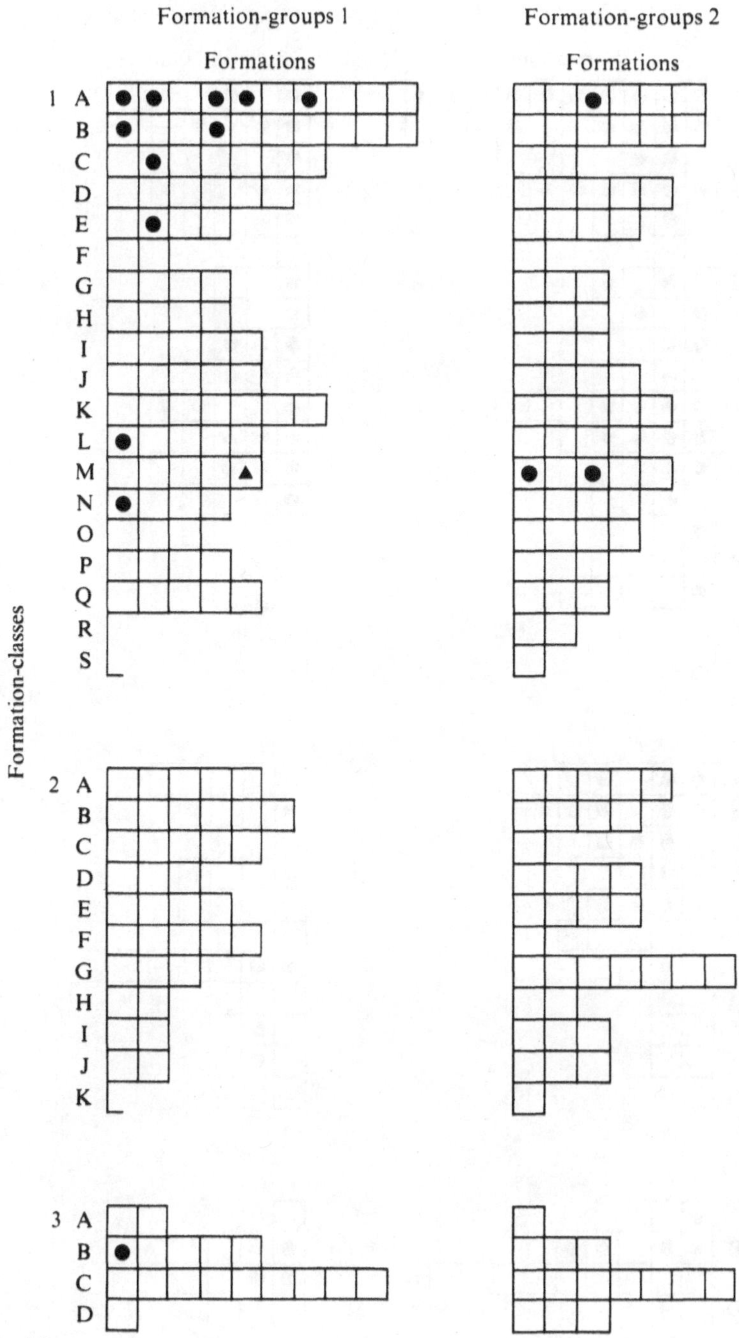

Fig. 4.5 *c*
For letters and symbols see Table 4.9 (p. 155) and Table 4.10 (p. 157).

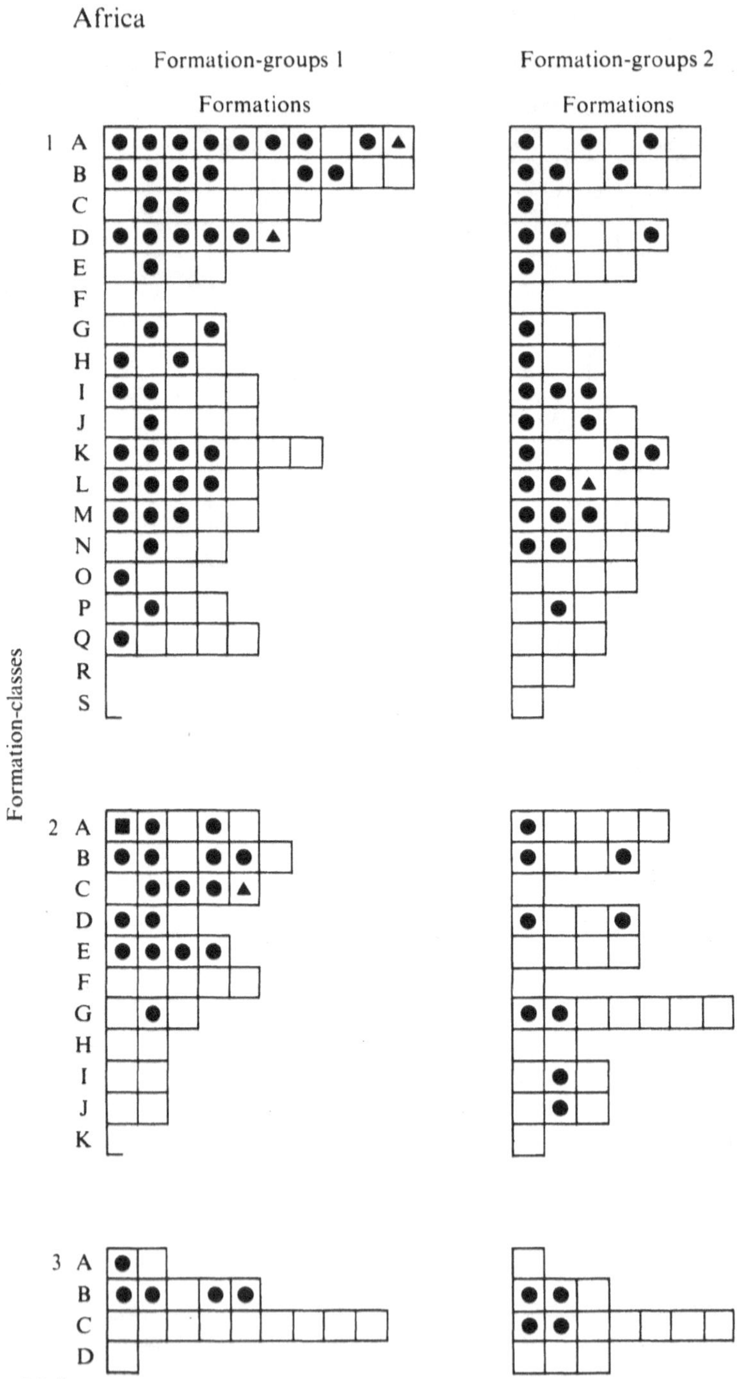

Fig. 4.5 *d*
For letters and symbols see Table 4.9 (p. 155) and Table 4.10 (p. 157).

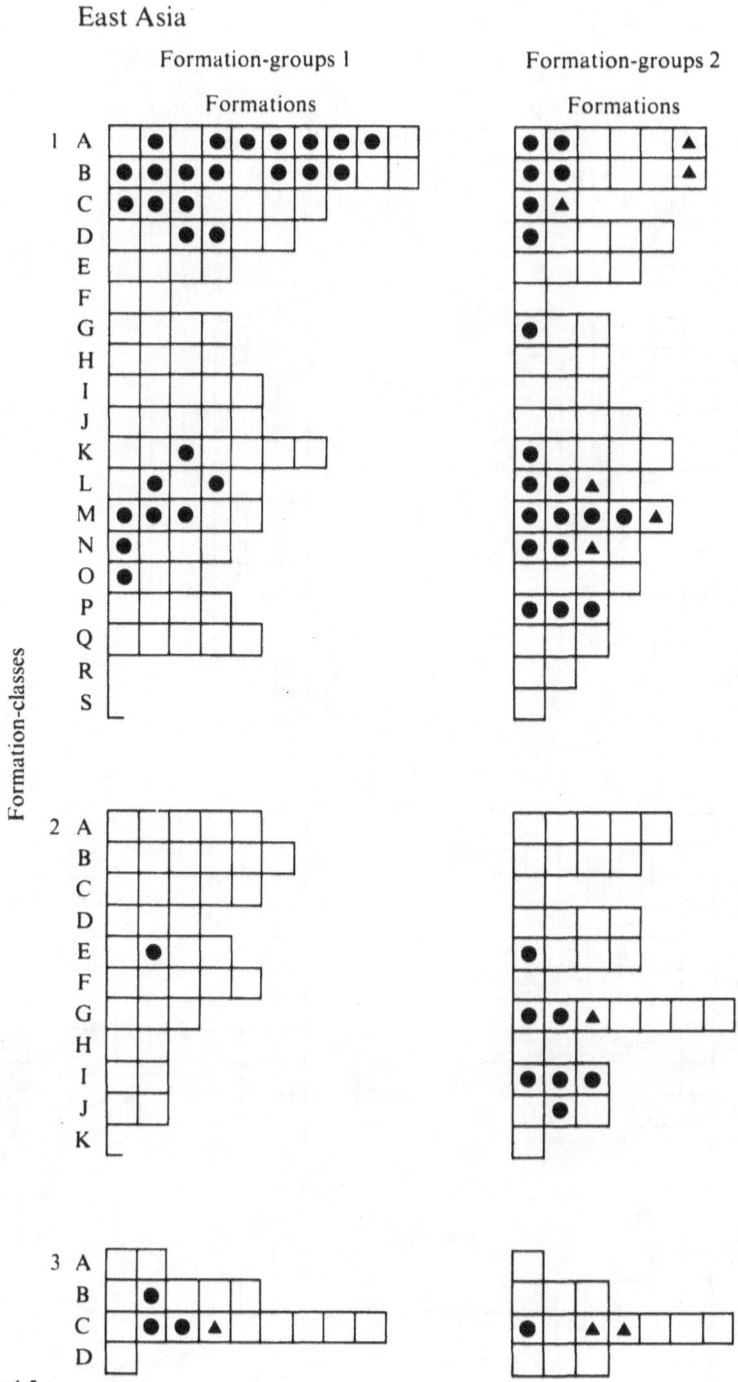

Fig. 4.5 *e*
For letters and symbols see Table 4.9 (p. 155) and Table 4.10 (p. 157).

South-East Asia

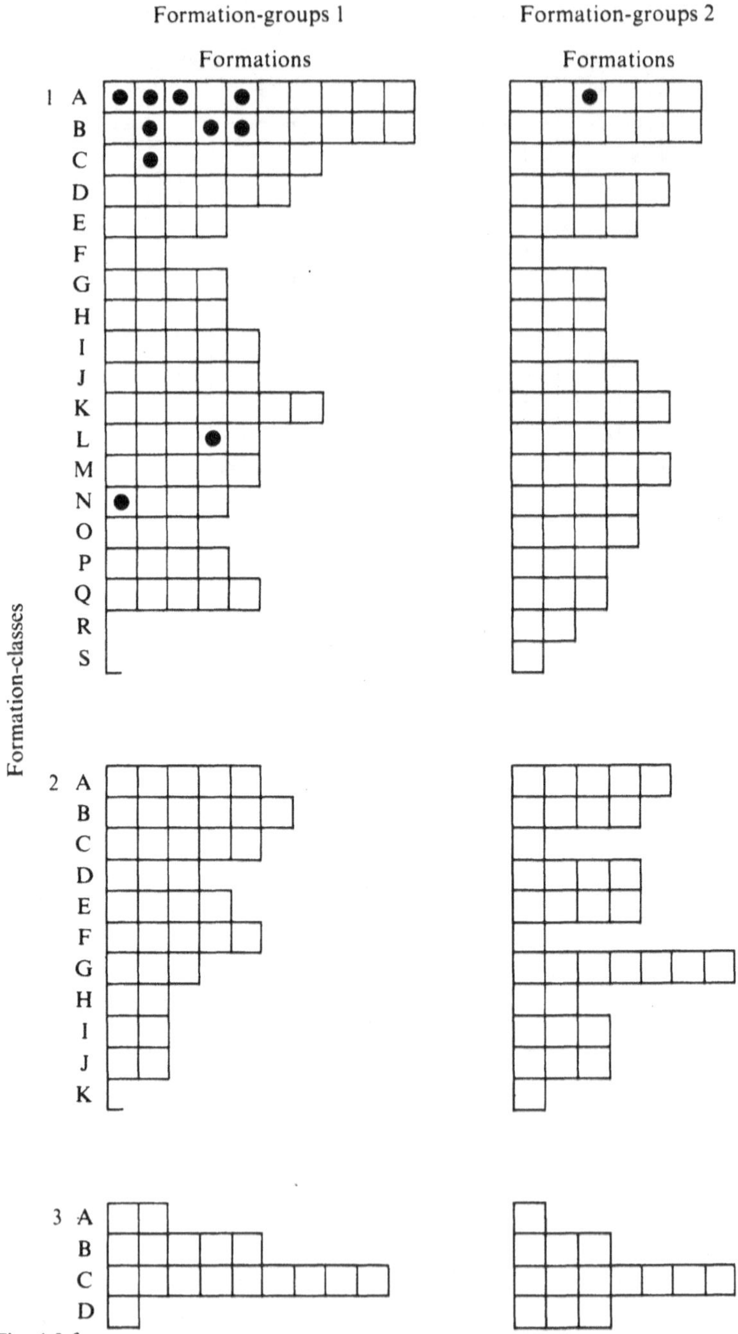

Fig. 4.5 *f*
For letters and symbols see Table 4.9 (p. 155) and Table 4.10 (p. 157).

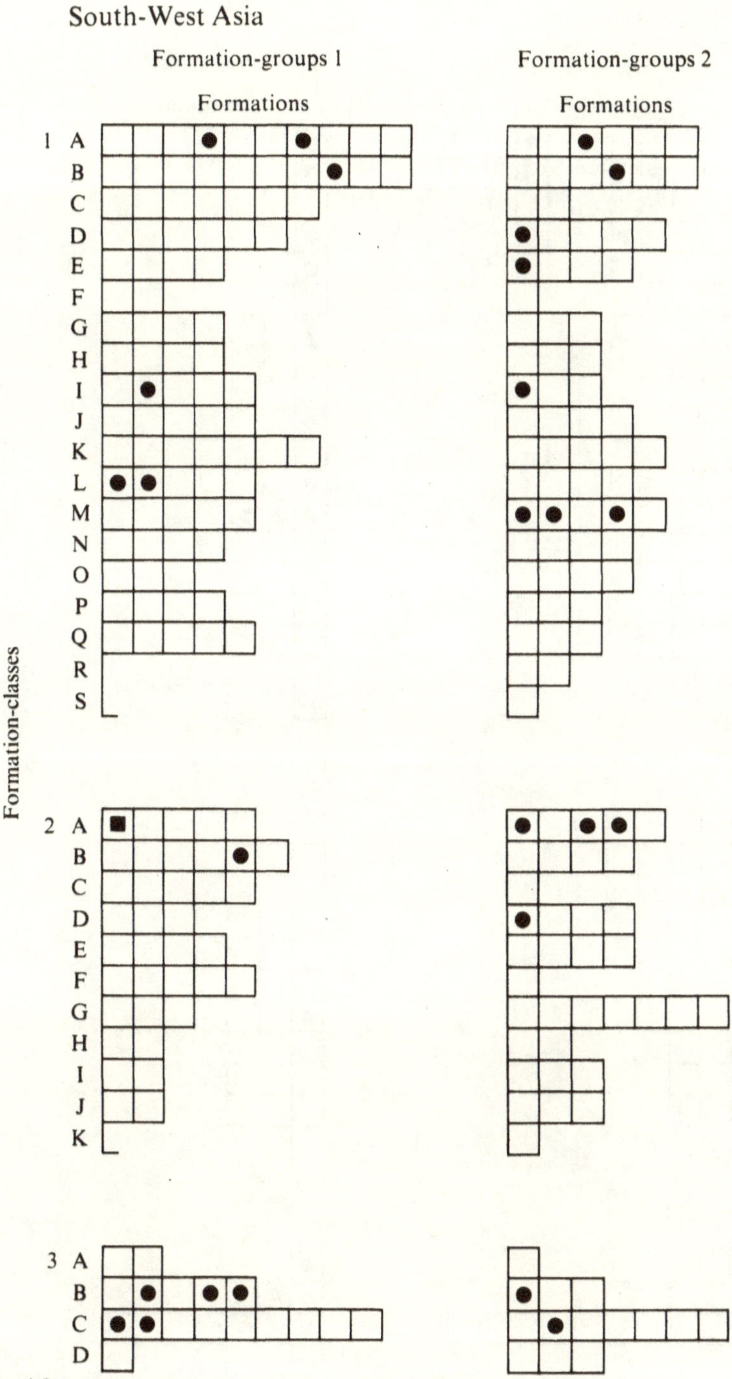

Fig. 4.5 *g*
For letters and symbols see Table 4.9 (p. 155) and Table 4.10 (p. 157).

Northern Asia

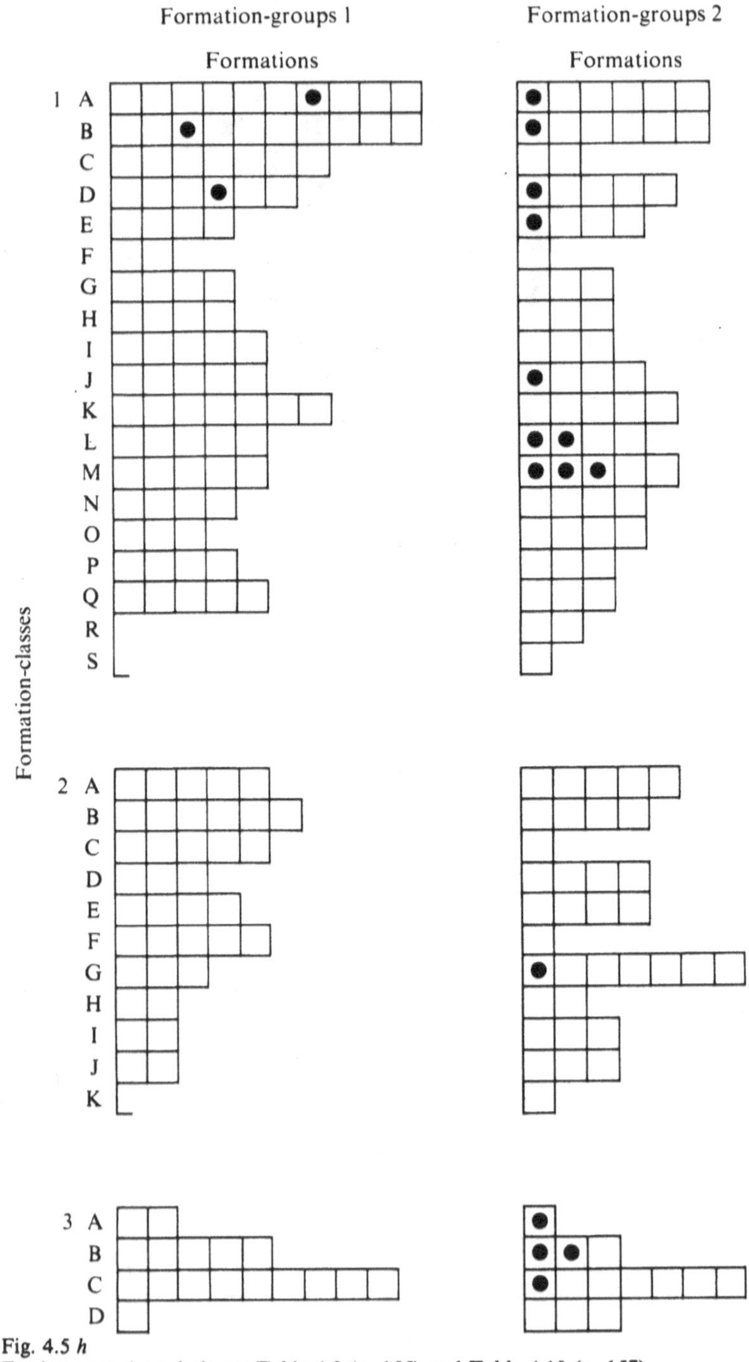

Fig. 4.5 *h*
For letters and symbols see Table 4.9 (p. 155) and Table 4.10 (p. 157).

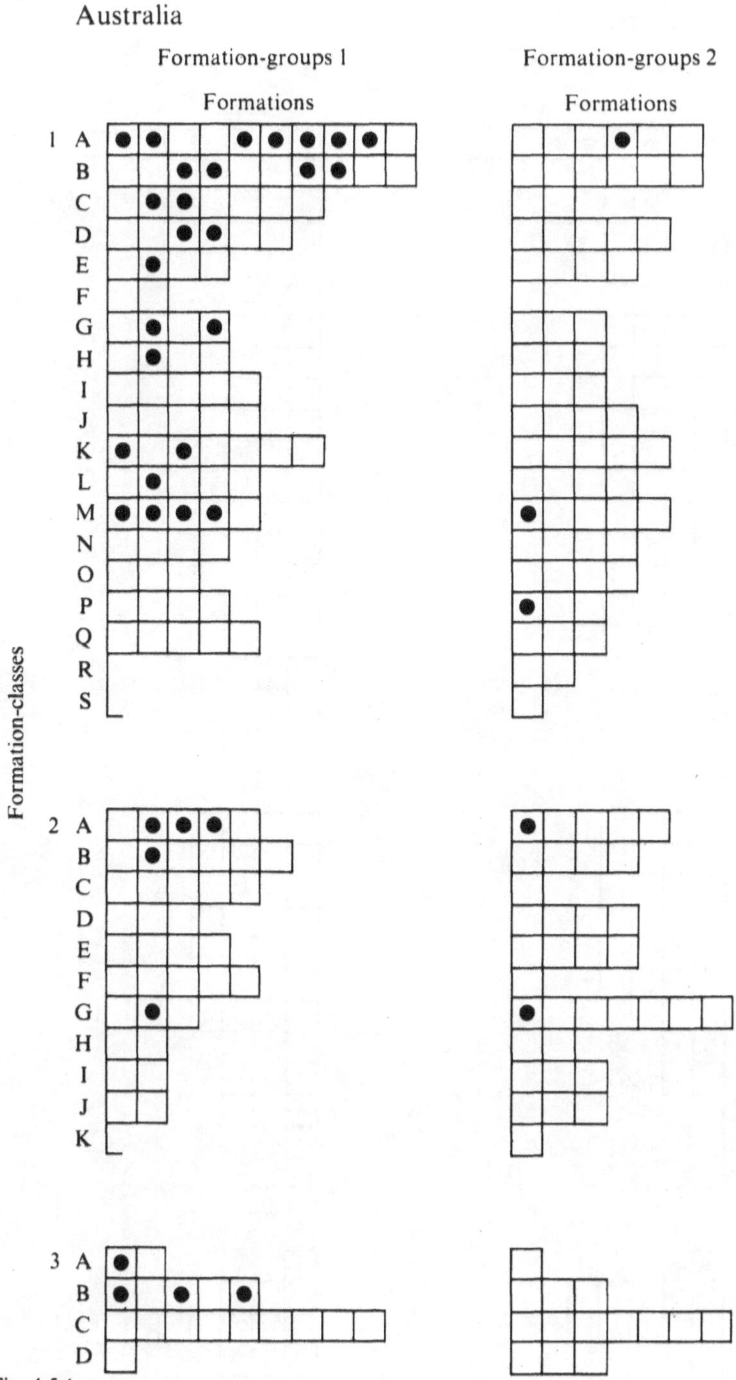

Fig. 4.5 *i*
For letters and symbols see Table 4.9 (p. 155) and Table 4.10 (p. 157).

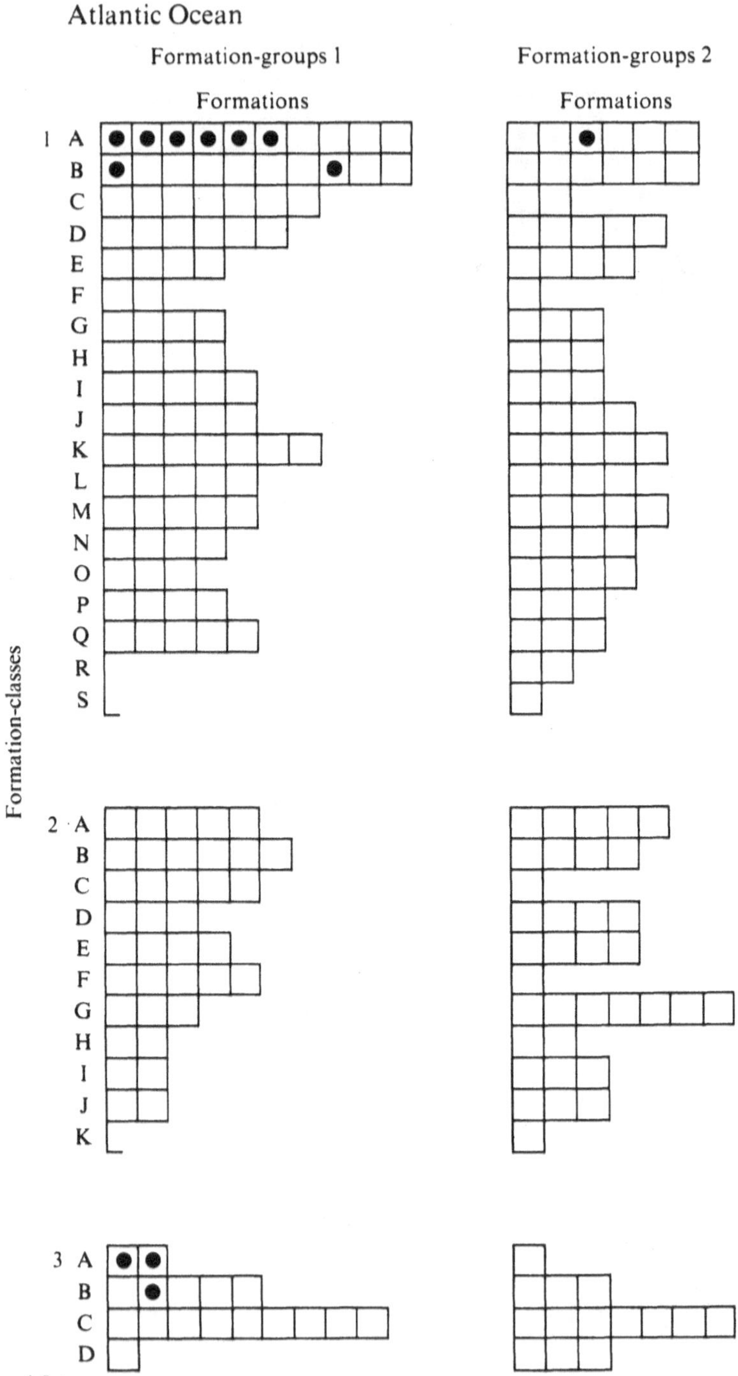

Fig. 4.5 *j*
For letters and symbols see Table 4.9 (p. 155) and Table 4.10 (p. 157).

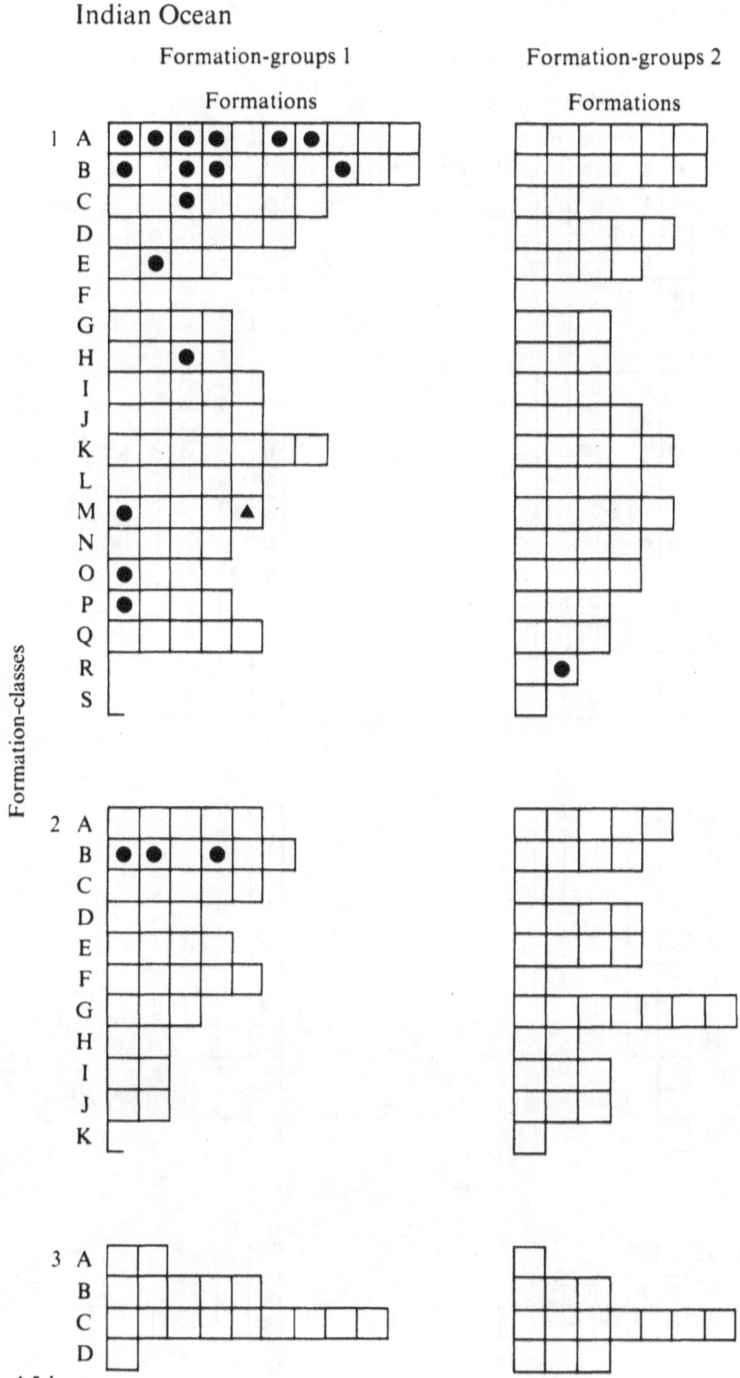

Fig. 4.5 *k*
For letters and symbols see Table 4.9 (p. 155) and Table 4.10 (p. 157).

Pacific Ocean

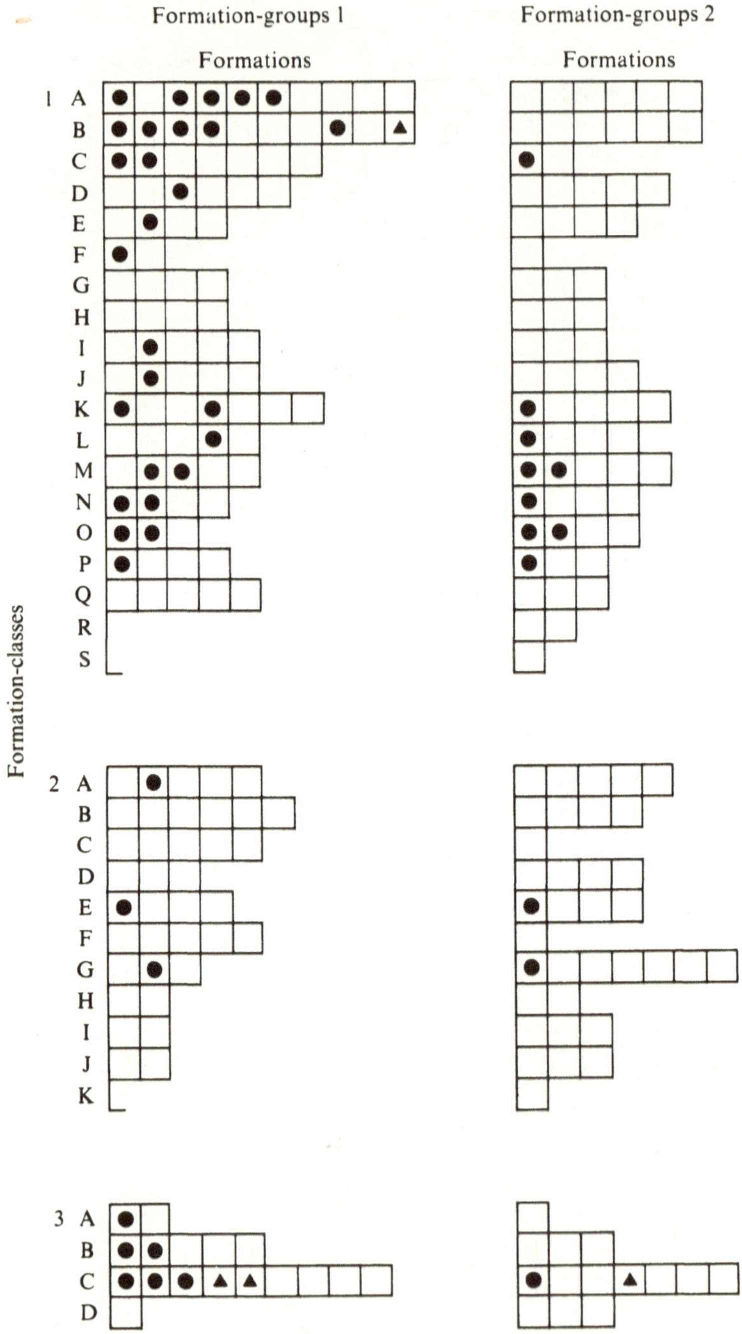

Fig. 4.5 *l*
For letters and symbols see Table 4.9 (p. 155) and Table 4.10 (p. 157).

the strikingly larger numbers from fewer sites in Africa (112 from 128 Areas) and Asia (113 from 181 Areas). L. J. Webb (personal communication) writes 'I made a very rough estimate of the number of Fosberg formations which I could recognize in Australia and got over eighty. The discrepancy means that 'conservation areas' sampled in Australia do not include many ecosystems, e.g., in Queensland none of the important grasslands of the west is in national parks. . . There is also unevenness in classificatory levels in tropical countries – many different rain forests are lumped under Fosberg's 1A1 1'. The Australian contribution to vegetational recording is discussed at some length in Section 4.3.3.

4.3.3 Coverage of vegetation-types in selected countries

Estimates of the coverage of vegetation-types in the check-sheet survey depend upon knowledge of the total number of different types present within the area under consideration. It was possible to state (p. 154) that about 80% of all Fosberg formations had been recorded on completed check-sheets because it now seems unlikely that any further substantial addition will be made to the present list of 261 formations occurring throughout the world. Comparable estimates of coverage for individual countries are possible only if it is known how many of these formations are present in each of them, and this information is available for regrettably few countries. There is a still smaller number for which we know, more or less precisely, the full list of the different plant communities occurring in them in terms of a floristic classification, which has also been used consistently in recording on check-sheets which of them are present within surveyed sites. It is therefore very rarely possible to answer the question 'What is the coverage of floristically-defined plant communities in completed check-sheets?' as part of an attempt to secure adequate representation in protected areas of the full range of vegetational diversity in a particular country or region. There are, nevertheless, some countries from which it is possible to gauge the success of the survey in providing answers to such questions, and examples are considered below.

Netherlands

A country for which much of the requisite information is available is the Netherlands. In contrast with the United Kingdom the Netherlands has a generally accepted comprehensive and detailed classification of plant communities based on floristic criteria, that of Westhoff & Den Held (1969), and this was used consistently both in the check-sheet survey and also in a recent independent examination of samples of natural vegetation throughout the country. All the floristically-defined plant communities have been allocated to their Fosberg formations (Appendix 2), so that it is possible to make estimates of the country-wide coverage in check-sheets of vegetation-types both of the purely structural and of the floristic kind. In the Final Report

(1974) of the Netherlands Committee for IBP, covering the years 1966–71, it is stated that ninety-two check-sheets had so far been completed at least at 'second level' in the survey of ten Class A, seventy-six Class B/C and six Class D sites having a combined area of 590 km², or about 1.5% of the total area of the Netherlands. These check-sheets recorded 92% of the eighty-eight 'alliances' and 73% of the 272 'associations' of the Westhoff–Den Held classification which have been listed for the Netherlands. It may be supposed that about 90% of the eighty-one Fosberg formations listed for the Netherlands have been recorded on check-sheets.

Australia

Attention has already been drawn in Section 2.2.4 to the decision of the Australian IBP/CT Committee to undertake a Conservation Survey of the whole of Australia and of Papua–New Guinea and to develop a classification of vegetation-types which would be applicable throughout the area to be surveyed. This classification is purely structural at the highest level, that of the 'structural formation', but floristic composition plays an increasingly large part in the definition of units at progressively lower levels in the hierarchy – 'alliances', 'associations' and 'societies'. For the most part the teams engaged in the survey recorded the different alliances they encountered but only occasionally recognized different associations.

In order to make an assessment of the coverage of vegetation-types in the Australian check-sheet returns we need to know the total numbers in each State and in the country as a whole as well as the numbers recorded on check-sheets; and for comparison with other countries we need this information for Fosberg formations as well as for the appropriate units of the Australian classification. Lists of structural formations and alliances may be found for each State in *Conservation of Major Plant Communities in Australia and Papua New Guinea* (Specht, Roe & Boughton, 1974), with the qualifying statement that 'Even though all available ecological information has been collated to make the lists as comprehensive as possible, ecological knowledge varies greatly from State to State.' It should be pointed out that the alliances of the Australian system are units distinctly finer than Fosberg formations, which are more nearly comparable with Australian structural formations. The still finer associations have not yet been mapped or listed in a comprehensive way.

Table 4.14 shows the numbers of 'important structural formations' and alliances, and of conservation areas of various kinds, listed for each State in the 1974 publication cited above, also the numbers of conservation areas from which check-lists were returned and of alliances and Fosberg formations recorded on the check-sheets. It will be seen that only in Queensland and South Australia were sheets completed for any appreciable proportion of the conservation areas, and it is therefore possible only for these two to make

Table 4.14. *Coverage of conservation areas (reserves) and vegetation-types in Australian check-sheet returns, based on information in Specht, Roe & Boughton (1974) and at the IBC/CT data-centre*

	Numbers of					
State	Important structural formations	Listed alliances	Listed reserves	Check-sheet returns	Alliances on check-sheets	Fosberg formations on check-sheets
New South Wales	22	202	106	3	25	10
Northern Territories	16	77	37	—	—	—
Queensland	22	153	261	261	58	25
South Australia	17	127	97	64	52	19
Tasmania	21	305	60	—	—	—
Victoria	23	125	97	2	—	9
Western Australia	23	218	81	1	—	7
Australia	31	*c.* 900	739	331	(135)	40

any useful estimates of coverage. For Queensland, the only State in which check-sheets were completed for all the conservation areas that were visited, fifty-eight alliances were recorded out of a listed total of 125, giving a coverage of 38%. For South Australia, with fifty-two recorded in check-sheets out of a listed total of 127, the coverage is 41%, despite the incomplete representation of conservation areas. Even fewer data are available for Fosberg formations, but for Queensland L. J. Webb (personal communication) estimates a total of sixty-four, so that twenty-five recorded on check-sheets constitute a coverage of 39%, very close to the corresponding figure for alliances of the Australian classification. For Australia as a whole the small set of completed check-sheets record forty Fosberg formations out of an estimated total of rather more than eighty, a coverage of something under 50%.

An objective of the Australian Conservation Survey was 'to enumerate all the plant communities...which were already adequately conserved in National Parks and Reserves', comparable with the aim of the check-sheet survey to provide information on a world basis on the extent to which plant formations were already adequately safeguarded. Table 4.15 summarizes data in the 1974 publication on the conservation status in each State of every alliance recorded for the State. This is assessed in the terms of a subjective classification:

Excellent — well conserved in several large reserves exhibiting ecological diversity and subject to little human influence

Table 4.15. *Conservation status of the alliances of the Australian vegetational classification which are listed for each State in Specht, Roe & Boughton (1974)*

State	Conservation status (%)		
	Adequate	Inadequate	Nil
New South Wales	32	33	35
Northern Territories	47	15	38
Queensland	12	42	46
South Australia	48	21	31
Tasmania	51	34	15
Victoria	22	58	20
Western Australia	24	36	40

Reasonable conserved in a large reserve or a number of small reserves; ecological diversity is usually present and human interference is minimal

Moderate conserved only in one or two small reserves often subject to human pressures

Poor conserved in only a few very small reserves subject to human activities; conservation thus precarious

Nil no reserve includes the plant community.

In Table 4.15 these grades have been reduced to three, the first two being combined as 'adequate' and the next two as 'inadequate'.

It will be seen that there is wide variation between States in the proportion of listed alliances regarded as at least reasonably well conserved, with an average close to one in three. At the other extreme there is about the same average proportion of alliances not included in any conservation area. The chief reason for this is the uneven distribution of National Parks, Forest Parks, Wildlife Reserves, Scenic Reserves, etc., which tend to be concentrated within easy reach of the large centres of population and to be very sparse in areas where land is in greatest demand for other uses, such as grazing. 'Of over 600 000 km^2 of tussock grassland only eight alliances out of 111 are adequately conserved' and 'out of fifty-four alliances of low shrubland only one has been included in that category' (Specht, Roe & Boughton, 1974). L. J. Webb (personal communication) confirms this for Queensland: 'most of the National Parks in Queensland (with the exception of the recently proclaimed Simpson's Desert in the interior and a couple of other western ones in rugged sandstone country) are along the mid-eastern and south-eastern coast. This excludes the extensive and diverse grasslands, grassy woodlands, grassy open woodlands, etc., of western Queensland, i.e., the savanna and steppe formation classes as well as the tropical mixed woodlands, often with palms and interspersed with orthophyll Indomalesian species, of Cape York Peninsula'. It had been noted at the IBP/CT data-centre that there were no records, on check-sheets from Australia, of the savanna and steppe formations of the Fosberg system, and it is now clear that the explanation lies in the location of conservation areas. There are too few completed check-sheets, except from Queensland, for assessing the proportion of Fosberg formations that are adequately conserved, but there is no reason to suppose it to differ greatly from the figures for alliances in Table 4.14.

The results of the Australian Conservation Survey are of considerable relevance to the purposes of this volume. The Australian IBP/CT Committee were faced with the problems that confronted a majority of the countries participating in the check-sheet survey, in particular those surveying as many as possible of the conservation areas already established and of estimating how far they included representatives of all communities and species meriting conservation. There were the further requirements of finding satisfactory systems of description and classification at least of plant communities and soils so as to convey more of the information essential for decisions on nature conservation than could be provided by the exclusive use of the Fosberg and Ball systems. The vegetational classification adopted by all the States has been explained above (pp. 39–41). Its 'alliances', which have been identified and listed with varying degrees of comprehensiveness for every State and for most of the conservation areas, are units distinctly finer than Fosberg formations. Units giving a still greater degree of discrimination and with a stronger floristic element will almost certainly be requisite for final decisions on nature conservation. This purpose might be served by the 'associations' of the system, but to extend the classification down to this level over the whole country will call for a large-scale mapping programme based on aerial photography and subsequent ground survey. Meanwhile it has become clear that the present conservation areas, numerous as they are, are not yet so located or of sufficient number and size to ensure the conservation of more than about a third even of the listed alliances.

Northern Tunisia

Reference has already been made (pp. 78–80) to the survey of IBP Areas in Tunisia carried out by Radford and Peterken (1971). Comprehensive ecological and vegetational studies had culminated in the production in 1966–7 of a series of five vegetational maps covering northern Tunisia at a scale of 1:200 000, *Carte Phyto-Écologique de la Tunisie Septentrionale* (M. Gounot & A. Schoenenberger), followed a little later (Le Houerou, 1969) by two for central and southern Tunisia at a scale of 1:500 000. At about the same time a number of lists was drawn up of sites of scientific importance in Tunisia, in particular:

(i) the *List of European and North African Wetlands of International Importance*, compiled by Olney (1965).
(ii) a list of sites of ecological and botanical importance prepared by A. Schoenenberger in 1967;
(iii) lists in contributions by A. Ben Aissa, A. Schoenenberger and others to the IBP/CT Conference at Hammamet in 1968.

These three sources provided between them a list of sixty-three sites judges to be worthy of conservation on scientific grounds. They fell into four geographical groups: (1) north of the Dorsale chain of mountain which extend parallel to the coast westwards to Cap Bon; (2) the Cap Bon peninsula; (3) the Dorsale chain, and (4) the area south of the Dorsale. As a result of discussions during the visit by Radford and Peterken, seven of these sites were provisionally graded as Class A, 'of outstanding scientific importance within Tunisia and possibly of international importance', and a further eighteen as Class B, 'sites of considerable scientific importance either for the conservation of representative stands of Tunisian vegetation or for the preservation of populations of particular plants and animals'. The remaining thirty-eight, placed in Class C, were judged as scientifically less valuable than the Class A and Class B sites.

The majority of these sixty-three sites are in the northern part of the country: only seventeen lie south of the Dorsale chain, most of them selected mainly for their relict plant communities or their bird populations. The continued steady advance of desert communities at the expense of steppe and scrub made it appear a matter of secondary importance to conserve samples of all of them for scientific purposes, and there is little doubt that the range of vegetational variation in the southern part of the country is not fully represented in the relatively small number of selected sites.

The extent to which the full range of vegetational diversity is represented in listed and surveyed sites can therefore be better assessed for the northern part of the country. The *Carte Phyto-Écologiques de la Tunisie Septentrionale* (M. Gounot & A. Schoenenberger), to which reference was made above, were produced by scientists from the Centre d'Études Phytosociologiques et Écologiques at Montpellier and cover, at a scale of 1:200 000, the whole area in which the forty-six northern sites are located. It is, therefore, possible to estimate what percentage of the total number of vegetational mapping-units found in the whole of northern Tunisia are present in the listed sites. The units in question, as far as natural and semi-natural vegetation is concerned, are the *groupements forestiers* (pp. 78–80) of each of the *étages bioclimatiques* (Emberger, 1960). For the area as a whole sixty-five *groupements* are mapped, and of these forty, or 64% of the total, are included in the selected sites. If only Class A and Class B sites are considered, the number of *groupements* falls to twenty-six, or 41% of the total and 65% of those recorded for all forty-six sites. The twenty-five *groupements* which have not been recorded for selected sites include seven about which there is uncertainty, because of imprecise boundaries, as to whether they are or are not present. There remain eighteen *groupements* which definitely do not occur within any of the sites. These constitute 27.7% or rather more than a quarter of the total number, but many are of very small extent or differ only very slightly from *groupements* that are included in one or more sites. Others again are of little ecological importance, but there are some significant omissions. Amongst these, for example, is the '*groupement à Quercus ilex, Acer monspessulanum, Lamium longiflorum*' which covers a very restricted area at the summit of Djebel Zaghouan but is of great phytogeographical interest as an eastward extension of the flora of the Atlas Mountains. It is hoped that extended or additional sites may be found where this *groupement* can be conserved.

As has already been explained (pp. 78–80) the *groupements forestiers* which were the basis for mapping the vegetation of northern Tunisia are neither structural nor simply floristic units. During their visit in 1969 Radford and Peterken were able to make on-the-spot surveys of fewer than half the sites. At these they recorded the vegetation in terms of the Fosberg system and by the major species present, including the ecological dominants where appropriate. For sites not visited by them, information obtained from the maps was interpreted and supplemented by Dr C. Floret, who knew most sites personally and was therefore able to report on the degree of openness and seasonal features of the communities essential for their Fosberg coding. Even so, some of the finer variations in physiognomy could not be recorded satisfactorily, and the level of variation in the sites in terms of the Fosberg system is no doubt underestimated.

Radford and Peterken state that, out of a total of sixty-five Fosberg formations reported for Tunisia up to 1969 (Fig. 4.6), all but five, or sixty, were recorded from the forty-six northern sites. These sites were believed to be 'reasonably representative', but the total number of formations in northern Tunisia is not known with any precision; nor can it be safely inferred from the recorded *groupements* for reasons discussed on pp. 79–80. It is of interest in this connection that fifty-four of these sixty formations, or 90%, were recorded from the twenty Class A and Class B sites which together constitute less than half the total number. The corresponding figure for *groupements* was only 65%, despite the fact that the total number of formations was not far short of the number of *groupements* – sixty against sixty-three. This emphasizes yet again the fundamental differences between these two modes of specifying types of vegetation. No simple equating of formations with *groupements* is possible, nor can any combination of the two procedures be envisaged as a practicable single system of vegetational recording.

The upshot of this assessment of coverage of vegetation-types in check-sheet

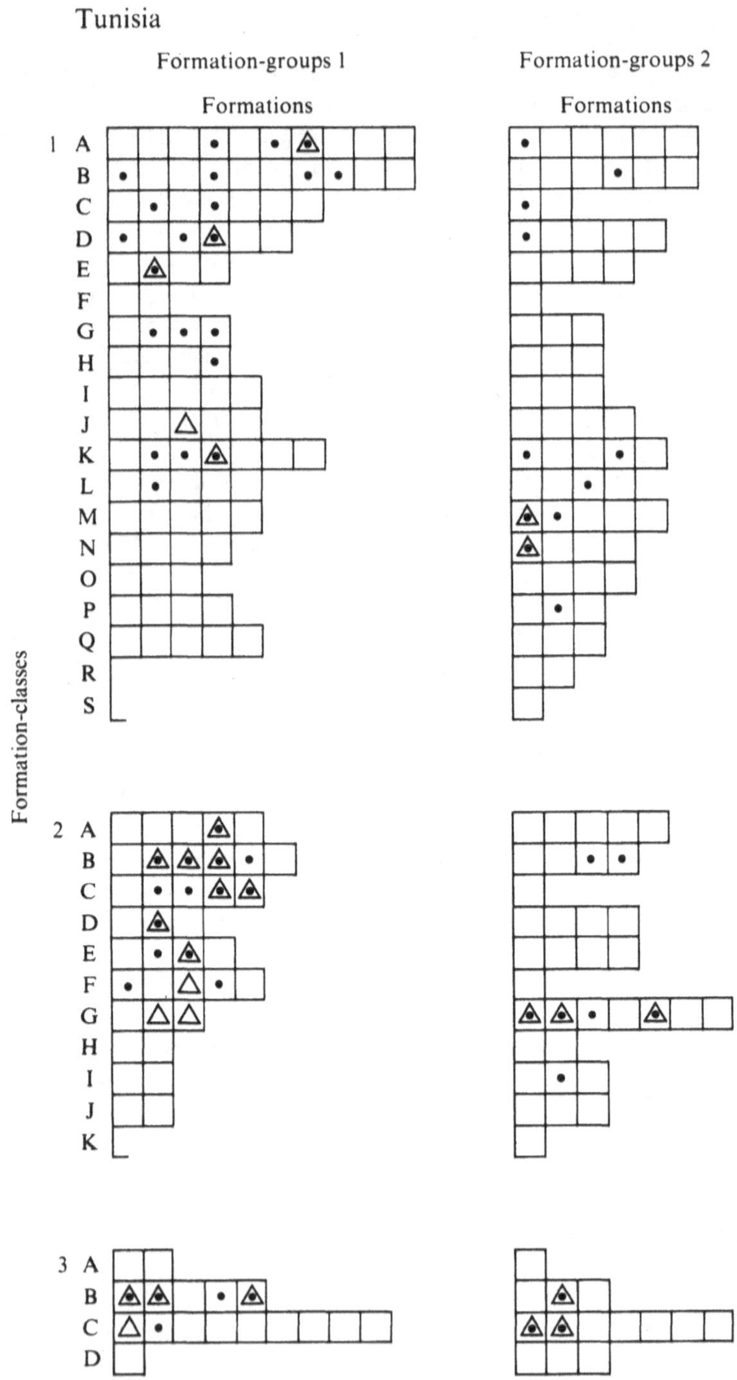

• North and Central Tunisia, from CEPE maps and field work
△ South Tunisia from Le Houerou's maps

Fig. 4.6. Fosberg formations present in Tunisia.

returns from northern Tunisia is that there is good reason to believe that quite a high proportion, perhaps as high as 90%, of the Fosberg formations of the whole area were recorded in the returns. The proportion of the *groupements forestiers* is lower but is still almost two-thirds of all those shown on the vegetation-maps at a scale of 1:200 000, and the remainder seem likely to include many that are relatively unimportant. This is an encouraging outcome both of the selection and of the check-sheet survey of these sites proposed for conservation.

4.4 Coverage of climates and factors of the physical environment

The questions in Section 6 of the check-sheet ask for the name and location of the climatological station nearest to the survey-area and whether there is a station of any kind within the area itself. It had previously been ascertained that only rarely is all the desirable meteorological information about an IBP Area available without recourse to a recording station outside the area and often at some considerable distance from it. On the other hand many such areas record certain data, especially temperature and rainfall, the surveyors were therefore invited to attach summaries of the findings to their completed sheets. Since data from the nearest full recording station could be obtained and inserted by the data-processing centre, it was thought advisable to relieve the surveyor of what might turn out to be a troublesome and time-consuming task. In the test-sample of 200 completed sheets it was found that 97% provided answers to the questions in this section and 18% appended meteorological data collected in the survey-areas, but only 6% of the areas had climatological stations.

An assessment of the coverage of climatic types in completed check-sheets must depend upon the criteria adopted for distinguishing and classifying different types. No attempt has so far been made to make any estimate of what proportion of the full range of types recognized in any of the widely accepted systems of classification is represented in the answers in Section 6. It may be inferred from the global distribution of surveyed IBP Areas (pp. 150–1) that the various types of temperate climate, however classified, are far better represented than are those of tropical and northern subtropical latitudes between 30° N and 10° S, and no surveyed area lies within the Antarctic continent. It would seem to be a useful exercise to explore this aspect of coverage as a means of obtaining a valuable pointer to gaps in the series of nature reserves and equivalent areas.

Section 9 of the check-sheet seeks information on the geomorphology of the survey-area, i.e., on its general type of landscape (whether flat, undulating, hilly or mountainous, how sharply or gently dissected and whether with incised or skeletonized relief) and on any special landscape features. Entries were made in 86% of the sample of 200 completed sheets and for the most part these were satisfactorily informative (Appendix 6). As with types of climate it is not possible to estimate the degree of check-sheet coverage of the full range of variation in physical landscape, but it seems certain that there is an overrepresentation of hilly or mountainous and sharply dissected landscapes, those less likely to have been taken over for agriculture or plantation-forestry than flat or gently undulating country.

The next sections, 10–13 deal with the nature of the coastline (if any) and the representation within the survey-area of the various forms of surface water: whether fresh, brackish or salt, whether running or standing, permanent or intermittent. Information on the trophic level of fresh water is also sought. Answers to these questions are considered in detail in Appendix 6. Analysis has not yet proceeded sufficiently far to yield information on the extent of check-sheet coverage of the full range of variation in these features.

4.5 Coverage of faunistic information

In two samples of completed check-sheets, each of 100 sheets, 52% recorded a positive entry for Question 14.2, which asked whether the survey area had outstanding faunal features, and 40% for 14.3, asking for names of threatened, endemic, relict and rare animal species found in the area. It is very difficult to assess the accuracy and completeness of the information supplied in these entries, but it is clear that the questions as worded might elicit no more than the surveyor's personal opinions in the light of his particular interests and training. The extent of the bias towards groups with which he was familiar cannot be judged unless a site is so well known that omissions can be detected: for sites being surveyed for the first time there is no means of judging. For most sites, records tend to be more complete for the higher than for lower groups of animals. Undoubtedly there is scope for some reconsideration of the form of this section of the check-sheet and of the wording of corresponding parts of the *Guide* so as to encourage a less subjective approach (pp. 198–9).

Reports from some participating countries show their check-sheet entries to have been based on detailed faunistic investigations. The Japanese Synthesis Volume 9, for instance, provides quantitative faunistic information for a number of areas. These comprise five 'Main areas', extending from the Mt. Daisetsu National Park in northern Hokkaido to the Mt. Kirishima Area in Kyushu, selected to cover the main climatic zones represented in the country as well as a large range in altitude; and sixteen 'Supplementary areas' which have 'special biotic communities of biological interest or offer special problems for nature conservation' and include the Mt. Fuji and Mt. Mino areas, Kinkasan Island, and the Ogasawara Islands National Park far to the south-east. Before the faunistic surveys of these areas were undertaken there were discussions on survey methods which resulted in a publication entitled *Research Methods for Describing the Fauna* (Kato, 1967). This dealt with all the main groups of animals and provided a common methodology which was followed in each of the Main areas, where teams of specialists covering the important animal groups took part in comprehensive surveys in a number of sites carrying different types of vegetation. For the most part these faunistic surveys took place in summer, but there were a few winter visits for observations on the larger mammals. The survey procedure was similar in many of the Supplementary areas, and the whole set of investigations provided a coherent body of faunistic information relevant to problems of nature conservation in Japan and ensured an unexceptionable basis for faunal entries in the check-sheet.

In many other countries too little precise information was available for a wholly satisfactory completion of the faunal parts of Question 14. The *Final Report* of the Netherlands Committee for the IBP (1974), for example, stated that ninety-two check-sheets had been completed for areas containing 72% of all the 'plant associations and equivalent communities known to exist in the country', but added that 'Data on fauna are incorporated in the check-sheets as far as they were available either in the literature or by personal communication.' The *Australian IBP Progress Report for 1967–1970* (published in Canberra, 1970) noted that 'Limited information on fauna (required for p. 8 of the...check-sheet) is available for most Parks and Reserves in Australia. An intensive faunal survey will need to be instituted at a later date to complete the check-sheets.' And in *Conservation of Major Plant Communities in Australia and Papua New Guinea* (Specht, Roe & Boughton, 1974) it is explained that although 'it would be impossible to tackle the conservation status of Australian fauna without a detailed faunal survey', it was hoped that 'most fauna would be conserved if a satisfactory network of reserves containing all major plant communities recorded in Australia was achieved'.

It must be concluded that little would have been gained by asking for more detailed faunistic information in the check-sheets. It is widely recognized, however, that present knowledge is inadequate for an effective conservation policy and it has become clear that faunistic survey is one of the fields into which more effort must be directed in the immediate future.

4.6 Coverage of types of human impact

Section 16 of the check-sheet deals with 'significant human impact in the survey-area and the *Guide* explains that information is sought on 'the actual human impact which is taking or has taken place within the IBP Area', in contrast with the aim of Section 17 which is concerned with legal restrictions on various kinds of human activity in the interest of nature conservation and also with management policies tending to promote conservation. The major forms of disturbance and modification are listed for the surveyor to check, and the *Guide* explains at some length what is implied by *significant* impact and considers each of the main types: cultivation, drainage, other soil disturbance, grazing, selective flora disturbance (extraction of commercially valuable trees, collection of orchids, reed-cutting, etc.), logging (removal of all or most trees without replacement), plantation, hunting, removal of predators, application of pesticides, introductions (of plant or animal species), fire, permanent habitation, recreation and research. Space is left for reference to other types of human impact.

It is, perhaps, not surprising that entries tend to emphasize current and quite recent impacts and seem often to have overlooked those of the more distant past. Thus the removal of predators, which must commonly have accompanied hunting in its various forms and is an impact of undoubted ecological significance, is rarely entered. On the other hand research, though its ecological effects cannot often have been significant in comparison with those due to others of the factors, is easily the most frequently recorded type of impact. It must be concluded that the completed check-sheets provide a satisfactory coverage of more recent forms of impact but that in many areas more information is desirable on earlier human influences whose consequences may well persist to the present day.

In Appendix 6 further points arising from entries to this section are considered in detail, including certain omissions (as of the use of fertilizers) and the desirability of seeking some assessment of levels of impact rather than mere presence or absence.

4.7 Coverage of rare and threatened species

Subsections 2–5 of Section 14 of the check-sheet ask for information as to the presence or absence in the survey-area of rare and of threatened/relict species of animals and plants and also of species of biogeographical interest, together with lists of such species. The *Guide* distinguishes the first two of these categories as 'rare in the region containing the IBP Area, but not necessarily elsewhere' and 'rare when considered on a world scale' respectively; and on 'Species of biogeographical interest' there is the comment 'e.g., on the edge of their range'. Problems arising from these headings and notes and the ways in which surveyors have reacted to them are considered in Appendix 6, where some suggestions are made for revision of the wording. Even after removal of all obscurities and ambiguities of wording, however, this set of questions would remain difficult for surveyors to answer in a satisfactorily objective manner, and virtually impossible in any but well explored regions. It is correspondingly difficult to assess the extent to which positive entries are accurate and comprehensive by the time of the survey. In only a few countries had a special effort been made to record the distribution of all higher plant species judged to fall into any of the above categories as a necessary preliminary to detailed planning for species-conservation. Whereas the basic information as to which mammals and birds were threatened was becoming available in the IUCN Red Data Books, only very recently have comparable plant data been assembled. Indeed, it will be some years before the majority of the estimated 25 000 plant species dangerously rare or under severe threat of extinction have been identified. (The comparative figure for higher animals is around 1000.)

Such a task can only be carried out nationally or internationally, as the full distribution of each species must be considered and use made of the knowledge of numerous experts, each covering a different group or area. Only when such a base list, either by country or by region, has been put together can field workers in general ascertain which species in their area are rare or threatened on a national or international basis and hence worthy of special protection.

One of the first such efforts culminated in *Conservation of Major Plant Communities in Australia and Papua New Guinea* (Specht, Roe & Boughton, 1974) where it is stated that

> As a logical development of the survey of the conservation status of major plant communities in each State, a list of endangered plant species has been prepared by the herbaria of each State... An attempt has been made to list the following categories of rare and endangered species:
>
> (1) Probably extinct.
>
> (2) Endangered (only small colonies remain, under adverse conditions).
>
> (3) Rare (population of adequate size, but needs constant monitoring).
>
> (4) Depleted (population originally widespread but now reduced in area; needs constant monitoring).
>
> (5) Species known only from the original collection (and more information needed on its distribution).
>
> (6) Species of geographical importance (with a disjunct or isolated distribution).

A list has also been made of Australian seed-plants 'which retain some primitive floral or morphological attribute and thus provide vital evolutionary evidence on the development of the flowering plants'.

These categories derive from those used by IUCN in the earlier Red Data Books. In the early 1970s, however, the categories were modified to the following:

Extinct (Ex)

Endangered (E)

Taxa in danger of extinction and whose survival is unlikely if the causal factors continue operating. Included are taxa whose numbers have been reduced to a critical level or whose habitats have been so drastically reduced that they are deemed to be in immediate danger of extinction.

Vulnerable (V)
Taxa believed likely to move into the 'Endangered' category in the near future if the causal factors continue operating. Included are taxa of which most or all the populations are decreasing because of over-exploitation, extensive destruction of habitat or other environmental disturbance; taxa with populations that have been seriously depleted and whose ultimate security is not yet assured; and taxa with populations that are still abundant but are under threat from serious adverse factors throughout their range.

Rare (R)
Taxa with small world populations that are not at present 'Endangered' or 'Vulnerable', but are at risk. These taxa are usually localized within restricted geographical areas or habitats or are thinly scattered over a more extensive range.

Indeterminate (I)
Taxa *known* to be 'Extinct', 'Endangered', 'Vulnerable' or 'Rare' but where there is not enough information to say which of the four categories is appropriate.

Out of danger (O)
Taxa formerly included in one of the above categories, but which are now considered relatively secure because effective conservation measures have been taken or the previous threat to their survival has been removed.

For plants an additional category is required:

Insufficiently known (K)
Taxa that are suspected but not definitely known to belong to any of the above categories, because of the lack of information. In practice, 'Endangered' and 'Vulnerable' categories may include, temporarily, taxa whose populations are beginning to recover as a result of remedial action, but whose recovery is insufficient to justify their transfer to another category.

For species which are neither rare nor threatened, the symbol 'nt' is used.

Although national lists of threatened plants will take many different forms, IUCN has expressed the hope that all will include the use of these categories as indicators of the degree of threat to individual species. The Specht lists duly appeared in 1974, but with the notable exception of Papua New Guinea. Its biota is described as 'one of the richest and most diverse in the world'. Here the problem for conservationists, as generally in regions where biological exploration is still incomplete, is the inadequate knowledge of what plant and animal species are present and how they are distributed: 'new distribution records and, in fact, new species are constantly being discovered...The greatest need...is for a greatly accelerated botanical survey combined with attempts to bring rare species into either refuges or botanic reserves'.

Since then, whereas lists of rare, threatened and endemic species have been or are being produced for most temperate and subtropical areas and for most island groups, little has been possible for the tropical floras, especially those of rain forests. Here, rather than trying to produce a list of threatened species, studies on centres of endemism, on generic endangerment and on sample threatened species from particularly noteworthy groups such as palms and orchids seem likely to provide the most useful data base for conservation, to complement surveys of principal and representative sites based on purely ecological grounds.

In those countries where a special effort has been made, the degree of check-sheet coverage of rare and threatened species will depend mainly on the representativeness of the series of areas for which sheets have been completed. They were completed for fewer than half the sites included in the Australian conservation survey, but where, as for example in the Netherlands, Japan and British Columbia, the representativeness of surveyed areas, and the proportion of these for which check-sheets have been returned, are both high, it may be inferred that the coverage of rare and threatened species is also very satisfactorily high.

A high check-sheet coverage of rare and threatened species should ensure that information on their distribution remains readily accessible not only to conservationists and planners but also to biological research workers. For it must be emphasized that rare species are scientifically important because of the light they may throw on factors determining biological success or failure. A species may be rare, nationally or globally, for one or more of a number of reasons. In one country it may be near the limit of its climatic tolerance, in which case it is likely to be less rare in certain climatically different countries. It might, on the other hand, be rare because its habitat requirements, apart from climate, are met only in a few localities within a given country or even in the world as a whole. And this limitation by habitat may have operated over indefinitely long periods or may alternatively be a consequence of past climatic and geographical changes or a recent phenomenon due to man's activity in draining wetlands, destroying woodland, keeping large numbers of grazing or browsing animals and, in many parts of the world, using fire to maintain types of vegetation suitable for their grazing, or, in more recent times, introducing chemical means of controlling weeds and pests and influencing the growth of his crops. Apart from this a species may be rare because it is young, having arisen too recently to have spread far from its point of origin; or because it is 'senescent', having lost genetic variability to the point of being no longer capable of sufficiently rapid adaptation to changing conditions in space and time. These are all situations demanding scientific exploration in order to enlarge our understanding of living organisms in their natural surroundings, and this consideration reinforces the arguments from rarity *per se* in favour of the conservation of as many as possible of these rare, threatened and relict species.

5. Conclusions and recommendations

A. R. CLAPHAM & G. L. RADFORD

5.1 Aims and methods of the check-sheet survey

The objectives of the programme of the CT Section of IBP are explained in the Introduction and Chapter 1 of this volume and very briefly may be summarized as:

(*a*) to provide for the selection and protection of the sites and species required for scientific study now and in the future, including scientifically adequate samples of the main types of ecosystems throughout the world; and for the storage in suitable and accessible form of the relevant information about these sites and species;

(*b*) to make a contribution to the general methodology of biological surveys on a world-wide scale.

It was decided that these objectives could best be achieved through a world-wide survey, all the surveyors being asked to complete the same questionnaire or check-sheet, thus ensuring the collection of certain common items of information from every site examined and the consequent comparability of the data. It was further decided that the completed check-sheets should be returned to a single centre where the data could be analysed and stored and where computerized processing could be undertaken.

5.2 Outcome of the survey

Of 13000 check-sheets distributed to all parts of the world, 3010 have been returned, and of these 2690 were more or less satisfactorily completed and have been processed (pp. 117–128). The completed sheets came from over 2600 IBP Areas in fifty-five 'countries', using the term in the broad sense of p. 140. More than 80% of the sites covered by completed sheets are in temperate regions, this concentration being evident in the map (Fig. 4.1, pp. 142–3).

Of the 979 National Parks and Equivalent Reserves in the United Nations List for 1973, 275, or 28% of the whole, have been covered by check-sheet returns, but from only 22% of the total number, or 25% of the total area, was information returned which was fully utilizable for comparative purposes. The reasons for these low proportions are that only 71% of all listed national parks and equivalent reserves are in countries taking part in the survey, and only about 40% of those were in fact surveyed.

These figures may be thought somewhat disappointing, but nevertheless a very useful start has been made towards the main aim of securing information about all sites already protected or deemed to merit protection because of their interest to conservationists and other biologists. It may confidently be expected that coverage will increase substantially by the inclusion in the near future of sites in participating countries which have so far been left unsurveyed.

5.2.1 Effectiveness as a source of information

A primary aim of the check-sheet survey was to collect information about survey-sites selected as representative of actual or potential conservation areas. The information was to be processed and stored so as to provide a firm basis for comparative studies and for planning future action in this field, both national and international. It is therefore most important to consider how far the information already collected is *reliable*, *consistent* as between different observers and different countries, *adequate* in factual content and sufficiently *comprehensive* in the sense of being based on a selection of sites that includes satisfactory samples of all vegetation-types or ecosystems of the geographical units for which returns have been made. These questions are considered in some detail in Chapters 1 and 2 of this volume, but it is appropriate to summarize the conclusions in this final section.

For the most part the information in check-sheets has been reliable wherever a genuine attempt has been made to answer questions fully. There have been some evident misunderstandings of questions such as those on types of vegetation and soil, and due care has been exercised in processing the affected sheets. There has been greater difficulty with some of the questions demanding judgement on the part of the surveyor, such as Question 14 inviting comment on 'outstanding floral and faunal features', and most of all with Question 16 on 'significant human impact', which depends to the greatest extent on the surveyor's own assessment. Omissions and detectable discrepancies in answers to these questions served, indeed, to give some guidance as to the general usefulness of the information supplied in a particular check-sheet. Another test of reliability was to compare selected check-sheets with published descriptions of the site in question. The general level of agreement was good when the necessary allowances were made for the form of presentation imposed by the questionnaire and for the fact that check-sheets for well-known areas were frequently completed from published records. For less-known sites the information is best regarded as reliable unless internal discrepancies are revealed: contradictory statements from other sources often prove unreliable and incomplete.

The consistency of information judged fully reliable varied with the form and wording of particular questions as well as with their subject-matter. The use of the simple Fosberg and Ball classifications for vegetation and soil respectively (Question 7) was highly consistent as between different surveyors and different countries, as were answers involving the selection of one from several alternatives, like those to Question 12 on salt and brackish water, to Questions 11.2 and 11.3 on freshwater, and to Question 4.3 on international class. Inconsistencies between free-style descriptions of plant communities and soils were to be expected, but it had been hoped that they would not be very significant at least between surveyors from the same country. Unfortunately, however, not all the check-sheets received, for example, from USA and Scandinavia, made use of the national systems for vegetational description. The Australian returns showed the highest degree of consistency, but this was at state rather than federal level: there has been no attempt to co-ordinate contributions from different states, and equivalent communities certainly appear under different names in lists from different states.

Problems of reliability and consistency interact more closely as questions give scope for differences of interpretation and invite a higher degree of subjectivity in the answers. Question 14, on 'outstanding floral and faunal features', proved to be open to differences in interpretation. One surveyor might regard a thinly but widely scattered species are rare, but a colleague might think otherwise, though they would probably agree on most other matters. The question on human impact (Question 16) left room for subjective treatment and it became difficult to assess both reliability and consistency. There is the further point that the value of the information may depend as much on negative as on affirmative answers. Thus it is important to know that certain ecosystems may occur in sites *not* affected by particular types of human impact, and in this respect the question differs from some others demanding subjective answers.

How far the information in completed check-sheets is adequate in factual content is considered later on pp. 196–9. How far it is sufficiently comprehensive, covering the whole range of ecological variation in a country, varies greatly from one country to another and is in part a matter of the effectiveness of the organization of the survey at national and international levels (pp. 194–6). Comprehensiveness can be assessed only by comparing the information in check-sheets with the results of overall national surveys carried out in a sufficiently detailed manner and using the same units of description. The outcome of such a survey may take the form either of maps or of inventories indicating not only the units – vegetational, pedological, etc. – recognized as distinct at a suitable scale, but also their location and areal extent. The Country Check-Sheet (Appendix 1) was designed and circulated to collect this information, but the response was disappointingly poor and attempts to assess national coverage have usually to be based on other sources. The following are examples of published classifications of vegetation-types used in check-sheet entries, some dealing only with forests:

Australia   Specht, Roe & Boughton (1974)
Canada (forests only)   Krajina (1969–70)
Finland (forests only)   Cajander (1916, 1917, etc.)
India (forests only)   Champion (1936)
Netherlands   Westhoff & Den Held (1969)
South Africa   Acocks (1953)
USA   Küchler (1964)
USA (forests only)   Society of American Foresters (1964)

Only the Westhoff–Den Held classification in the Netherlands and the Krajina forest classification for British Columbia were used consistently as a basis for check-sheet entries, though most Australian returns made use of the system devised for the Australian Conservation Survey.

There was inadequate information for assessing the coverage of other features of ecosystems, such as soil-types or land-forms.

5.2.2 Success in provision for conservation

The first aim of the survey was to provide for the selection and *protection* of the sites and species required for scientific study, now and in the future, and for the storage of relevant information about them. Participating countries have varied greatly in the extent to which the survey was used as an opportunity for increasing the number of protected sites. Some were content to collaborate in the collection and storage of information about areas already protected or meriting protection but did not proceed with any immediate expansion of their conservation programmes.

These included countries which had already made considerable progress in nature conservation and others in which little or nothing had so far been achieved. At the other extreme were countries in which the survey occasioned a notable extension of interest in conservation and a consequent designation of many new national nature reserves or sites to be protected in other ways. Noteworthy amongst these was Canada (pp. 151–2) and particularly British Columbia, where the survey was treated very much as a reconnaissance of vegetation-types and of sites that might constitute a representative series of 'ecological reserves'. Plant communities were assigned to their Fosberg formations in accordance with the instructions in the *Guide* but were also placed in the appropriate classes of Krajina's phytosociological classification of forest types. Canada had participated in IBP since 1964, and in 1968 the Province of British Columbia authorized the formation of an Ecological Committee under the chairmanship of the Deputy Minister of Lands and instructed it to advise on the selection of potential ecological reserves. In 1971 the provincial government approved an *Ecological Reserves Act* allowing the establishment of reserves on Crown land. Up to June 1975 there were 235 applications for establishment, 65 of them, totalling 45 067 ha, being authorized. Similar Acts were subsequently passed in Quebec (1974) and in New Brunswick (1975).

The Fennoscandian IBP/CT projects were designed to increase knowledge, by means of inventory and mapping, of the most valuable areas for further nature conservation. The proposals from the Central IBP/CT Committee were closely examined and discussed, and it was decided to participate in the conservation survey. There were, however, reservations about the suitability of the check-sheet for Scandinavian conditions, especially for recording the plant communities of survey-sites, and attempts were made to secure agreement on a uniform vegetational classification for use both in mapping and for completing check-sheets (pp. 75–6). In the event there were some differences in practice between Norway, Sweden and Finland in the procedures adopted and therefore in comparability of the resulting data, but within the three countries the effort stimulated by the survey resulted in substantial increases in the number and size of protected areas. In Finland there was published in 1973 a plan for thirty-five new national parks and the enlargement of eight existing ones, and for eleven new nature reserves and the enlargement of four already in existence. Apart from this there was co-operation between the IBP/CT Sections in the Fennoscandian countries in undertaking inventories of peat-lands for the international project Telma and of forests as a contribution to project Silva, both aimed at protecting areas of scientific importance. There were other IBP/CT projects which were restricted to single countries, like the surveys and inventories leading to the establishment of nature reserves on the coast of the Gulf of Bothnia and in important habitats for wildfowl (project MAR, Finland); and the studies of survey techniques for assessing and recording conservational interests in alpine and subalpine areas of the Jotunheimen mountains, and of the population-size and biology of polar bears in Svalbard (Norway). While not parts of the check-sheet survey these additional studies indicate the extent of the conservational activity resulting from the work of the IBP/CT organization in general.

These examples, to which many more could have been added, show that the survey has certainly led to increased provision for nature conservation, though by no means to the same extent over all participating countries.

5.2.3 Success in furthering the methodology of biological surveys

The check-sheet survey probably constitutes the most broadly based biological survey so far undertaken and the most extensive trial of a check-sheet as a means of collecting ecological information. The reasons for adopting a check-sheet are explained in Chapter 1 and the outcome of the survey in Chapter 4. The content and wording of certain of the questions in the check-sheet were found to some degree to be unsatisfactory, and suggestions have been made for improving them. On the whole, however, the results have been encouraging and have justified the use of a check-sheet. Where expectations have not been fulfilled the main causes have been shortcomings in the organization of the survey, especially at national level, and deficiencies in recording procedures and in the knowledge requisite for remedying them. These are discussed below.

5.3 Problems of planning and execution

The experience gained during the planning and execution of the check-sheet survey has already brought to light a number of defects and difficulties, and others emerge from the examination of completed sheets. Together they are responsible for the failure of the survey to achieve quite as much as was hoped of it. Clearly it is important, both for any continuation of the present survey and for future biological surveys of the same general kind, that these defects should be identified and remedied as soon as possible.

Three major aspects of the undertaking have been concerned:

(*a*) the organization of the survey at international and national levels;
(*b*) the prerequisites for effective national participation;
(*c*) the adequacy of adopted procedures and of instructions to participants.

5.3.1 Organization of the survey

It has become clear that success in an enterprise such as the check-sheet survey demands effective organization at the international and also at regional and national levels. The requirement at international level is for an advisory group of specialists to give encouragement and guidance to those carrying out the work at regional and national levels. This international advisory committee must be composed, largely or exclusively, of specialists in relevant scientific fields and certainly must include competent ecologists and phytosociologists. They should be prepared to explain the objectives of the survey and the principles underlying the adopted methodology, and to organize discussions on any questions requiring clarification at the international level. They should also be given responsibility for co-ordinating regional and national contributions and ensuring the necessary degree of standardization of the whole procedure. Proposals for modifying the check-sheet should require to be approved and their adoption synchronized by the International Advisory Committee, and this should apply also to changes in the instructions to surveyors.

The organization of the survey has suffered in many participating countries because members of the national CT committees had insufficient interest in nature conservation and too little awareness of its objectives and problems. Organization and co-operation at the national level were most effective in those countries, such as Australia, Canada, the Netherlands, Japan and Poland, where strong CT Committees included leading representatives of governmental bodies responsible for nature conservation as well as experts from a range of relevant disciplines, and it must be concluded that this was an important factor in their success. Ideally the international advisory committee should be given the opportunity, within their terms of reference, to approach the most appropriate organizations within a country or region, whether governmental or not, in order to secure the best foundation for a co-ordinated contribution to any international biological survey comparable with the check-sheet survey, whether its aims are concerned exclusively with nature conservation or with wider aspects of land-use. What is required is a working group of specialists deeply sympathetic to the aims of the survey and willing and able to assume responsibility for the organization and co-ordination of the whole project at national or regional level and for maintenance of effective contact with the international committee.

5.3.2 Prerequisites for effective national participation

The first task of the co-ordinating body for a biological survey comparable with the check-sheet survey is the adoption of uniform procedures for recording the required information, especially about features of vegetation and soils. This is essential for taking decisions on potential land-use in general and on nature conservation in particular. There are problems concerning the availability of suitable methods, and the matter is considered in Chapter 2 and in 5.3.4 below. The same recording procedures should certainly be used by all the surveyors in any one country, and the aim must be to extend uniformity to the whole of large regions and ultimately to achieve world-wide agreement, this being a main concern of the international organization.

The next requirement is a national or regional inventory of all the natural and semi-natural ecosystems, at an appropriate level of discrimination, that occur in the area to be surveyed under the aegis of any one local co-ordinating body. Their recognition and discrimination, using the agreed uniform recording systems of the foregoing paragraph, would be likely to require comprehensive aerial photography followed by ground survey and the presentation of the results in the form of maps. The scale most appropriate for the mapping would vary from one country to another, but uniformity may be expected to become feasible in due course. The desirability of a comprehensive cover of vegetation maps, and a consideration of problems of preparation and scale, are discussed on pp. 28–34.

The adoption of agreed recording procedures and the preparation of comprehensive maps or inventories provide the necessary basis for the detailed planning of the selection of sites for survey. The check-sheet survey undoubtedly has suffered from the failure of many countries to formulate and follow a definite and adequate plan for site-selection. This has commonly, though not invariably, been an outcome of insufficient knowledge of the variety of vegetation-types present in the country and of their location and area. Survey-sites accordingly have not been fully representative of the vegetational diversity of the country as a whole, and there is a likelihood of serious omissions in the resulting proposals for nature reserves and other protected areas. This state of affairs might be remedied to some extent by more detailed guidance on site-selection from the central organization of future surveys (pp. 18–19), but this will be ineffective unless the requisite overall survey has been undertaken and the resulting maps made available, and that must be the responsibility of national bodies.

Survey-sites having been selected, local co-ordinating bodies should decide on such procedural matters as the number and timing of visits and the mode of sampling within sites, in order to provide adequate information about vegetation, fauna and soils and about such special features as the presence of rare species and the consequences of human activity. If the main interest is nature conservation they should ensure that sites selected primarily on zoological grounds should nevertheless be visited and described in the standard way with the help of suitably qualified recorders of the geomorphology, vegetation and soils and, conversely, that zoologically-trained recorders should visit sites of primarily botanical interest.

Local co-ordinating bodies should also be responsible for the selection and training of surveyors, ensuring that all will be capable of following a common set of recording instructions. Consideration will need to be given by the international advisory committee to the special problems of countries with inadequate resources in trained personnel for surveys of the kind required.

5.3.3 Adequacy of adopted procedures and of instructions to participants

The check-sheet

Attention has been drawn in Chapter 1 to the basic problems of using check-sheets in biological surveys, and the Introduction records the gradual evolution of the questionnaire into the present form. Appendix 6 analyses the adequacy of replies to all the questions, those dealing with types of vegetation and soil also being considered at some length in Chapter 2. There can be no doubt that these last were the ones causing most difficulty to surveyors, primarily because of the lack of either internationally agreed schemes for description and designation or, for the most part, of schemes consistently adopted even within individual participating countries or regions. To repair this deficiency must become an urgent task for all concerned with the future of this and comparable biological surveys (see pp. 205–9).

A major criticism of the remaining questions was that greater precision should have been sought and achieved. There were two main reasons for the vulnerability of the check-sheet to this kind of criticism. In the first place it was designed to serve a number of different though related functions:

(*a*) to provide, for purposes of inventory and comparison, a standard documentation-system for areas already protected or deemed to merit protection;

(*b*) to provide a factual basis for new or enlarged conservation-programmes, especially in parts of the world where little relevant information was so far available and where there might be considerable difficulty in collecting it; and

(*c*) to provide ecological base-line information on many areas of special interest as sites for biological observations and experiments of longer or shorter duration, for monitoring programmes and the like, but not necessarily recommended for conservation on the usual grounds.

A second reason for failure to seek and achieve a uniformly high standard of precision was that the survey lacked a sufficiently strong national and international organization (pp. 194–6) to ensure a high level either of planning or of training and supervision of field-workers, and therefore of the co-ordination of replies to exacting questions. It was recognized that no rigorous selection and sampling of sites, and no satisfactory standardization of replies to certain types of questions, would be possible except through the most active co-operative national committees and organizers. Questionnaires were sent out to many countries, all of which promised to participate, but it was impossible to predict which would participate actively and what would be the scope and quality of their contributions. That being so, both the questionnaire on the one hand, and the systems for data-handling and retrieval on the other, were necessarily kept at a fairly general level.

Question 16, on 'significant human impact' was one for which modifications were suggested in order to increase the amount and precision of the information elicited. The change proposed by the Scandinavian CT Committee, and adopted by the French Committee, sought an indication of the intensity of each type of impact and also of its extent within the particular site. This made it possible to record, for example, that hunting had once been practised over the whole of the site, that it had now been discontinued and that its persistent effect was very slight; or that mineral extraction was currently having a strong impact on a small part of the site. But the design of the question after this modification becomes inappropriately complicated in relation to the reliability and level of detail secured, and there is still no information as to how the impact is recognized and its intensity assessed. To restructure the question so that it asks for the vegetational evidence for impacts of different kinds and intensities would place it outside the range of what is generally feasible at the present time, but this is the kind of change that should be considered for a future period when there has been a wider experience of more detailed field-work.

Question 14, on 'outstanding floral and faunal features', is another that might at some future date be modified to provide more detailed information. At present 14.2 invites no more than a tick in the appropriate box to indicate that, amongst representatives of the major classes of conspicuous animals (Mammalia, Aves, Reptilia, Amphibia, Pisces and Insecta), there is species diversity, abundance or superabundance of individuals, rare species, threatened or relict species, species of biogeographical interest, exceptional associations, breeding or nesting populations, migrating populations or wintering populations. The accompanying notes state, for this and for the corresponding questions (14.4) about plants, that 'Only the presence of outstanding features should be noted by checking the appropriate box. No other information is required here: we do not want, for example, the number of bird species present inserted under "Aves – species diversity", because this is not in itself an indication that this number is outstanding. Columns have been left vacant for additional types of outstanding features, and additional taxonomic groups may be added in the vacant rows. The vacant rows may also be used to give more precise data for the groups listed, e.g., if the outstanding interest centres on the Carnivora of the Mammalia, "Carnivora" may be inserted in a vacant row.'

This example illustrates and emphasizes the intention of those who framed the questions not to expect and therefore not to demand that surveyors should be sufficiently expert over the whole field of local natural history to be able to provide detailed information about individual species of plants and animals in answers to 14.2 and 14.4, while at the same time giving scope for recording additional facts where they were known. This was a deliberate compromise that took account of the frequent deficiencies in numbers and training of field-workers but nevertheless sought to collect a preliminary body of basic data about the areas surveyed. There can be no doubt that the general value of the data-bank would be much enhanced by the addition for each site of quantitative estimates of species-diversity and of the abundance of named species, and also lists of breeding or nesting, migrating and wintering populations, again with estimates of numbers: at present only the names of the main threatened, endemic, relict and rare species are asked for in 14.3 and 14.5.

With regard to rare species it would be valuable to invite an indication of the presence of 'globally' rare species with estimates of their numbers and their distribution within the site, e.g., 10 000 breeding pairs restricted to a stated part of the site. A full list of 'regionally' or 'nationally' rare species should also be invited, again with some indications of numbers present and distribution within the site; and similarly for relict and endemic species and other 'biogeographically interesting' species. It will be readily appreciated, however, that such additional information will be of real value only if its accuracy is assured; and it will be many years before all sites of long-term biological interest can be surveyed by sufficiently expert teams. Meanwhile the aim should be to invite, but not to insist on, additional information of a more detailed kind than is required for the present check-sheet, and also to press on with the training of surveyors all over the world.

The Guide to the check-sheet

The *Guide to the Check-Sheet for IBP Areas* (Peterken, 1967) was intended primarily to assist surveyors actively engaged in the check-sheet survey. The first three of its four sections deal respectively with 'General information' about IBP as a whole and the CT Section in particular, the UN List of National Parks and Equivalent Reserves, the objectives of the check-sheet survey, the development of the check-sheet and of the data-handling scheme; with the 'Selection of sites' for survey and guidance on the subdivision of certain sites, and with the 'Contents of the check-sheet', explaining the significance of each question in turn. The last section consists of 'Field instructions', and there follow appendixes setting out the recommended 'Classification of vegetation for general purposes' by F. R. Fosberg, the 'Classification of soils' by D. F. Ball, and the 'Geocode', explained by F. H. Perring, all of which have to be used in completing a check-sheet. A 'Glossary' of technical terms accompanies the vegetational classification.

The *Guide* has proved an extremely valuable handbook to the check-sheet, but it may reasonably be criticized as unduly lengthy for use in the field. This criticism could readily be met by dividing its two main functions between two separate publications, one a source of information on the purpose and organization of the survey, the other giving guidance on answering the more difficult questions and including the keys for recognizing the chief types of vegetation and soil, with a glossary of technical terms used in the keys, and an explanation of the geocode.

The first of these booklets would concentrate on providing a background to facilitate the planning of a regional or national contribution to the survey without need for detailed consultation with the international advisory committee. The first three sections of the present *Guide* might be amplified by adding examples of the kinds of answers expected and of ways in which participants could benefit from well-planned and carefully executed surveys. The second would be a booklet readily carried into the field. Its size could further be reduced by making the questions more self-explanatory than at present and by having the field instructions restricted to the backs of the pages of the check-sheet and not repeated as now in the booklet. The aim should be that the combination of check-sheet and field booklet should be just sufficient to enable a surveyor to work independently in normal circumstances.

The present *Guide* includes a glossary provided by Dr Fosberg to explain terms appearing in the keys to his classification of vegetation-types and in the brief descriptions of the different types. This glossary contains strictly technical terms such as epiphyte, graminoid, pneumatophore and sclerophyll, but also many that are more generally familiar, like bog, desert, ephemeral, mangrove, marsh, meadow, savanna, shrub, steppe and swamp. In general these are defined so as to give them a rather more precise meaning than in common usage. Many of the terms in the second group, however, have been used by British plant ecologists and others in a still more restricted way than is recommended in the glossary, this being particularly true of some terms for wet habitats and for types of grassland. A. G. Tansley established the use of 'marsh', 'fen' and 'bog' for different, though intergrading, types of wet habitat in which the water-table is below the surface of the substratum for most of the year, and experience has justified this conferment of precision on popular words used in a technical scientific sense. Ecologists seem prepared to adopt them as readily as newly-coined technical terms and misunderstandings are infrequent. Similarly 'meadow' and 'pasture' are often limited to their original meanings of mown and grazed grassland respectively, and the distinction is of real ecological value even though many 'meadows' are customarily grazed after an initial moving.

These considerations are important for an attempt to establish an agreed vocabulary for habitat-descriptions in check-sheets and elsewhere, not only in relation to the designation of vegetation-types but also for recording the habitats of all kinds of animals, including, for instance, the sites where birds sing, nest or flock. For such purposes it would clearly be an advantage for the field-surveyor to have available a precise though non-technical terminology for different kinds of plants and plant-assemblages, and it might well be thought desirable to include in such an extended glossary terms for features of the physical landscale and for use in descriptions of soil-morphology. Its content would depend on the various classifications or alternative procedures that are eventually adopted for dealing with vegetation, soils, geomorphology, etc., but there seems much to be said in favour of including in the *Guide* a general glossary of this kind in place of the present addendum to the section on vegetational classification. One employing English words in common use would raise the problem of translation into such other languages as may be commonly used in completing check-sheets, but this should not present insuperable difficulties.

The data-handling scheme

Early versions of the check-sheet consisted of simple questions about the main features of IBP Areas designed to enable punched cards to be used for the storage and retrieval of the data. The third version was tried in the field in several countries and the results were considered at a technical meeting held at the Monks Wood Experimental Station of the Nature Conservancy in 1966. They showed that the simplified approach did not sufficiently take into account the interdependence of biological and physical features, and there was a consequent attempt to frame more comprehensive questions, the answers to which would convey something of the individual ecological characteristics of each site.

Two strong influences upon the discussions at this meeting came from the Centre d'Etudes Phytosociologiques et Ecologiques at Montpellier and from the Smithsonian Institution. The ideas then being developed at Montpellier centred upon a system of site-recording involving the examination of a wide range of features contributing to the general ecology of a site. The data-processing was to be entirely card-based but with a capacity for computer-analysis in specific detailed studies (CEPE, 1968). At the Smithsonian a system for indexing and labelling museum specimens was being developed and the experience gained in this project led to an offer of help in designing a computer-based retrieval system for the check-sheet survey. Both systems gave increased flexibility in data-handling and, in view of the dissatisfaction with earlier versions of the check-sheet, it was decided that ecological considerations involved in the design of questions should, for the time being, outweigh concern for ease of data-handling. It was only in the final stages of design that care was taken to present the questions in a form that would ease data-processing without distorting their balance. The punched-card machinery at the Biological Records Centre was incapable of meeting all the requirements considered necessary, so that the availability of computing facilities and expertise fairly close at hand led to a decision in favour of a computer-based system.

The check-sheet itself, as a tool in a new approach to ecological inventory, stimulated the design of a data-handling system tailored to meet all the foreseen needs. It also incorporated a degree of flexibility to accommodate change and to meet unforeseen needs. This gave the system a great potential for the retrieval and analysis of information from the data-base, but there were many constraints within which it had to operate. The most important of these was the budget, which was tight for such an undertaking, although extra financial support was forthcoming from IBP upon any reasonable request. Another constraint was the type of computing facility available at the Computer-Aided Design Centre in Cambridge, which is mainly concerned with engineering design and related display techniques. There is a high proportion of work requiring lengthy computation time or a large amount of store. With a priority of access third of four and a large amount of input needing attention at all stages of editing, processing and storing, the IBP/CT project was at a serious disadvantage in a system catering primarily for work of a very different nature and suffered, moreover, from a shortage of staff time. A further problem was a tendency for files to become misshapen on transfer from disc to magnetic tape, a condition detectable only by means of an extremely time-consuming check on each file after transfer.

5 Conclusions and recommendations

In addition to the above constraints there were limitations on data-processing arising from intrinsic features of the design and mode of operation of the data-base.

Limitations on the interpretation of the information in check-sheets

There are five main ways in which the design of the data-base may be said to limit interpretation of information in check-sheets. Four of these stem directly from the complexity of the questionnaire, limiting the extent to which the data-base could be enabled to cater for any deviation from a normally completed set of entries.

The first example of such a limitation was seen in 30% of a sample of 200 completed check-sheets that were subjected to preliminary analysis and arose from a tendency of surveyors to show two or more plant communities as sharing a common area. This practice, which was unfortunately adopted in the completed check-sheet printed in the *Guide*, could not be accommodated in the processing without unreasonable complexity.

Secondly there was the lack of a system for handling synonyms, though Willis' *Dictionary of Flowering Plants and Ferns* was used for reference in framing enquiries concerning the generic names for dominants of plant communities and for rare plants.

Third was the freedom given to surveyors in entering descriptions of plant communities, leading to difficulties in retrieving the information in a later enquiry because any search for a particular community had to be framed very loosely so as not to omit examples described in different ways. Individual inspection of all possibly relevant entries had to be undertaken after clearly irrelevant descriptions had been rejected by machine.

The fourth limitation concerned the use of single descriptors in the data-files for multiple attributes. The clearest example was the coding for 'exceptional interest of the area' in which entries were classified into one or more of eight categories, each having further subcategories to a maximum of five. In the case of zoological interest the subcategories were: rare species, other biogeographical interest, important breeding populations, high species-richness or other interest. Only one of these could be indicated, and this led to difficulty where an outstanding interest included examples of two or more subcategories. Such cases were infrequent and did not justify a more comprehensive system of coding, but there was a regrettable loss of information for those particular sites.

The fifth stems from the lack of any context for the site-files in the data-base. All synthesis involving the presentation of site-data in a national or regional context was done after presentation of the relevant site-data by the computer. Insufficient numbers of completed Country Check-Sheets ruled out the possibility of developing this capability in the data-base.

The second of these limitations, the treatment of synonyms, could have been given closer attention at the input stage, but resources were constantly stretched to process the material and it was decided that since a matching routine could be incorporated into the output stage, if required, the extra checking was not justified.

Scope for modifications, future or retrospective, in the mode of specification of features showing continuous variation

It was recognized in designing the data-base that experience might call for changes in the questionnaire. The blueprint governing the form in which the answers to all questions could be accepted for processing was therefore made flexible (Chapter 3). Additional questions could be included and existing questions could be altered or expanded within fairly generous limits. The greatest likelihood of change was foreseen as being the treatment of continuous variation in those questions where the establishment of classes was the most difficult because of problems with definitions.

The treatment of a continuous variable must depend on the scale of survey and the detail which can be recorded at that scale. It is not always desirable and certainly not always practicable to collect precise values, but a truly parametric approach is usually possible in an intensive rather than an extensive survey. An alternative is to use a number of classes for each variable, chosen on a mathematical scale or with cut-off values of known or suspected significance. In the system adopted at Montpellier, for example, variables are coded on a logarithmic scale where a geometric progression is appropriate and on a parabolic scale for percentages. The number of classes can be varied to suit the possible or desired accuracy of measurement of any particular variable. But once such a step is taken the analytical die is cast, and even though the number of classes may make precise values necessary, the sensitivity is subsequently determined by the class and not by the value.

For these reasons the check-sheet asks for specific values of continuous values whenever it is reasonable to do so, though leaving the choice of unit to the surveyor, areas being given in acres, hectares, square miles, etc., and tidal range in centimetres, metres, inches or feet as convenient. The limits to classes in a scale convey the order of accuracy required, and the surveyor can judge whether his information is or is not within the tolerance. For more complex situations continuous variables are classed rather broadly. Thus in the Fosberg vegetational classification there are three classes for plant-cover, four for the height of woody layers and two for herbaceous layers. The length of coastline is asked for in one of three classes: less than 1 km, between 1 and 10 km, and over 10 km.

The data-base was designed to handle continuous variables exactly as they were treated in the check-sheet itself. Their analysis cannot therefore be more sensitive than the original classification, but it can be made coarser. It was nevertheless recognized that more sensitive classes might be needed at some later stage, and there was accordingly provision for changing scales at any time. The limiting factor would then become the earlier style of framing the questions, since all check-sheets up to that time will have used fewer classes and comparisons across the board will be possible only at the original sensitivity.

The greatest difficulties with continuous variation have been in handling data relating to interdependent variables in complex systems. In the early technical meetings of IBP/CT there was much discussion of continuous and discontinuous variation in vegetation and of the relative merits of ordination and classification as the principles behind numerical analysis. Discontinuities frequently arise from fairly abrupt changes in combinations of characters that may nevertheless show continuous variation over their total range. The abruptness varies with the environmental factors mainly involved, transitions due to changes in soil-type, for example, tending to be more abrupt than those caused primarily by climatic differences.

Of particular interest here are the limitations imposed by the data-handling system itself as an additional source of distortion in the recorded representation of ecosystems. The structural characters used in vegetational classification commonly show continuous variation, this being true of height and degree of openness and of the size, shape and texture of leaves. The Fosberg classification uses different combinations of broad classes of these characters to define its 'noda', but recognizes that intermediates and mosaics occur and must be represented by suitable coding. The limitation introduced by the data-base is upon the flexibility with which this can be done. In the original blueprint provision was made for a single full code for each plant community, but many surveyors entered compound Fosberg codes to classify certain vegetation-types. The data-base was unable to handle these and was, therefore, altering the mode of representation of the communities in question. It had then to be decided whether to change the blueprint, expanding an already complicated section and adding to the time taken for processing each check-sheet, or to tolerate the distortion already present in sheets processed earlier. In the event the change was made, but only after considering a further very important general point: how far provision should be made for any future modifications to meet the tendency of surveyors to evade or elaborate the simple answers they are invited to make to complex questions. This tendency is readily intelligible and some provision is desirable, but it must not go so far as to jeopardize the whole system. An example is the frequent bracketing of two or more plant communities to indicate that they are in some way linked, spatially or ecologically. Unfortunately information of this kind cannot be handled by the data-base without lengthy notation.

Restrictions on the length of entries

A less important source of distortion due to the data-handling procedure was the convention adopted for limiting the length of free-style answers. Lengthy answers were shortened wherever possible: in community descriptions, for example, the number of named species was limited to four. This involved rejection of a considerable part of the information supplied in certain check-sheets with lengthy species-lists, but in general it has proved adequately generous.

The conclusions must be that the operation of the data-handling scheme has imposed few limitations on the interpretation of the data, mainly because the data-base was specifically designed for this check-sheet, which was planned to secure information in a form suitable for the appropriate kinds of interpretation. In some instances the wording of questions led to deviations from the expected form of answer and therefore to loss of information in processing which could perhaps be avoided by reframing the questions. There was also some restriction on the scope for later modification or expansion of the check-sheet, but more flexibility could be introduced at the expense of the ease of operation of the system. The same is true of the small element of misinterpretation and misrepresentation arising from the editing of free-style entries, and it may well prove better to reduce the number of such entries rather than to complicate the data-handling.

It should be added finally that there has been rapid recent progress in the flexibility and power of small computers. This calls for a reconsideration of the systems of data-handling available for future biological surveys, and some changes are provisionally suggested in the recommendations on p. 214.

5.4 Requirements for future research and agreement

5.4.1 Recording procedures for vegetation, soil, etc.

International agreement on standard methods for recording types of vegetation would greatly assist all engaged in biological field-work and would much facilitate the selection of sites for nature conservation. The problems are formidable, but every effort must be made to ensure that site-descriptions convey, in as much detail as is requisite for the task in hand, internationally intelligible information so that comparisons can be made with other sites, whether in the same country or further afield. This would best be achieved by the adoption of common recording procedures. The problems, considered in Chapter 2, derive in large part from present deficiencies in our knowledge of both the ecology and taxonomy of natural vegetation, particularly in many equatorial and subtropical regions, and also in the numbers of trained experts in these fields. The two main questions immediately raised are, first, what categories of information about vegetation would be most useful as a basis for decisions on the objectives of world-wide surveys of the kind envisaged; and, second, how far is it currently possible, or likely to be possible in the foreseeable future, to collect this information.

It was suggested in Chapter 2 that the physiognomic information implicit in the Fosberg codes is of consideration value in site-assessment and has the further advantage that it can be recorded by field-surveyors without specialized training. This argues in favour of continuing the use of the Fosberg system in future surveys. On the other hand the 'formations' and 'subformations' of the system are, in general, too coarse for many of the purposes of a biological survey, including the final selection of areas for nature conservation. Experience in north-eastern Australia shows, nevertheless, that a level of discrimination much superior to that given by Fosberg's units is possible by the use of a wider range of structural and physiognomic features of rain-forest vegetation together with certain environmental data, and still without recourse to species-lists, though the sensitivity of site-assessment might be further enhanced if they were available. The Queensland workers also propose the use of a *pro forma* to secure answers to specific questions about climate, topography and soil and about the form and structure of the vegetation rather than to invite the surveyor to assign the class or code of any recognized system of vegetational classification. The interpretation of the answers would then be the task of the data-base associated with the survey.

These studies suggest that it would be useful to explore the feasibility of extending comparable procedures to the recording of natural vegetation of all kinds and in all parts of the world. This would entail the co-ordination of widespread investigations similar to those undertaken in Queensland. It is conceivable that separate *pro formas* for the climatic formation-types, like equatorial rain-forest, desert scrub and tundra, might lead to site-assessments showing a reasonable sensitivity, but there might well be a need for variants for each of the geographical regions where the formation-type is represented, South American, African and Indo-Malayan rain-forests requiring to be handled separately. The eventual outcome could be an agreed first stage of vegetational description and interpretation complementing or replacing the use of the Fosberg scheme while retaining many of its advantages.

Success in these investigations would not dispense with the need for species-lists for certain purposes and in particular for selecting areas for nature conservation. Floristically-based systems of vegetational description and classification have been brought into general use in many parts of the world and are commonly forms of the Braun-Blanquet system. The standard procedure is to record selected environmental as well as vegetational information and there is, therefore, a steadily growing volume of knowledge about the local ecological significance of both plant associations and individual plant species. Full species-lists are drawn up, so that vegetational recording by a surveyor who is both familiar with the system and expert in local taxonomy may be expected to provide all the information that will serve as an adequate basis for decisions on nature conservation as well as other types of land-use. The fact must be faced, however, that an individual plant association or plant species has a limited geographical range and the ecological significance of its presence or absence in a particular site can be established only through much detailed local research. Present deficiencies in taxonomic knowledge and in numbers of experienced taxonomists are unlikely to be remedied rapidly. It follows that efforts must be made to find some half-way house towards the ultimate goal of a biological data-bank storing the information needed to supply a trustworthy assessment of any site for which standard survey-data are provided.

The Fosberg scheme goes some of the way towards this goal, and a global extension of the Queensland procedure holds some promise of further progress, but neither supplies the necessary indication of floristic composition unless experienced local taxonomists are members of the survey-teams and can provide it. Some indication of floristic composition might be inferred from knowledge of the usual species content of otherwise similar sites within the same general area. This would depend on the further knowledge that there really are extensive areas over which there is a more or less unchanging relationship between environmental factors and species-list. This is claimed provisionally for the 'biotic provinces' of Dasmann and others and the 'biogeographic provinces' of Udvardy (pp. 88–9), but a great deal of further investigation is required to demonstrate that they could serve as a basis for inferences of adequate reliability for decisions on nature conservation. It would be of great value if systematic investigations of the whole question of defining and classifying natural regions of this kind could be initiated on a global scale, as proposed by Dasmann (1972–3; IUCN, 1974).

The national and international bodies concerned with world-wide biological surveys must, therefore, assume responsibility for a number of important decisions on methods of description and designation in order to ensure success. These decisions fall into two main groups: those on the standardization of procedures for collecting information about the vegetation of survey-sites and also about their fauna, environmental features, etc.; and those on the organization and co-ordination of further data-collection and studies designed to support and improve procedures already available.

(i) The world-wide standardization of procedures for recording the types of vegetation present in survey-sites is faced by the problems that there is still disagreement, both within and between countries, as to what procedures should be adopted; and that in many parts of the world it would be impossible at the present time to adopt all the procedures widely recognized as desirable, in particular those demanding greater taxonomic knowledge and a much larger number of experienced taxonomists than currently available. The outcome of the check-sheet survey suggests that all field-surveyors should be asked either to follow revised instructions for assigning Fosberg codes or, as an alternative or in addition, to record certain non-floristic features selected as increasing the amount of relevant information above that derivable from the Fosberg coding alone. As far as floristic data are concerned, it is to be hoped that steps will be taken to bring about the world-wide adoption of the Braun-Blanquet system, though perhaps with local modifications. It is recognized, however, that this will be a long-term matter in some species-rich and little-explored areas. If these decisions were agreed internationally, vegetational recording would normally become a two-stage procedure but with the second stage dependent for the time being on adequate taxonomic knowledge and personnel.

(ii) Every country or regional group of countries should publish, as soon as possible, details of their standardized procedures for vegetational recording. They should also proceed with the publication of comprehensive lists of all the vegetation-types, at an appropriate level of discrimination, present in the country or region, as well as lists of all component species of higher plants and of the more important animal groups occurring in individual sites representative of each type. The lists might be prepared and published in instalments to avoid lengthy delays. This information would provide a firm interim basis for decisions on site-selection for survey for various purposes if accompanied by maps showing the location and extent of all the types of vegetation. It would also enable a suitable international body to explore the feasibility and value of delimiting and defining biogeographic provinces (pp. 88–9) and thus, perhaps, of helping still further the search for decisions about areas from which full species-lists are as yet unobtainable.

Much of what has been proposed for the recording of types of vegetation is also applicable to types of natural soils, as has been discussed in Section 2.3 and especially on pp. 111–3. There is the same need for world-wide standardization of procedures for collecting the information essential for achieving the objectives of biological surveys, the same uncertainty as to what information really is essential, and the same shortage in many countries of surveyors with the requisite training and experience. In the immediate future, too, it may well be more valuable to invite the recording of a predetermined set of readily observable facts about each soil rather than to ask that it be allocated to its class in a particular classification but, as with vegetational recording, there is need for a great deal of further investigation.

5.4.2 The need for maps of vegetation and soil

It would be of considerable value to biologists of all kinds, including nature conservationists, if vegetation maps were available for the entire land-surface of the world at a range of scales from 1:1 million to at least 1:50 000 in areas of high diversity (pp. 52–4). Agreement on standardized schemes for vegetational recording and designation is a prerequisite for the ready intelligibility of such maps and for comparisons between one area and another. Agreement having been reached, recent advances in aerial photography and remote-sensing techniques would enable maps at the larger scales to provide good resolution of vegetation-types whose floristic composition could then be established from the ground survey of selected sample areas. Mapping of this kind would greatly facilitate the identification of areas meriting close examination as possible candidates for protection or for investigation on other grounds. It might well be combined with drawing up species-lists from recognized types of vegetation within a country or region, and also with research into the delimitation of biogeographic provinces.

Maps showing the distribution of soil-types, at a level of distinctness agreed to be appropriate for the purposes of biological surveys, would also be of considerable value.

5.4.3 The choice of sites for protection

The primary criterion for selection of a site for survey should be that it is known or believed to include typical examples of plant communities, with their characteristic animals, occurring within the country or region. The aim must be that the surveyed sites as a whole cover the complete range of different communities represented, at an appropriate level of distinctness. To ensure this will normally entail some prior knowledge of the distribution of vegetation-types throughout the country, and this is most readily secured through vegetation-maps at a reasonably large scale and based on a standardized system of description and designation, as has been emphasized earlier (pp. 52–4).

Survey-data having been collected from all the chosen sites, there must next be a second selection of those, not already conserved, that are to be recommended for future protection of some kind. The basis for this choice should be both an assessment of the scientific value of each, in the light of what has already been protected, and a consideration of other factors affecting the probability of successful conservation.

The scientific value of a site depends upon the number, extent and representativeness of the different natural communities and habitats it includes, how far it displays relevant special features, and whether there are full and reliable records of its past history. The special features include rare or threatened species of plants and animals, unusual habitats or local climates, long freedom from disturbance, etc. Amongst other factors for consideration are the size and shape of the whole site and the presence or absence of surrounding land that could act as a buffer zone against disturbance of various kinds. These last points are important aspects of the problem of *scale*. The probability that a site will survive catastrophies such as fires, incursions of grazing or browsing animals, pollution, and other consequences of human activity along its margins, will increase with its size and compactness and with the effectiveness of its buffer zone. And this is true also of size acting indirectly through the numbers of individuals of all component species, whose chances of surviving, for example, periods of exceptionally unfavourable weather or epidemic disease will increase with increasing population-sizes. Another relevant factor is the degree of isolation from other areas carrying the same communities and the same rare species, though this can affect the scientific value of a site in two opposing ways. The greater the isolation the smaller the likelihood of natural replacement of lost species, but inadequate isolation may reduce the value of a site for certain kinds of ecological research and as a source of unique genetic material.

Many other considerations inevitably influence decisions between sites closely comparable as candidates for conservation. The presence, for instance, of rare or unusual species in an example of a community may result from minimal past disturbance, the species in question being characteristic of the community in its wholly natural state but being sensitive to certain kinds of change and not readily returning after being displaced. Alternatively it may indicate that the example is not truly representative owing to atypical features of local climate, substratum, topography or past treatment. If the latter is judged the correct explanation, then it might seem advisable to conserve both the atypical and a more typical site; but if the former, then the site with rare species would be chosen rather than one lacking them. The proximity of an example of a community meriting conservation to areas easy of access and having a high potential for any kind of economic development may be an important argument in favour of looking elsewhere for a less vulnerable site. Yet another consideration is that stands of a given community will in general be most representative near the centre of the range of the community rather than towards its limits.

The present check-sheet asks for information on many of these points but the thoroughness of the recording often seems to reflect the extent to which their significance was appreciated. There is need for a much greater effort in future, at both national and international levels, to ensure that criteria for the selection of sites, for survey in the first instance as well as for subsequent protection, are better understood and applied. There is also need for a review of sites already protected in order to assess the validity of criteria used for their selection and to see what additional factors should be taken into account.

5.5 Recommendations

The following recommendations are addressed to any organization undertaking the continuation or extension of the check-sheet survey initiated by the IBP/CT Committee. They should also interest any international bodies concerned with extensive biological surveys of areas of more or less natural vegetation aimed at providing information for decisions on land-use in general and on nature conservation in particular.

5.5.1 Organization and procedure

(1) That an international advisory group or committee be established to support a continuation of the check-sheet survey or any comparable future survey, such groups or committees to include specialists in all the appropriate scientific fields.

(2) That the international group should prepare a clear statement of the general aims of the survey and of the consequent practical objectives, and should indicate the methodology deemed appropriate for attaining the objectives. This statement would serve as a guide to national or regional organizing bodies and assist in the international co-ordination of the survey, but it should be reinforced whenever necessary by detailed discussion of any points requiring clarification or modification.

(3) That the international advisory group should be enabled to approach the most appropriate organizations within any participating country or region, whether governmental or otherwise, in order to secure the best foundation for co-ordinating the local contribution to any world-wide or extensive regional survey.

(4) That the national or regional co-ordinating bodies should include experts in nature conservation as well as in a range of relevant disciplines such as ecology, biogeography, geomorphology, pedology, biometrics and cartography, with channels for consultation and collaboration.

(5) That the national or regional co-ordinating bodies should assume responsibility for drawing up inventories of the different assemblages of living organisms (ecosystems, associations, etc.) present in the area to be surveyed. For this purpose there would need to be agreement on the scale of the inventories and on a uniform scheme for recording, and subsequently for describing, designating and classifying the assemblages, one accepted and used at least throughout the country or region involved (pp. 91–4). The international advisory body should be consulted before a decision is taken on the recording scheme to be adopted.

(6) That the results of the inventory should be set out in such a way, as by mapping, that the location and extent of the various assemblages, recognized as different at the adopted scale, can serve as a basis for the

selection of survey-sites, this selection being a responsibility of the local co-ordinating body, as should decisions on subsequent procedures for securing the requisite information from selected sites (pp. 195–6).

(7) That local co-ordinating bodies should undertake the training of surveyors to complete check-sheets or other types of record in a manner agreed internationally and locally; and that all completed records should be scrutinized by the local co-ordinating body before being processed.

(8) That for participating countries with inadequate resources in personnel for surveys of the required kind, the international advisory committee should be enabled to organize the training of both senior scientists and of field-surveyors or, in appropriate circumstances, to send advisory groups or task-forces to a country needing such assistance.

5.5.2 The collection of data from the survey-sites

The value for a biological survey of a questionnaire or check-sheet is that it favours the collection of comparable data from all survey-sites and therefore eases the tasks of processing and interpretation. Comparability involves the use of standardized procedures both for obtaining and recording the data. Procedures for obtaining the data can, if necessary, be laid down in field instructions for surveyors, but those for recording will in general be conveyed by the wording of the questions, and standardization will then be dependent on their form, clarity and freedom from ambiguity. The recommendations below follow from these considerations.

(9) That questions should as far as possible demand answers involving a simple choice between alternatives, i.e., two-state or yes/no answers, which greatly simplify data-manipulation. Where this is impossible or inappropriate, the invited choice should be one from a set of stated and readily intelligible possibilities.

(10) That questions involving lengthy free-style answers should be reduced to a minimum or avoided entirely.

(11) That any question requiring lengthy explanation should be regarded as doubtfully valuable for the purposes of a general survey and should either be made more straightforward or the topic should be approached in a different manner.

(12) That information about non-floristic features of the vegetation or about the soils of a survey-site should be recorded in answers to a number of direct questions rather than solely through the assignment by the field-surveyor of classes or codings appropriate to particular systems of classification.

(13) That information about the species-composition of different types of vegetation is best provided through lists of species, and data on their abundance or cover, drawn up in accordance with a standardized procedure

which is in general use in the country or region and full details of which have been made accessible in the scientific literature (see Recommendation (5) above).

(14) That any terms in a questionnaire which might be misunderstood should be defined, carefully and simply, in a glossary included in the field instructions.

5.5.3 The guide to the check-sheet or questionnaire

The *Guide to the Check-Sheet for IBP Areas* (Peterken, 1967) is a book of 133 pages and may reasonably be deemed too lengthy for convenient use in the field. The following recommendations are therefore made, irrespective of whether the present check-sheet is retained more or less unchanged or is replaced by some alternative questionnaire.

(15) That two separate booklets replace the present *Guide*. One of these should be a general guide for members of national or regional organizing bodies as well as for field-surveyors, giving information on the purposes of the survey, on agreed basic principles of site-selection, sampling and recording that must guide all participants, and on the organization, at both national and international levels, designed to ensure success in realizing the stated aims. It should also list, explain and give guidance on any preliminary requirements without which useful participation would be difficult or impossible (see Recommendations 5 and 6 above). The other booklet should be a field guide concerned solely with assisting field-surveyors to make the requisite records and providing an explanatory background to the more difficult parts of the questionnaire, including those dealing with vegetation and soils, and with a glossary if necessary.

(16) That field instructions covering all the straightforward questions should accompany the check-sheet or questionnaire itself and need not be reproduced in either booklet.

(17) That a glossary of technical terms used unavoidably in the check-sheet or questionnaire should also accompany the field instructions and need not appear in either booklet. Technical terms requiring definition should be reduced to a minimum, and it must be made clear that they are to be interpreted strictly as laid down in the glossary. A separate glossary may be necessary in the explanatory booklet dealing with the recording of vegetation-types and soils (see Recommendation (15)). The purpose of this would be to establish an agreed vocabulary for habitat-descriptions, not only for designating vegetation-types but also for recording the habitats of animals of various kinds and the features of the physical environment (pp. 199–200).

(18) That even if the present check-sheet is retained more or less unchanged, the contents of the *Guide*, and therefore of the two booklets replacing it, should be revised in the light of experience derived from the outcome of the survey

(pp. 199–200) and in relation to changes in organization which require placing greater emphasis on national and international co-ordination.

5.5.4 The data-handling system

The objectives of the data-handling system adopted for the check-sheet survey are stated in Chapter 3 and some of its limitations are considered in Section 5.2.4 above. It has been emphasized that the whole of the information on completed sheets was processed in order to provide a general service, and that this resulted in a slower throughput of data and a greater complexity in programming than the ultimate use of the data justified. Limits to use stemmed very much from the way in which information was collected (pp. 202–5), and some changes have, therefore, been recommended. Recent progress in the design especially of small computers, with a consequent increase in their power and flexibility, favours a fresh approach to the system of data-handling, and the following provisional recommendations indicate one such approach.

(19) That the computer used in data-handling should be small and should embody disc files. The PDP 11/10 is of this type. It has a core size of 16K comprising 24-bit words with exchangeable disc packs and capable of storing 1.25 million words. The tape cassette units are available for backing storage and archive, and a fast paper-tape reader and card reader provide input independent of the keyboard terminal.

(Random-access disc files are of great value as directly addressable arrays requiring no core-storage. They may be stored permanently on disc packs which are relatively inexpensive, and provide a medium for data-storage which is effectively an extension of core.)

(20) That the language employed should be BASIC, as being particularly suitable for this kind of application, where the data-items are heterogeneous but relatively simple to code and largely independent of each other. BASIC is capable of handling long alpha-numeric strings as individual records and of segmenting these as appropriate for general sorting procedures or computation based on the numeric components.

(A small computer with random-access disc files and using BASIC provides most of the flexibility of punched-card systems with increased potential for correlation between fields, the added capability of computation and the absence of laborious handling. The recommended combination has many other advantages over punched-card systems, so that comparisons weigh heavily in its favour despite limitations in output and in flexibility of program control.)

5.5.5 Methods of description and designation

The most difficult matters facing the organizers of extensive biological surveys concern the information to be collected about the vegetation and soils of survey-sites, matters considered at some length in Chapter 2 and also briefly on pp. 205–9. Most questions bearing on potential land-use, including use for nature conservation, should be answerable from records of present vegetation and soil together with some information about past use. The important problems are those of deciding precisely what information should be collected and in what form. The following recommendations take into account the present deficiencies in our knowledge of the natural vegetation and soils, and of their significance for land-utilization, in many parts of the world. They also seek to remedy the current lack of generally agreed world-wide systems for describing and designating types of vegetation and soil at a suitable level of discrimination to meet the purposes of extensive surveys.

(21) That in view of its considerable though incomplete success, Fosberg's *Classification of Vegetation for General Purposes* should continue for the time being to be used in biological surveys. Meanwhile international investigations should be undertaken, on as extensive a scale as possible, into the potential use of readily observed and recorded non-floristic features of natural vegetation as a more adequate basis for site-assessment and decisions on land-utilization (pp. 91–3).

(22) That from the outcome of these investigations questionnaires or *pro formas* should be drawn up to elicit the relevant information from survey-sites in natural vegetation of all kinds. Decisions must then be reached on the number of different *pro formas* which are essential or desirable in order to cover the whole range of vegetational physiognomy and structure and which will need to be incorporated in future check-sheets (pp. 92–3).

(23) That records of the floristic composition of the natural vegetation should also be drawn up for each survey-site in a biological survey, preferably following procedures agreed and generally adopted in groups of countries or whole regions and used in preparing the inventories of Recommendations (5) and (6) above. These procedures might most usefully be those of the Braun-Blanquet system, but surveyors should aim at producing reliable and complete species-lists, with estimates of abundance and cover, rather than merely recording the presence of previously named communities.

(24) That international investigations be initiated as soon as possible into the extent to which the recognition and careful demarcation of *biotic or biogeographic provinces* would allow usefully detailed inferences, for the purposes of an extensive biological survey, concerning the species of plants and animals present in areas for the vegetation types of which no full species-lists are available (pp. 88–9).

5 *Conclusions and recommendations*

(25) That consideration be given on an international scale to devising a procedure for recording soils in terms of features readily observable in the field and retaining the advantages in simplicity of Ball's classification but allowing a greater degree of discrimination, especially of soils of the more arid tropical and subtropical zones (pp. 111–3).

Appendix 1. The check-sheet and the country check-sheet

INTERNATIONAL BIOLOGICAL PROGRAMME

SECTION CT: CONSERVATION OF TERRESTRIAL BIOLOGICAL COMMUNITIES

CHECK SHEET (Mark VII) FOR SURVEY OF IBP AREAS*

To be completed with reference to the GUIDE TO THE CHECK SHEET

Serial Number | | | | | | |

For Data Centre Use only

1. 1. Name of surveyor ....
   2. Address of surveyor ....
   ....
   ....
   3. Check Sheet completed (a) on site .... (b) from records ....
   4. Date Check Sheet completed ....

2. 1. Name of IBP Area ....
   2. Name of IBP Subdivision (or serial letter) ....
   3. Map of IBP Area* showing boundaries attached? Yes .... No ....
   4. Sketch map of IBP Area*. Please mark direction of north, the scale and grid numbers where applicable.

* For " IBP Area ", read IBP Area and/or IBP Subdivision.

		For Data Centre Use only
3.	**Location of IBP Area*** 1. Latitude ....° .... N/S Longitude ....° .... E/W 2. Country .... State or Province .... County .... (State or Province .... County ....)	
4.	**Administration** **National** 1. Official category .... 2. Address of administration .... **International Class** 3.	
5.	**Characteristics of IBP Area*** 1. Surface area (state units of measurement) .... 2. Altitude (state units of measurement) Maximum .... Minimum	
6.	**Climate** Nearest climatological station : 1. Name .... 2. Climatological station on IBP Area*? Yes .... No .... 3. If (2) not, distance from edge of IBP Area* (state units) .... 4. Direction from IBP Area* .... 5. Additional data sheet attached? Yes .... No	

Included in U.N. List	Rejected from U.N. List	Area with formal conservation status	No formal cons. status
(A)	(B)	(C)	(D)

For Data Centre Use only

7. **Vegetation and Soil**

1 **Vegetation**

Community Reference Number	Vegetation Code: Primary Structural Group	Vegetation Code: Class	Vegetation Code: Group	Vegetation Code: Formation	Vegetation Code: Sub-Formation	Plant communities (give usual name using full Latin names of a species where applicable)	Area (state units)
1							
2							
3							
4							
5							
6							
7							
8							
9							
10							
11							
12							
13							
14							
15							
16							
17							
18							
19							
20							

Please give information about further communities on a separate sheet.

		For Data Centre Use only
7. (cont.)	2 **Soil**	

Community Reference Number	Soil type	Other notes
1		
2		
3		
4		
5		
6		
7		
8		
9		
10		
11		
12		
13		
14		
15		
16		
17		
18		
19		
20		

For Data Centre Use only

B. **Similar Communities in Country (or State)**

Community Reference Number	Protected					Protected and Unprotected				
	Abundant	Infrequent	None known	Decreasing	Increasing	Abundant	Infrequent	None known	Decreasing	Increasing
1										
2										
3										
4										
5										
6										
7										
8										
9										
10										
11										
12										
13										
14										
15										
16										
17										
18										
19										
20										

		For Data Centre Use only

9. Landscape

1. General Landscape (give brief description) ....

....

....

....

2. Relief Type

	Flat	Undulating (0)-200 m.	Hilly 200-1000 m.	Mountainous > 1000 m.	%
Sharply dissected					
Gently dissected					
Incised					
Skeletonised					
%					100%

3. Special landscape features (list) ....

....

....

10. Coastline of IBP Area*

1. Protected bays and/or inlets Many ☐ Few ☐ None ☐
2. Substratum. % of coast

Rock	Boulder Beach	Shingle Beach	Sand Beach	Shell Beach	Mud	Coral	Ice

3. Physiography. % of coast

Cliffed	Sloping	Flat

4. Special Coastal Features (list) ....

....

....

5. Tide. Maximum range (state units of measurement) ....
6. Total length of coastline :

Less than 1 km. ☐ 1-10 km. ☐ Above 10 km. ☐

		For Data Centre Use only

11. **Freshwater within IBP Area***

1.

	Permanent	Intermittent
General		
Standing		
Running		

2. Standing Water

	Permanent	Intermittent	Unproductive	Productive
Swamps				
Ponds				
Lakes				

3. Running Water

	Permanent	Intermittent
Springs, cold		
Springs, hot		
Streams		
Rivers		

4. Special freshwater features ....

....

12. **Salt and Brackish Water within IBP Area***

Salt Lakes		Lagoon		....	
Estuaries		Salt pools		....	

13. **Adjacent Water Bodies (not within IBP Area*)**

1. Fresh ☐ Lake ☐ River ☐ Stream ☐

2. Salt and Brackish

Estuary	Salt lake	Salt pool	Lagoon	Ocean		

		For Data Centre Use only

14. **Outstanding Floral and Faunal Features**

1. None ....

2. Fauna

	Species diversity	Abundance of individuals	Superabundance of individuals	Rare species	Threatened/Relict species	Spp. of biogeographical interest	Exceptional Associations	Breeding or Nesting Populations	Migrating Populations	Wintering Populations		
Mammalia												
Aves												
Reptilia												
Amphibia												
Pisces												
Insecta												

3. Names of main threatened, endemic, relict and rare species

....

....

....

....

....

....

....

For Data Centre Use only

4. **Flora**

	Species diversity	Abundance of particular species	Rare species	Threatened/relict species	Spp. of biogeographical interest	Exceptional associations	Outstanding specimens				
Angiospermae :											
trees											
shrubs											
herbs											
grass											
Gymnospermae											
Pteridophyta											
Bryophyta											
Lichens and Algae											

5. Names of main threatened, endemic, relict and rare species

....

....

....

15. **Exceptional Interest of IBP Area***

....

....

....

....

....

For Data Centre Use only

16 **Significant Human Impact**

1. General : None in entire IBP Area* ....

None in part of IBP Area* ....

Impact on entire IBP Area* ....

2. Particular

	Past impact	Present impact	Trend: Increasing	Trend: Decreasing	Trend: No change	Trend: No information
Cultivation						
Drainage						
Other soil disturbance						
Grazing						
Selective flora disturbance						
Logging						
Plantation						
Hunting						
Removal of predators						
Pesticides						
Introductions — plants						
Introductions — animals						
Fire						
Permanent habitation						
Recreation and tourism						
Research						

3. Additional details on each type of impact attached?

Yes .... No ....

For Data Centre Use only

17. **Conservation Status**

	Protection			Utilisation			Conservation Management			Permitted Research		
	none	partial	total	none	controlled	uncontrolled	none	to alter status	to maintain status	experimental	observational	prohibited
Flora												
Fauna												
Non-living												

18. **References**

1. List major biological/geographical references for the IBP Area.

   Sheet attached? Yes .... No ....

2. List main maps available for the IBP Area.

   List attached? Yes .... No ....

3. Aerial photographs for the IBP Area available?

   For whole area .... For part of area .... None ....

19. **Other Relevant Information**

Signed ....
(Surveyor)

Appendix 1

GUIDE TO THE CHECK SHEET

by G. F. Peterken

PART FOUR

FIELD INSTRUCTIONS

This part is designed to assist the surveyor to fill in the Check Sheet, and thereby facilitate the task of the Data Centre in transferring the contents of each Check Sheet to the computer tape. It contains all definitions and instructions necessary for completing the Check Sheet, except the classifications of plant formations and soils, which are presented in Appendices 1 and 2 respectively. Together with these appendices, it can be used in isolation from the remainder of the Guide, and is therefore suitable for translation in those countries where it is not possible to translate the entire Guide. Previous parts explain the purpose and objectives of the survey (Part 1), the selection of sites (Part 2) and the meaning and purpose of each question on the Check Sheet (Part 3). Following this part are four appendices dealing with the classification of Plant Formations, classification of soils, the Geocode and an example of a completed Check Sheet

Incomplete Information

It is likely that for many IBP Areas* the surveyor will not have enough information to complete every question To a limited extent this does not matter, for even incomplete returns will contain valuable information. Nevertheless, there is a minimum number of sections which must be completed before a returned Check Sheet can be accepted as adequate : **Sections 1, 2, 3, 4, 5, and 7(1) must be completed before it is worth sending in a Check Sheet to the Data Centre.**

A returned Check Sheet containing only the bare minimum of information will possess only limited worth. In practice it is expected that for most IBP Areas* much more information will be available : any ecologist reasonably familiar with an IBP Area* should have no difficulty in answering Sections 6, 7(2), 9, 10, 11, 12 and 13 in addition to those listed above. The remaining Sections — 8, 14, 15, 16, 17 and 18 — ask for more detailed information which may not be readily available Since these later sections largely correspond with the conservation content of the Check Sheet, it is hoped that surveyors will make every effort to obtain the additional information necessary to complete the Check Sheet As the number of unanswered questions increases, so does the value of the survey decrease.

IBP Area and IBP Subdivision

IBP Area : An IBP Area is a site of class A, B, C or D as defined below under 4(3).

IBP Subdivision : An IBP Subdivision is part of an IBP Area. It is an area, variable in extent, which is of interest to conservationists and biologists, and which is of such size and uniformity that its features can be meaningfully set out on a single Check Sheet.

Notes on Sections

In the paragraphs below, the numbers correspond with the section (question) numbers on the Check Sheet

General rules :

(a) Where quantitative information is requested (e g area) this should be given as accurately as possible Estimates are acceptable in the absence of accurate values.

(b) In general only positive statements should be made (i.e. **presence** of a particular feature) but when a feature is known **with certainty** to be absent, this may be stated.

1(1) **Name of surveyor**

(2) **Address of surveyor**

(3) **Check Sheet completed on site/from records.** Check (i.e. ✓) one or both as applicable.

(4) **Date Check Sheet completed.**

2(1) **Name of IBP Area.** If the IBP Area is Class A, B or C (see 4(3) below), insert the name as it appears in the U.N List (A and B), or in national lists of protected sites (B and C). For Class D IBP Areas insert the name by which the IBP Area is generally known. If the U.N List is not available for Classes A and B, fill in the name by which the IBP Area is generally known.

N B * IBP Areas and IBP Subdivisions

(2) **Name of IBP Subdivision.** To be used **only** when the IBP Area is divided into two or more IBP Subdivisions. IBP Subdivisions for which there is no suitable name should be given a reference letter (a, b, c . . . etc.), thus distinguishing them from other IBP Subdivisions in the same IBP Area. **This question should only be left blank if the Check Sheet refers to an IBP Area.**

(3) **Map of IBP Area* showing boundaries attached?** Yes/No. Check

(4) **Sketch map of IBP Area*.** This should show
— the shape of the IBP Area*
— its relation to compass directions
— boundaries common with the boundary of the IBP Area (for IBP Subdivisions only)
— major features of the land form and vegetation (e.g. peaks, rivers, woods, etc.)
— sites of field stations and other permanent habitations.

3(1) **Latitude and Longitude.** Delete the N or S, E or W which does not apply.

(2) **Country, State or Province, County.** Insert names of administrative areas in which the IBP Area* is situated. The following levels are recognised :
— National or Territorial, embracing the whole contiguous area under one political sovereignty **(Country)**
— Regional or Provincial units intermediate between national and local levels **(State or Province)**
— Local, e g **county,** parish, commune, gemeinde, etc.
Spaces are provided for IBP Areas* which overlap "Province" or "County" boundaries.

4(1) **National Category,** e.g. National Park, Strict Nature Reserve, etc.

(2) **Address of Administration** responsible for the IBP Area*. Full postal address.

(3) **International Class.** The following four classes have been adopted. Check under the appropriate class.
Class A : Included in U.N. List
Class B : Considered for inclusion in U.N. List, but rejected. These sites are mentioned in Chapter V of the U.N. List.
Class C : Other sites at present protected.
Class D : Unprotected sites of interest to conservationists and biologists.

5(1) **Surface area,** may be inserted in any units, but please state units.

(2) **Altitude. Maximum and Minimum.** Please state units used.

6(1) **Name of Nearest Climatological Station.** As used in publications of national climatological organisations.

(2) **Climatological Station on IBP Area* Yes/No.** Check.

(3) **Distance from edge of IBP Area* if outside.** State units.

(4) **Direction from IBP Area*.** Insert compass direction from centre of IBP Area*. Use 16-point compass notation (N, NNE, NE, . . . NNW) or degrees (0°, 10°, . . . 350°).

(5) **Additional data sheet attached? Yes/No.** Check.

7(1) **Vegetation**

Plant Communities. List these by their usual names, using "Latin" names for all species mentioned. Space is provided for 20 Communities : further Communities should be listed on a separate sheet. There is no restriction on the methods by which Communities may be defined, so long as the Communities so formed can be easily recognised by local scientists. Community Reference Numbers are provided to facilitate cross reference between 7(1), 7(2) and 8

Vegetation Code. The Formation (and sub-formation) to which each Community belongs should be entered. These Formations (and sub-formations) may be identified in Appendix 1. A key is provided to facilitate identification. Enter only the code numbers for each Formation (and sub-formation), placing one digit in each square.

Area of each Community should be entered to maximum available accuracy.

7(2) **Soil**

Soil Type. Enter the code number for the soil type which occurs under each Community. These can be identified in Appendix 2. Where more than one soil type occurs under one Community, either the definition of the Community should be revised, or an explanatory note should be added under ' Other notes '.

Other Notes. Sub-types present should be mentioned, together with short descriptions of significant features, e.g., colour, humus content, depth.

8 **Similar Communities in Country (or State)**

This Section will normally refer to the entire Country, but in the case of large countries (Australia, Brazil, Canada, China, India, U.S.A., U.S.S.R.) it should refer to states or provinces (primary administrative subdivisions). All Communities should be considered here -- in exactly the same order as in 7, using the Community Reference Number for cross-reference.

Insert up to four checks in each row.

Protected : refers to sites of A, B and C (see 4(3) above).

Protected and Unprotected : refers to all sites within the Country (or State).

None known : The Community does not occur elsewhere in the country/state.

Infrequent : Other examples of the Community exist in the country/state, but the loss of any one of them would be a grave depletion of its type.

Abundant : Other examples of the Community are sufficiently common and widespread that the loss of any one of them would not be a significant depletion of its type.

Decreasing/Increasing : Insert a check only when the change observed appears to be leading to a permanent change in the status of the Community.

9(1) **General Landscape.** Describe in less than 50 words. Confine description to geomorphological features. It is permissible to consider land outside the IBP Area* (see Part 3).

(2) **Relief Type.** Check off type(s) present. It is possible to consider land outside the IBP Area* (see Part 3).

Altitudinal range : divided into four classes of which the lowest is ' flat ', in which there is very little variation in altitude.

Erosion Types may be illustrated as follows :

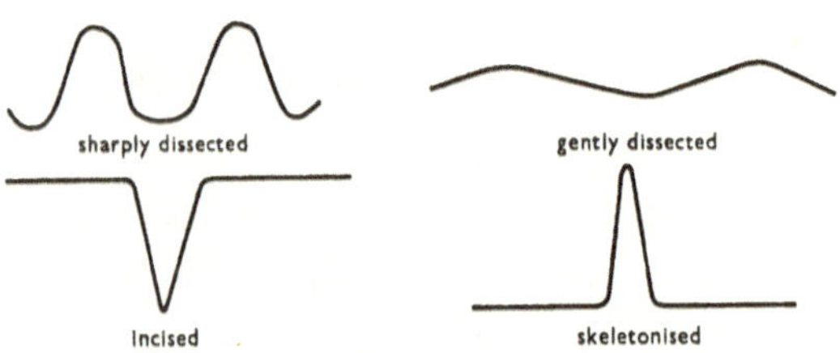

(3) **Special Landscape Features** should be listed according to widely known terms (e.g. cliff, ice fields, dunes, recent vulcanism . . .). Interpret ' special ' liberally.

10(1) **Protected Bays and Inlets. Many/Few/None.** Check.

(2) **Substratum.** Insert approximate percentage value for the length of coast occupied by each type of substratum. It is possible for the total to exceed 100%. Definitions are as follows :

Rock : Fixed, stable, unweathered rock.

Beach : Mobile, or potentially mobile, material, of which the particle size ranges from very large (boulder) to minute (mud).

(3) **Physiography.** Insert approximate percentage value for the length of coast occupied by each type. These values should total 100%.

Definitions are as follows :

Cliffed : Wholly or partially ' vertical ' with at least some part inaccessible to land animals.

Sloping : Cliffed coastlines in which no part is inaccessible to land animals.
Flat : Coastlines which lack cliffs and sloping 'cliffs'.

(4) **Special coastal features** should be listed accordingly to widely terms (e.g., reefs, sand bars . . .)

(5) **Tide. Maximum Range.** State units.

(6) **Total length of coastline.** Check appropriate value.

11 **Freshwater within IBP Area*.**

(1), (2) and (3) Check in the spaces the features which are present. Surveyors may insert indications of abundance, e.g., many, few, etc., provided it is clear which features are present and which absent.

Definitions :

General :	All types of freshwater.
Standing :	Water not flowing continuously in a definite direction.
Running :	Water flowing in a definite direction.
Swamp :	A lake, pond or other site of such small depth that it is occupied ± completely by emergent vegetation.
Pond :	A body of standing water whose area of open water is less than 10,000 m^2.
Lake :	A body of standing water whose area of open water is greater than 10,000 m^2.
Spring :	A site at which water is issuing through a natural opening in such quantity as to form an appreciable current. A hot spring has an average temperature more than 10°C above the yearly mean for the surrounding air.
Stream :	A watercourse or part of a watercourse whose mean width is less than 5 m.
River :	A watercourse or part of a watercourse whose mean width is greater than 5 m.
Permanent :	Never, or very rarely disappears. All other situations are regarded as 'Intermittent'.
Productive :	Eutrophic waters, and those with relatively high biological productivity which are morphometrically oligotrophic.
Unproductive :	Other oligotrophic waters, and those of relatively low biological productivity.

(4) **Special freshwater features** should be listed according to widely known terms (e.g., rapids, geysers, seasonally inundated land . . .).

12 **Salt and Brackish Water within IBP Area*.** Check.

13 **Adjacent water bodies,** i.e. those whose margins form part or all of the boundary of the IBP Area*, which are therefore not within the IBP Area*

Definitions as follows :

Freshwater :	Salinity generally within the range 15-300 p.p.m.
Salt and Brackish water :	Salinity above the normal range of freshwater.
Ocean :	Should only be used for the interconnected oceans.
Salt Lake :	A body of standing salt water whose area of open water is greater than 10,000 m^2.
Salt Pool :	A body of standing salt or brackish water whose area of open water is less than 10,000 m^2.
Lagoon :	Shallow lake formed in association with coral.
Estuary :	Tidal portion of a river mouth.

14(1) **Outstanding Floral and Faunal Features.** Check if none known.

(2) and (4) Only the presence of outstanding features should be noted by checking the appropriate box. No other information is required here: we do not want, for example, the number of bird species present inserted under "Aves — species diversity", because this is not in itself an indication that this number is outstanding. Columns have been left vacant for additional types of outstanding feature, and additional taxonomic groups may be added in the vacant rows. The vacant rows may also be used to give more precise data for the groups listed, e.g., if the outstanding interest centres on the Carnivora of the Mammalia, "Carnivora" may be inserted in a vacant row Always designate taxonomic groups by their "Latin" name.

(3) and (5) **Names of main threatened, endemic, relict and rare species.** List the species by their "Latin" names. Vernacular names in addition are welcome but not obligatory.

15 **Exceptional Interest of IBP Area*.** List items and salient facts (e.g., botanical, ornithological, teaching area, site of classic research since 1930 . . .).

16(1) **Significant Human Impact. General.** Check one line

(2) **Particular** types of significant human impact. Types of human impact additional to the 16 types listed should be entered in the vacant rows. Where the impact does not operate today, but has operated in the past, check 'past'. Where it does operate now but did not operate before 1900, check 'Present' only. Where a present-day impact operated before 1900, check both 'past' and 'present'. For all types of present impact check off the trend. Only check 'increasing' or 'decreasing' if this is certain, otherwise check 'no certain change'.

(3) **Additional details on each type of impact attached? Yes/No.** Check.

17 **Conservation Status.** Refers to human influence on material objects within the IBP Area*. This influence may be 'partial' in space, time or manner.

Protection (from exploitation): Refers to current legal position regarding deleterious influence of man. If practice falls significantly short of theory, this fact should be noted in 19.

Utilisation: Restrained exploitation, to take a long term 'crop'. The extent and period of utilisation may be legally limited ("Controlled") or not ("Uncontrolled").

Conservation Management: Utilisation with the primary object of maintaining, restoring or creating an ecosystem which has some special interest to biologists. 'Status' refers to biological status, which may be equated with vegetation type for the purposes of this survey.

Permitted Research: 'Observational' research does not interfere with the ecosystem, 'Experimental' research usually involves interference of some sort.

18(1) **List major biological/geographical references for the IBP Area*.** Attach list and check.

(2) **List main maps available for the IBP Area*.** Attach list and check.

(3) **Aerial photographs for the IBP Area* available?** Check one space.

19 **Other relevant information.** Can also be used when there is insufficient space for the answer to another question.

Additional Information

In a number of sections surveyors are asked to attach additional information, when this is available, on separate sheets. These sections are:

2(4) Map of IBP Area*.

6(5) Climatological Data.

16(3) Significant Human Impact. Explanatory notes.

18(1) Major biological/geographical references.

(2) List of main maps available.

Data Centre

Completed Check Sheets should be returned to the national organiser, or direct to the Data Centre whose address is:

IBP/CT Survey,
Biological Records Centre,
The Nature Conservancy,
Monks Wood Experimental Station,
Abbots Ripton,
Huntingdon, England.

INTERNATIONAL BIOLOGICAL PROGRAMME

SECTION CT: CONSERVATION OF TERRESTRIAL COMMUNITIES

COUNTRY CHECK SHEET

To be completed for a Country, Province, State or other political unit

Serial Number ☐☐☐☐☐☐☐

1.	1. Country .... 2. Province .... 3. State .... 4. Geocode	
2.	Area of country, province or state (state units)	
3.	Conservation Administrations. Names and addresses with classes of reserves administered.	
4.	Extent of knowledge of the vegetation. Is the occurrence and distribution of the vegetation formation classes ☐ completely known ☐ completely known with minor local exceptions ☐ completely known for the greater part of the country, province etc. ☐ completely known only for certain regions ☐ known only in imprecise or partial terms ☐ still in the course of active investigation Check as appropriate	

Appendix 1

5	6	7	8	9	10
Fosberg formations	Usual name of community	Area occupied (see key)	Distribution (see key)	Protection status (see key)	Check Sheets completed (see key)

Please enter further information on a separate sheet

11. Give the location and area of all "best examples" listed under Category 9.7
Indicate which occur in areas currently proposed or under negotiation as protected areas

Name of area	Location Latitude/Longitude	Area (State units)	Proposed Reserve

Please enter further information on a separate sheet

12. **Areas of special interest. To be completed for those areas which exhibit features of outstanding biological interest**

Name of Area	Location Lat/Long	Area (State units)	Special Interest	Protection status (see key)	Check Sheets completed (see key)

Please enter further information on a separate sheet

13.	What other problems of conservation not dealt with above are currently serious in their incidence in the country or province. (Based on Q.16 of Mark VII Check Sheet).
14.	Give main references on the vegetation and fauna of major regions.

KEY

7. Where available information permits, indicate which of the units listed in 5 occupies within the country/province:

- 7.1 more than 1 million sq. km.
- 7.2 100,000 - 1 million sq. km.
- 7.3 10,000 - 100,000 sq. km.
- 7.4 1,000 - 10,000 sq. km.
- 7.5 100 - 1,000 sq. km.
- 7.6 10 - 100 sq. km.
- 7.7 under 10 sq. km.

---

8. Where available information permits, indicate which units are:

- 8.1 generally distributed throughout the country
- 8.2 generally distributed throughout one or more major named regions
- 8.3 irregularly or patchily distributed throughout the country
- 8.4 localised within certain limited geographic sections (specify)
- 8.5 localised within certain ecological limits (specify)

---

9. Name the reserve in which the best example of each Fosberg unit is protected and give its category. If the unit is not protected, name the site of its best example and enter category 9.7.

Categories:

- 9.1 National Park
- 9.2 Equivalent Reserve
- 9.3 Other nature reserve or protected area
- 9.4 Forest refuge
- 9.5 Wildlife Refuge or Game Reserve
- 9.6 Conservation area
- 9.7 Area with no conservation status

---

10. As all examples used in Questions 9 and 12 qualify as IBP Areas, check whether special Check Sheets (Mark VII) have:

- 10.1 already been completed and sent in
- 10.2 are in course of completion
- 10.3 still need to be arranged for

Appendix 2. A classification of vegetation for general purposes

Introduction

Vegetation is the term used to designate the total plant cover of a region, area, or place. It should not be used when speaking of individual plants, kinds of plants, or lists of kinds that occupy an area. For the latter the term *flora* is used. Vegetation is generally made up of one or more plant communities or aggregations of plants, usually forming a mosaic or complex. It is a geographic feature of great importance, as it determines the appearance and general character of most land areas of the earth.

The study of vegetation is often regarded as the special province of the science of plant sociology (phytosociology, phytocoenology) but in practice it is studied in more or less detail by many whose main purposes are not plant sociology. Workers in many other scientific aspects of botany and ecology, zoologists, pedologists, geologists, geographers, and those interested in a host of practical fields where vegetation has a bearing, or, indeed, may be the principal material concerned, as in forestry and agriculture, have a legitimate concern with vegetation and often include valuable descriptions of it in their papers. Information from such sources is about all that is available on the vegetation of many areas, as any regional vegetation bibliography shows.

Vegetation studies by plant sociologists usually either follow one of the recognised schemes or, especially in the case of British and American workers, *ad hoc* descriptive systems especially constructed for the particular areas studied, and particularly suited to these areas or to the particular plants of the studies. Descriptions and maps of vegetation by those who are not special students of the subject are very diverse. Sometimes they follow or attempt to follow existing classifications. More usually the authors either evolve special classifications or do not attempt to classify, because a general classification well adapted to their purposes is lacking.

While a great deal of valuable work has been done and expressed in these *ad hoc* arrangements, there is difficulty in relating information from one area

* As recorded in the *IBP/CT Progress Report: 1971*, the field trials in various parts of the world have led to the formulation of an additional number of formations and subformations. These further categories are listed in an annex at the back of this reprint.

with that from another and in making comparisons. It is felt that if a generally acceptable classification can be devised for use on a world-wide basis such comparison will be greatly facilitated. It would then, also, be possible to organise the vegetation information available for the entire world into a co-ordinated body so that any part of it would be readily available in usable and comparable form.

Such a classification, to be serviceable to a maximum extent must possess the following features: (1) It must be possible to apply it even with the minimum of information commonly available. (2) It must be applicable to all macroscopic vegetation, or at least be capable of expansion to accommodate all types that cover mappable areas at ordinary scales. (3) It must convey or be adaptable to convey the kinds of information useful to a wide range of users of vegetation data. (4) It must be capable of refinement to utilise and convey detailed and quantitative information when this is available. (5) Its terminology must, in itself, convey a substantial amount of information about the vegetation. (6) It must be based entirely on the features of the plants, themselves, without involving features of the physical or animal environment.

With this background statement there can now be presented some general remarks on the uses of classification in mapping vegetation, the kinds of classifications now available and the kinds of information that may be used in classifying vegetation.

General remarks on vegetation classification

Classification of vegetation has as primary objectives to facilitate recording of information in an orderly manner, to aid in storage and prompt recovery of such information, to make possible intelligent discussion of vegetation at various levels of abstraction, to aid in understanding the phenomenon itself, and to enable us to communicate information on vegetation easily and unambiguously. The storage and communication functions are of most concern in vegetation mapping.

Mapping is done in terms of classificatory units which can be used to characterise and designate areas, which are then outlined on a map. Taken together, these areas form patterns. One of the principal uses of maps is to compare patterns representing different features and factors, to detect correlations. Another, of course is to present information in a graphic, readily comprehensible form, to save the reader's time.

As is well known, there are many systems of classification of vegetation, or systems that purport to be classifications of vegetation. These are all of value, serving different purposes and interests. The desire, often expressed, for a single uniform classification of vegetation is based on the idea that all purposes can be best served by a single classification, that if we only could formulate the right classification we would not need any others. This is clearly illusory.

It seems inherently impossible to serve, adequately, all of the multifarious interests in vegetation with any one classification. Therefore there will continue to be as many different ones as are needed.

Most of the important vegetation classifications are based not only on the features of the vegetation itself but also on various factors in the environment, such as climate, ground water, soil, and biotic influences. Thus, they should more properly be called classifications of ecosystems.* These are important and necessary, to develop proper understanding of ecosystems including their vegetation components. However, in the use of maps, great care must be taken to avoid circular reasoning. If a vegetation map that is really a map of an ecosystem including climate as one of its bases is compared with a climatic map there will certainly be a correlation, but it may be a false one. Therefore it seems essential that, for mapping purposes, at least, a classification be devised that is based only on the characteristics of the vegetation itself.

Of the important existing systems of classification, only that of Braun-Blanquet seems to satisfy fully this requirement. Some of its other features, however, limit its usefulness for mapping, especially small scale mapping. The system of Rübel also comes fairly close, and his main larger categories are based strictly on vegetation characters, but if I interpret it correctly, the attempt breaks down in the more extreme situations. The other systems frankly include environmental criteria.

There are two radically different points of departure employed in vegetation classification, perhaps in all classification. These are (1) the subdivision of collective phenomena into successively smaller units on the bases of their differences, and (2) the isolation of the basic smallest units, comparison and sorting into kinds with like characteristics, and the aggregation of these into successively larger units on the bases of their similarities. These are both logically acceptable techniques, each useful for certain purposes, but there is no reason to assume that they would necessarily produce similar arrangements. The Schimper's scheme and that of Rübel exemplify the first technique, that of Braun-Blanquet, the second.

For small scale mapping it seems clear that the subdivision technique is greatly to be preferred over the aggregation one, as it yields a general picture promptly which may be refined later. The other method, while ideally good because of the avoidance of a series of states of approximation, involves so much work and so much time that a map of a large area could not be constructed on this basis within a single lifetime, and the vegetation would

* Ecosystem is the term used for the sum total of vegetation, animals and physical environment in whatever size segment of the world is chosen for study.

likely be so changed in the time elapsed that the map would be obsolete long before it was finished.

Information used in classification of vegetation

The information considered by students of vegetation and from which the criteria for classification may be selected can be roughly and arbitrarily arranged in seven categories and commented on as follows:

A. **Physiognomy** is the appearance, especially the external appearance, of the vegetation, partly resulting from, but not to be confused with, *structure* and *function*, which are much more exact and objective categories. It is the roughest and least precise of the classes of information, but is very useful in separating such large categories as forest, grassland, scrub, desert, savanna and others. Gross compositional features, luxuriance, seasonality, biotic influences, and relative xeromorphy and the like show up here. Classifications based on physiognomy are about the easiest to agree on and also the easiest to use in cartography. Their categories, however, are usually extremely broad and their significance is often highly debatable. They are also hard to refine in a quantitative manner. The systems of Schimper and Rübel are essentially physiognomic.

B. **Structure** is here defined as the arrangement in space of the components of vegetation. Earlier definitions have varied according to different authors, some having restricted it to stratification and spacing, others including data on life-forms, growth-forms, leaf characters, functional adaptations of various sorts, and even dispersal mechanisms. Others have confused structure and physiognomy.

It seems that clarity may be served by separation of characteristics pertaining to the spatial distribution of the biomass, or material making up the vegetation, from functional or supposedly functional adaptations, considered below under *function*. We are left with the phenomena of height of plants, branching habit, size of stems, size of crowns, density of crowns, thickness and density of canopy, layering or stratification, and depth, density, spacing and stratification of root systems. We then have a more logical concept of structure, dealing only with spacing and size phenomena. This is perhaps useful by itself for certain special classifications, but must perhaps be combined at least with *function* for a general classification.

C. **Function,** as used here, includes features that seem to suggest special adaptation to environmental situations, either present or past. It has, by some authors, been restricted to periodic phenomena and dispersal mechanisms, but there seems to be little basis for this restriction, as these are only specific categories of adaptation. Here may be grouped such 'epharmonic' features as periodicity of any sort, life forms (in the Raunkiaer sense), growth forms, protective mechanisms, means of resistance to fire, xeromorphy, halomorphy, hydromorphy, and dispersal mechanisms. Another consideration that should come in this category is effectiveness in using the resources of the environment, but since at present no very practical means of measuring this are available it is difficult to use it in classification. Dansereau has used many of the features grouped here under structure and function to divide the world's climax vegetation into 15 major categories and is presently preparing a world vegetation map on a structural-functional basis. This is awaited with great interest.

D. Composition, the list of species making up the vegetation, is without doubt the most significant type of information about vegetation. However, the number of species involved, the complexity of the combinations in which they occur, and the inadequate state of their taxonomy combine to defeat effective use of composition as a criterion for classification except on a very local scale, in extreme habitats, or in regions that are very well known, such as Europe. There the several schools of phytosociology, especially that of Braun-Blanquet, have made use of floristic composition in a valuable way.

The difference between simple floristic composition, giving equal weight to all species present in whatever abundance or rarity, and proportional composition is important. No consistent scheme is known to me using proportional composition in its full sense, though a scale of abundance is an accompanying feature of the Braun-Blanquet method. However, species dominance, which is an aspect of proportional composition, plays an important part in many, if not most, vegetation classifications.

In the higher categories in a general classification composition data can be used to advantage only to the extent of some indication of dominance and of relative numbers of species present. The introduction, at this level, of more detailed floristic data is likely to obscure, rather than clarify. In the lower levels of most classifications floristic peculiarities are the principal means of separating units.

E. Dynamics or successional phenomena are of the utmost importance in the understanding of vegetation, which cannot be successfully regarded as a static concept. It is desirable to recognise dynamic status in any classification, and indeed, classifications based on climax types are only partially successful because of their failure to accommodate the widespread secondary types. However, except for attempts to arrange vegetation in vast 'cliseres' and the more successful attempts to interpret it in terms of hydroseres, xeroseres, and haloseres there has been little progress made in constructing world classifications which recognise the dynamic status of their units. Again, this sort of information appears to be too complex to be introduced effectively into a general scheme. Further, much of the evidence for dynamic status is inferential, derived from other data, hence its introduction into classification may give double weight to certain features.

F. Habitat or environmental relations are certainly essential data for the understanding of vegetation. Most classifications make full or partial use of such information. However, as noted above, such schemes are more appropriately termed classifications of ecosystems, or perhaps of landscape types. Environmental features will not be considered further here.

G. History is one of the most underemphasised categories of information on vegetation. That today's vegetation is as much shaped by its history as by its present environment shows up in its structure, its composition, its dynamic status, and even in its purely functional features. However, it is rarely possible to be very sure of historical influences, except by inference from other characteristics. This makes history a dangerous thing to be allowed a place in a classification, as it is too hard to stick to facts. When inferences are brought into the classification, the whole scheme may be weakened as indicated above for inferences about dynamic status. Therefore, it seems undesirable, as well as difficult, to bring historical features into a general classification.

Proposed general classification

With this background an attempt may be made to formulate a general classification of vegetation for use in mapping and for recording and presenting vegetation data. To achieve maximum objectivity, suitability for comparison with other features, and simplicity, a combination of features grouped above under *structure* and *function* appear to be the most satisfactory criteria available. This formulation has been carried down to the lowest levels at which these categories of information provide clear separations of units. The ultimate units, below this, must clearly be floristic,* and will correspond generally to the associations recognised by most systems. Above this, floristic criteria have been largely disregarded, as they do not permit broad enough correlations, at least in the present state of our knowledge.

The selection of the classes of criteria on which to base sub-divisions at the several levels recognised has been rather arbitrary, as the relative importance of different features is a subjective matter, depending on one's ultimate interest. An attempt has been made, however, to make the criteria consistent,

* Based on species composition.

in so far as the diversity of vegetation will permit. For this reason, the usual primary subdivision into woodland (or forest), grassland, and desert has been abandoned, as the basis of desert is spacing while that of woodland and grassland is habit or growth form.

Spacing as the primary basis for division, though not fitting any general scheme of vernacular terms, would seem to be of almost universal applicability, and also of great dynamic significance. There seems to be no available general term for primary categories based on spacing, at least in English, so *primary structural group* is here proposed. Three such groups are recognised—closed vegetation, open vegetation, and sparse vegetation or desert. In closed vegetation, in at least one layer or in all of them taken together the plants are predominantly touching or overlapping. In open vegetation they are separated by not more than twice their diameters. In sparse vegetation there is more space between the plants of the most complete layer.

It is realised that these three groups are based only on the visible portions of the vegetation, that the root systems in open vegetation may be, indeed in some cases are known to be, in contact. There is no necessary correlation between the spacing of shoots and root systems, just as there is no perfect correlation between stature of shoot and depth to which roots penetrate. However, with present means of observation, the study of underground aspects of vegetation is slow and laborious and the details are known for very few types. It is not practical, at present, to include information on root systems in a general classification, as this information will be lacking for a long time to come for most vegetation. However, when structural, and possibly even functional, information on the rhizosphere becomes more plentiful, revision of some aspects of this classification will doubtless become necessary.

At the next level, for which the old term *formation class* is adopted, habit and stature seem to provide a satisfactory and generally accepted set of criteria. Here the well-known terms denoting physiognomic types could be used, but it seems better to avoid those with 'land' as a suffix, because of a possible confusion with landscape types. Hence forest is used rather than woodland, scrub rather than shrubland, and grass rather than grassland.

The next rank, for which *formation group* may not be the best possible term but is at least less confusing than 'formation type', is characterised by whether or not its dominant layer is evergreen versus deciduous or otherwise visibly seasonally dormant.

The **formation** category has been preserved at somewhat its traditional physiognomic level, but based on dominant growth form, with emphasis on leaf texture and on such epharmonic features as thorniness, succulence, and graminoid habit.

The **subformation** category is used where further subdivision, on much the same general type of characters of growth form or epharmony, seems desirable. These could, with some justification, be regarded as independent formations, though a more logical arrangement seems to be effected by placing them as subdivisions.

These categories, both those on the same level and those on different levels, are not in all cases sharply separated. Intermediate terms such as scrub-forest and semi-open forest may be introduced where necessary. Generally it may suffice to include qualifications in any characterisation that accompanies the use of a term, in order to avoid undue multiplication of terms.

Descriptive diagnosis

It is intended that, where this classification is used on a general level, a diagnosis formed by combining the terms used for the applicable hierarchy of units will be used to characterise and refer to a particular vegetation. Where enough information is available to speak of a particular association by a name taken from some of its components, it may be placed in the system by such a diagnosis as just mentioned. Even without an accompanying description, such a diagnosis conveys a very considerable amount of structural and functional information. Where a vernacular term is available it may be used and the diagnosis merely added to characterise it. If no such term is available, a given vegetation type can be referred to by its diagnosis.

Refinement

In the form outlined below the classification is adapted for use where a minimum of information is available. This results in such anomalies as a 6 metre stand of young scrub pine in Virginia or a subarctic spruce taiga falling into the same formation with a 70 metre redwood or Douglas fir forest of northwest America. Such things are scarcely avoidable as long as insufficient information is available. However, when quantitative data, such as heights and trunk diameters of trees, measurements of spacing and coverage, stem counts, and any other significant information that can be applied to examples of formations or subformations are known, they can be worked into the proposed classification in a very simple manner without distorting or straining the system at all. Such information can be added to the diagnoses of the formations as modifiers to separate them into subunits in any manner desired, with the augmented diagnoses conveying the additional information that is regarded as important. Such units may easily be further refined or augmented by the addition of names of dominant genera or species, or in any other way that does not involve bringing in environmental characteristics.

When to use this classification

It is well recognised that this is a highly artificial classification, and that the selection of criteria has been rather arbitrary. It should not displace existing classifications. For many purposes special classifications, differing widely from this, will be preferable. For organising vegetational data for general purposes, however, for mapping, and for comparing vegetation patterns with environmental or even chronological patterns this one may be useful. If this or something similar can be widely accepted and used, much information that might otherwise be presented in very diverse fashion will become available in comparable form, making communication of such data more effective. It will, further, make possible the comparison of vegetation maps with climatic, soil, and other environmental maps without risk of circular reasoning.

How to use this classification

Since vegetation generally forms a continuum it is obvious that the categories in this, or any other, classification do not form discrete and clearly separated units but are simply convenient reference points in the continuum. They are chosen because of outstanding, describable characteristics and frequently because the types characterised occupy large areas. It must be understood that many examples of vegetation will not fit precisely any of the categories listed but will fall between them. Such a stand will have to be referred to in terms of the two or more units that it lies between or else placed in the one that it resembles most. The latter will be the normal practice in mapping, the former in description. For this survey, examples of vegetation which fall between two types may be indicated in the manner described in Part 3, Vegetation.

Scope

It is hoped that most or all of the existing *formation groups* have been accounted for. In some cases where the author does not know of the existence of a particular category, but where there seems no inherent reason why it does not exist, it has been included between brackets which may be removed if an example is pointed out or discovered. There is no doubt that many formations exist that are not listed. These may be readily inserted in their proper places. Where a given category should have the rank of formation or subformation may be a rather subjective matter in some cases. The arrangement in this respect may be readily altered. For the purposes of this survey, where a plant community is found which appears to justify a new formation, etc., this community should be fully described (see Part 3, Vegetation).

Appendix 2

Examples

An attempt has been made to provide one or more examples of associations or cover types for each formation or subformation. It is not claimed that these are necessarily the best available examples. They are merely ones that the author is personally familiar with or that he has seen adequately described or illustrated. Suggestions of additional or better examples are solicited and it is hoped that after discussion of this version with any colleagues who are interested a third and much more adequate approximation may be prepared.

Further remarks

The question of peat may cause some argument. Soil scientists will regard peat as part of the soil. It is, however, in my opinion, as much a part of the vegetation as is the dead outer portion of the bark or the dead heartwood of trees. In a bog it is difficult to separate the living from the dead part of the floating peat mat or the raised *Sphagnum* accumulation. Therefore, for purposes of this classification, peat is considered a part of the vegetation.

Finally, since the application of this classification depends upon the meaning conveyed by the terms used in the phrase-names and in the explanatory notes, a glossary of the words that may be possibly ambiguous is provided at the end.

Key to formation classes

1 Closed vegetation. Crowns or peripheries of plants mostly touching or overlapping. 2
Open and sparse vegetation. Crowns or peripheries of plants mostly not touching. 17

2 Trees present. 3
Trees absent, or nearly so. 7

3 Tree canopy closed. *1A Forests*
Tree crowns mostly not touching. 4

4 Shrub layer closed. 5
Shrub layer absent or open. 6

5 Tree crowns separated by less than their diameters.
1D Open Forest with closed lower layers
Trees scattered, separated by more than their crown diameters.
1E Closed scrub with scattered trees

6 Closed layer of dwarf shrubs. *1F Dwarf scrub with scattered trees*
Closed layer of tall grasses, etc. *1I Tall savanna*
Closed layer of short grasses, etc. *1J Low savanna*

7 Shrubs and dwarf shrubs present. 8
Shrubs and dwarf shrubs absent, or nearly so. 12

8 Shrubs present. 9
Shrubs absent, or nearly so; dwarf shrubs present. 11

9 Shrub layer closed. *1B Scrub*
Shrub layer open or sparse. 10

10 Shrubs separated by less than their diameters.
1G Open scrub with closed ground cover
Shrubs scattered, separated by more than their diameters.
1K Shrub savanna

11 Dwarf shrub layer closed. *1C Dwarf scrub*
Dwarf shrub layer open.
1H Open dwarf scrub with closed ground cover

12 Dominated by plants the leaves of which are adapted to lengthy or permanent submersion or to floating. 13
Dominated by grasses, etc., herbs, bryophytes or lichens whose leaves are not adapted to lengthy or permanent submersion or to floating. 14

13 Plants rooted. *1P Submerged meadows*
Plants floating, not rooted on bottom. *1Q Floating meadows*

14 Dominated by short and/or tall grasses, etc. 15
Dominated by broad-leaved herbs, bryophytes or lichens. 16

15 Dominated by graminoid plants exceeding 1 m in height.
1L Tall grass
Dominated by graminoid plants less than 1 m tall.
1M Short grass

16 Dominated by broad-leaved herbs.
1N Broad-leaved herb vegetation
Dominated by bryophytes and/or lichens.
1O Closed bryoid vegetation

17 Plants or tufts of plants not touching, but crowns not separated by more than their diameter; plants, not substratum, dominating landscape. 18
Plants so scattered that substratum dominates landscape. 27

18 Trees present. 19
Trees absent or nearly so. 20

19 Tree layer open; lower layers may be open or sparse.
2A Steppe forest
Trees scattered, not forming a well-defined layer.
2D Steppe savanna

20 Shrubs and dwarf shrubs present. 21
Shrubs and dwarf shrubs absent, or nearly so. 27

21 Shrubs present. 22
Shrubs absent, or nearly so: dwarf shrubs present. 23

22 Shrub layer open. *2B Steppe scrub*
Shrubs sparse, not forming a distinct layer.
2E Shrub steppe savanna

23 Dwarf shrub layer open. *2C Dwarf steppe scrub*
Dwarf shrubs sparse, not forming a distinct layer.
2F Dwarf shrub steppe savanna

24 Dominated by plants, the leaves of which are adapted to lengthy or permanent submersion or to floating. 25
Dominated by herbaceous vegetation whose leaves are not adapted to lengthy or permanent submersion or to floating. 26

25 Plants rooted. *2I Open submerged meadows*
Plants floating, not rooted on bottom. *2J Open floating meadows*

26 Dominated by grasses, etc., and/or broad-leaved herbs.
2G Steppe
Dominated by bryophytes and/or lichens. *2H Bryoid steppe*

27 Trees present. *3A Desert forest*
Trees absent, or nearly so. 28

28 Shrubs and dwarf shrubs present. *3B Desert scrub*
Shrubs and dwarf shrubs absent or nearly so. 29

29 Herbaceous plants predominate. *3C Desert herb vegetation*
Plants with leaves adapted to lengthy or permanent submergence or to floating predominate. *3D Sparse submerged meadows*

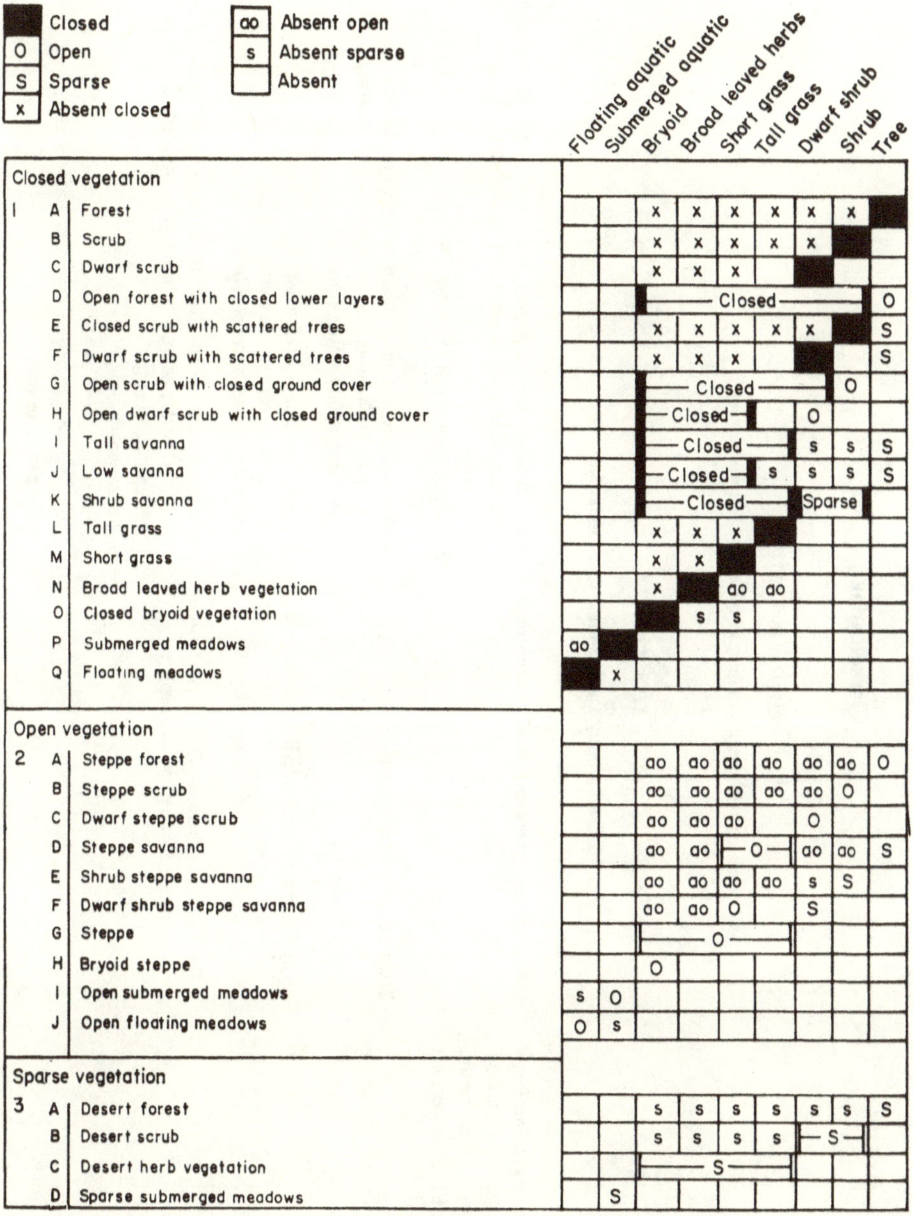
Closed
O Open
S Sparse
x Absent closed
ao Absent open
s Absent sparse
Absent
Floating aquatic
Submerged aquatic
Bryoid
Broad leaved herbs
Short grass
Tall grass
Dwarf shrub
Shrub
Tree
Closed vegetation
1 A Forest
B Scrub
C Dwarf scrub
D Open forest with closed lower layers
E Closed scrub with scattered trees
F Dwarf scrub with scattered trees
G Open scrub with closed ground cover
H Open dwarf scrub with closed ground cover
I Tall savanna
J Low savanna
K Shrub savanna
L Tall grass
M Short grass
N Broad leaved herb vegetation
O Closed bryoid vegetation
P Submerged meadows
Q Floating meadows
Open vegetation
2 A Steppe forest
B Steppe scrub
C Dwarf steppe scrub
D Steppe savanna
E Shrub steppe savanna
F Dwarf shrub steppe savanna
G Steppe
H Bryoid steppe
I Open submerged meadows
J Open floating meadows
Sparse vegetation
3 A Desert forest
B Desert scrub
C Desert herb vegetation
D Sparse submerged meadows

Fig. 2.

Classification

1 Closed Vegetation

(Crowns or peripheries of plants touching or overlapping)

1A *Forest*

(Closed woody vegetation 5 m or more tall)

1A1 Evergreen forest (at least the canopy layer with no significant leafless period)	1 Multistratal evergreen forest (rainforest)	(Tall, multistratal, orthophyllous or top of canopy sclerophyllous; epiphytes and lianas usually common) e. Dipterocarp forest (Malaya, Borneo)
	2 Evergreen swamp forest	(Bases of trees or root systems adapted to lengthy or permanent submergence; multistratal or unistratal; peat development often notable)
	(a) Evergreen orthophyll swamp forest	(Leaves of ordinary thickness) e. *Barringtonia racemosa* swamp (Guam) Most freshwater swamp types
	(b) Evergreen broad- sclerophyll swamp forest (Mangrove swamp)	(Sclerophyllous; usually unistratal; trees commonly provided with pneumatophores or prop roots) e. *Rhizophora* swamp *Avicennia* swamp Mixed mangrove swamps Most salt or strongly brackish water swamps, if trees over 5 m tall

(c) Evergreen megaphyllous swamp forest	(Usually composed of palms, more rarely *Pandanus*) e. *Mauritia* swamp (Colombia) Sago (*Metroxylon*) swamps, if over 5 m tall *Nypa* swamps, if over 5 m tall
3 Gnarled evergreen forest	(Low, trunks and branches tending to be twisted or gnarled, usually with one tree layer, usually more or less sclerophyllous)
(a) Gnarled evergreen mossy forest	(Bryophytes and other epiphytes abundant; generally mixed sclerophyllous and orthphyllous) e. *Metrosideros-Eugenia-Cheirodendron* forest (Hawaiian Islands). Most 'cloud forest' types
(b) Gnarled evergreen sclerophyll forest	(Moss and epiphytes not abundant; strongly sclerophyllous) e. Montane forest (Ceylon)
4 Evergreen hard-wood orthophyll forest	(One or two woody layers, irregular canopy, medium to low stature, much-branched trees) e. Dry evergreen forest (Northern Thailand) Dry oak forest (Ishigaki Island) Guava (*Psidium*) forest (Hawaii) Intermediate evergreen forest (Ceylon)
5 Evergreen soft-wood orthophyll forest	(Usually unistratal, of fast-growing trees, tangled with lianas). e. Belukar (Malaya) Most secondary forests in the humid tropics
6 Evergreen broad sclerophyll forest	(Leaves hard, stiff or coriaceous)

(a) Mesophyllous evergreen broad sclerophyll forest	(Leaves medium in size) e. Dry evergreen forest (Ceylon) *Dispyros-Osmanthus* forest (Hawaiian Islands) *Arbutus-Umbellularia* forest (California) *Laurus* forest (Canary Islands)
(b) Megaphyllous evergreen broad sclerophyll forest	(Leaves very large) e. Coconut (*Cocos*) groves Palm forest (Peruvian Amazonian drainage)
7 Evergreen narrow sclerophyll forest (needle leafed forest)	(Leaves (or equivalent) linear or scale-like, hard)
(a) Resinous evergreen narrow sclerophyll forest	(Dominantly coniferous) e. *Pinus* forest *Picea* forest
(b) Non-resinous evergreen narrow sclerophyll forest	(Dominantly non-coniferous) e. *Casuarina equisetifolia* forest
8 Evergreen bamboo forest	(Dominant layer of large bamboo or giant reed) e. *Schizostachyum* forest (Tahiti) *Gynerium* brake (Amazon)
9 Microphyllous evergreen forest	(Frequently but not always of compound-leafed trees, sometimes thorny) e. *Leucaena leucocephala* forest (Palau Is.) *Prosopis* forest (Hawaiian Is.) *Leptospermum* (*Manuka*) forest, taller aspects (New Zealand) *Pemphis acidula* forest (Pacific Is.)

1A2 Deciduous forest (at least canopy layer bare of leaves for a period during cold or dry season)	1 Winter-deciduous orthophyll forest (hardwood forest)	e. Beech-Maple (*Fagus-Acer*) (Eastern North America) Oak-Hickory (*Quercus-Carya*) (Eastern North America) Beech (*Fagus*) forest (Europe) Oak (*Quercus*) forest (Europe)
	2 Deciduous swamp forest	e. *Larix* swamp (Northern North America) *Taxodium* swamp (South-eastern U.S.) *Nyssa* swamp (South-eastern U.S.) *Acer rubrum* swamp (Eastern U.S.)
	3 Dry-season deciduous forest	(Orthophyllous; thick herb or shrub layer) e. Deciduous dipterocarp forest Most monsoon forests
	4 Microphyllous unarmed deciduous forest	(Trees mostly thornless or spineless) e. *Albizia lebbek* forest (Saipan)
	5 Deciduous thorn forest	(Trees mostly armed with thorns, spines or prickles, usually microphyllous) e. Caatinga (Ceara, Brazil)

1B *Scrub*

(Closed woody vegetation 5 m or less tall)

1B1 Evergreen scrub	1 Evergreen orthophyll scrub	
	(a) Broad-leaf evergreen orthophyll scrub	e. *Scaevola* scrub (dry coral islands) Coastal sagebrush when dominated by *Salvia* (California)

(b) Evergreen orthophyll vine scrub	(Shrubby vegetation tangled with climbers) e. Early stages of secondary vegetation in humid tropics
2 Evergreen bamboo or reed brake	(Dominant layer of dwarf bamboo, cane or woody reeds, may be strictly erect or tangled) e. *Chusquea* brake (Andes) *Pleioblastus* brake (Ryukyu Is.)
3 Evergreen swamp scrub	(Composed of shrubs with adaptations to stand lengthy or permanent submergence of root systems; peat development may be notable)
(a) Evergreen orthophyll swamp scrub	e. *Hibiscus tiliaceus* swamp (Pacific Is.)
(b) Evergreen reed swamp	(Composed of canes or reeds) e. *Phragmites karka* swamp (Pacific Is.)
(c) Evergreen broad sclerophyll swamp scrub	e. Low phase of mangrove swamp *Avicennia* swamp (New Zealand, Australia) *Rhododendron* swamp (Eastern North America)
(d) Evergreen narrow sclerophyll swamp	e. *Picea mariana* swamp or bog, low phases (Canada)
4 Evergreen broad sclerophyll scrub	(Stiff shrubs, leaves generally rather small)
(a) Mesophyllous evergreen broad sclerophyll scrub	e. Secondary scrub (Colombian Andes) Maquis (Mediterranean) *Ceanothus* chaparral (California) *Quercus dumosa* chaparral (California) *Rhododendron maximum* scrub (Eastern U.S.)
(b) Megaphyllous evergreen broad sclerophyll scrub	e. Low palmetto scrub (Florida)

	5 Mossy evergreen sclerophyll scrub	(Stiff shrubs, leaves generally rather small; with abundant epiphytes—scrub equivalent of mossy forest) e. Stiff scrub of wet mountain crests (Pacific Is.)
	6 Gnarled evergreen narrow sclerophyll scrub (krummholz)	e. *Pinus flexilis* krummholz (Rocky Mts.)
	7 Straight evergreen narrow sclerophyll scrub	(Not especially gnarled) e. *Pinus pumila* scrub (Japan) *Pinus mugo* scrub (Tyrol)? *Adenostemma* chaparral (California)
	8 Microphyllous evergreen scrub	(Often thorny)
	(a) Green microphyllous evergreen scrub	e. *Acacia farnesiana* scrub (Tropics) *Leucaena leucocephala* scrub (Pacific Is.) African heaths? *Ilex vomitoria* scrub, or yaupon (South-eastern U.S.) Some phases of *Ceanothus* chaparral (South-western U.S.)
	(b) Evergreen thorn scrub	e. *Acacia farnesiana* scrub (Tropics)
	(c) Gray microphyllous evergreen scrub	e. Sage brush (*Artemisia tridentata*) (Western U.S.) California sage when dominantly of *Artemisia californica* (California)
1B2 Deciduous scrub (shrubs periodically bare of leaves, usually in dry season or winter)	1 Deciduous orthophyll scrub	
	(a) Mesophyllous deciduous orthophyll scrub	e. *Salix* scrub, taller phases (Arctic and Subarctic) *Spiraea* scrub (Eastern U.S.) *Crataegus* thicket (Eastern U.S., Europe)

	(b) Microphyllous deciduous orthophyll scrub	e. *Leucaena leucocephala* scrub (E. Java)
	2 Deciduous swamp scrub	(Root systems adapted to prolonged submergence)
	(a) Mesophyllous deciduous orthophyll swamp scrub	e. Alder (*Alnus*) swamp (Eastern U.S.)
	(b) Microphyllous deciduous swamp scrub	e. *Larix* swamp low phases (North-eastern North America)
	(c) Deciduous broad sclerophyll swamp scrub	e. Mangrove swamps (Guajira Peninsula, Colombia) *Excoecaria* swamps (Malaysia and S.E. Asia)
	3 [Deciduous sclerophyll scrub]	
	4 Deciduous thorn scrub (Thornbush)	(Usually microphyllous, cacti often abundant. Rare in this closed form) e. Caatinga, low phases (North-eastern Brazil)

1C *Dwarf scrub*

(Closed predominantly woody vegetation less than 0·5 m tall)

1C1 Evergreen dwarf scrub	1 Evergreen orthophyll dwarf scrub	e. *Rhododendron* mat (Eastern Himalaya)?
	2 Evergreen broad sclerophyll dwarf scrub	
	(a) Mesophyllous broad sclerophyll dwarf scrub	e. *Arctostaphylos uva-ursi* mat (Northern temperate region)
	(b) Microphyllous evergreen dwarf scrub	(Without significant peat accumulation) e. Coastal *Osteomeles* scrub (Miyako Island) *Calluna* heath without peat (Western Europe)

	(c) Microphyllous evergreen dwarf heath	(With peat accumulation) e. *Empetrum* heath (Arctic and Subarctic) *Loiseleuria* heath (Arctic)
	3 Evergreen dwarf shrub bog	(Dwarf shrub with significant peat accumulation, root systems of plants adapted to constant immersion) e. Mountain bogs, more closed phases (Hawaii) *Chamaedaphne* bog (Eastern North America)
1C2 Deciduous dwarf scrub	1 Deciduous orthophyll dwarf scrub	
	(a) Deciduous orthophyll dwarf scrub	(Without significant peat accumulation) e. Low bush *Vaccinium* scrub (North temperate and subarctic regions)
	(b) Deciduous orthophyll dwarf heath	(With peat accumulation) e. *Vaccinium myrtillus* heath (Subarctic regions)

1D *Open forest with closed lower layers*

(Trees with crowns not touching, crowns mostly not separated by more than their diameters)

1D1 Evergreen open forest with closed lower layers	1 Open evergreen orthophyll forest	e. Denser phases of niaouli (*Melaleuca*) vegetation New Caledonia Foret-claire (Indo China)
	2 Open evergreen swamp	
	(a) Open narrow sclerophyll swamp	e. Open spruce muskeg
	(b) Open orthophyll swamp	e. Peat swamp (North Borneo) *Barringtonia racemosa* swamp (Micronesia)
	(c) Open broad sclerophyll swamp	e. Open phases of mangrove swamp with closed shrub layer (Tropics)

	3 Open evergreen broad sclerophyll forest	
	(a) Megaphyllous open evergreen broad sclerophyll forest	(Open palm or pandan forests) e. *Mauritia* groves (Colombian llanos) *Sabal* forest (Central Eastern Mexico)
	(b) Mesophyllous open evergreen broad sclerophyll forest	e. Denser phases of Orinocoan "savanna" Open oak forest (Northern Thailand) Live oak (*Quercus*) woodland (California)
	4 Open evergreen narrow sclerophyll forest	
	(a) Resinous open evergreen narrow sclerophyll forest	e. Open *Pinus* forests (Philippines, Mexico, Ryukyu Is.) Open aspects of pine barrens (New Jersey) Denser aspects of pinyon and pinyon-juniper (*Pinus-Juniperus*) with herbaceous or shrubby lower layers (Rocky Mts. and east slope of Sierra Nevada, Western U.S.)
	(b) Non-resinous open evergreen narrow sclerophyll forest	e. Some open phases of *Casuarina* forests (Mariana Is.)
	5 Open microphyllous evergreen forest	(May be somewhat thorny) e. Open phases of *Prosopis* forest (Hawaii)
1D2 Open deciduous forest with closed lower layers	1 Open deciduous orthophyll forest	e. Deciduous dipterocarp forest (Thailand)
	2 Open deciduous swamp	
	(a) Open broad orthophyll swamp	e. Tupelo (*Nyssa*) swamp, open phases (Southeastern U.S.)
	(b) Open broad sclerophyll swamp	e. Mangrove swamp, open phases where deciduous (Ryukyu Is.)

	(c) Open narrow sclerophyll swamp	e. Open *Larix* muskeg (Central Canada) Open *Taxodium* swamp (South-eastern U.S.)
	3 Open deciduous narrow sclerophyll forest	e. Open *Larix* forest (Central Canada)
	4 [Open deciduous broad sclerophyll forest]	(Possibly does not exist)
	5 Open microphyllous deciduous forest	e. Miombo forest (South Central Africa)

1E *Closed scrub with scattered trees*

1E1 Closed evergreen scrub with scattered trees (at least shrub layer evergreen)	1 Evergreen orthophyll scrub with trees	e. *Scaevola* scrub with scattered trees (Pacific coral islands)
	2 Evergreen sclerophyll scrub with trees	e. Scrub oak with scattered live oaks (*Quercus*) (California) Larger leafed phases of chaparral with scattered *Pinus* (California) Maquis with scattered *Pinus* (Mediterranean) *Kalmia latifolia* with scattered *Quercus* and *Pinus* (Virginia mountains)
	3 Microphyllous evergreen scrub with trees	e. *Purshia tridentata* with *Pinus* (Western U.S.) *Artemisia tridentata* with *Pinus* (Western U.S.) *Artemisia californica* with *Quercus* (California)
	4 Megaphyllous evergreen broad sclerophyll scrub with trees	e. Saw palmetto scrub with *Pinus* (Florida)

1E2 Closed deciduous scrub with scattered trees	1 Deciduous orthophyll scrub with trees	e. *Quercus ilicifolia* with *Pinus rigida* (Appalachians)
	2 [Deciduous sclerophyll scrub with trees]	(Perhaps does not exist)
	3 [Microphyllous deciduous scrub with trees]	(Perhaps does not exist)

1F *Dwarf scrub with scattered trees*

1F1 Evergreen dwarf scrub with scattered trees	1 Microphyllous evergreen dwarf scrub with trees	(Without significant peat formation) e. *Calluna* heath with *Pinus* (England)
	2 Microphyllous evergreen heath with trees	(With peat accumulation) e. *Empetrum* phase of heath birch *Betula* forest (Lappland)
1F2 Deciduous dwarf scrub with trees	1 Deciduous heath with trees	(With significant peat accumulation) e. *Vaccinium* phases of heath birch (Betula) forest (Lappland)

1G *Open scrub with closed ground cover*

1G1 Open evergreen scrub with closed ground cover	1 Open evergreen orthophyll scrub	e. Low phases of campo cerrado (Brazil)
	2 Open evergreen broad sclerophyll scrub	
	(a) Megaphyllous open evergreen broad sclerophyll scrub	e. Saw palmetto (*Serenoa*) prairie (Florida) *Eugeissonia* scrub (Malaya) *Chamaerops humilis* scrub (Algeria)

	(b) Mesophyllous open evergreen broad sclerophyll scrub	e. Low phases of *Diospyros-Osmanthus* forest (Lanai, Hawaiian Is.)
	3 Open gnarled evergreen narrow sclerophyll scrub or open krummholz	e. Open *Pinus flexilis* krummholz (Rocky Mts.) Open *Picea* krummholz (White Mts. N.H.)
	4 Open microphyllous evergreen scrub	e. Open phases of *Sarothamnus* scrub (N.W. Europe)
1G2 Open deciduous scrub with closed ground cover	1 Open deciduous orthophyll scrub with closed ground cover	e. Open phases of *Salix* scrub and *Betula* scrub (Subarctic)

1H *Open dwarf scrub with closed ground cover*

1H1 Open evergreen dwarf scrub with closed ground cover	1 Open evergreen orthophyll dwarf scrub	e. *Sida-Heliotropium* dwarf scrub (Christmas Island, Pacific)
	2 Open evergreen shrub bog	e. Open phases of mountain bogs (Hawaii)
	3 Open evergreen microphyllous dwarf scrub	e. Open *Calluna* and *Erica* heath lower phases (Western Europe)
1H2 Open deciduous dwarf scrub with closed ground cover	1 Open deciduous orthophyll dwarf scrub	(Without significant peat accumulation) e. Open phases of low-bush *Vaccinium* scrub (Eastern U.S.)
	2 Open deciduous orthophyll heath	(With significant accumulation of peat) e. Open phases of *Vaccinium myrtillus* heath (Subarctic)

1I *Tall savanna*

(Closed grass or other herbaceous vegetation 1 m tall or more with scattered trees)

1I1 Evergreen savanna (trees evergreen)	1 Evergreen orthophyll savanna	e. Fern savanna (Palau Is.)
	2 Evergreen broad sclerophyll savanna	
	(a) Mesophyllous evergreen sclerophyll savannna	e. *Curatella-Byrosonima* savanna (northern South America)
	(b) Megaphyllous evergreen sclerophyll savanna	e. Various palm savannas (tropical America and Africa) *Pandanus* savanna (Guam)
	3 Evergreen sclerophyll swamp savanna	
	(a) Evergreen narrow leaf swamp savanna	e. Saw-grass everglades with pines (Florida)
	(b) Evergreen broad sclerophyll swamp savanna	e. Coastal swamp savanna (West Africa)
	4 Evergreen narrow sclerophyll savanna	e. *Pinus* savanna (Luzon) *Casuarina equisetifolia* savanna (Guam)
	5 Microphyllous evergreen savanna	e. Closed erosion scar community with ferns, grasses and *Myrtella* (Guam)
1I2 Deciduous tall savanna (trees deciduous)	1 Deciduous orthophyll savanna	e. *Adansonia* savanna (Sudan)
	2 Deciduous microphyll savanna	e. Less thorny phases of *Acacia* tall grass savanna (Africa)
	3 Deciduous thorn savanna	e. Thorny phases of *Acacia* tall grass savanna (Africa)

1J *Low savanna*

(Herbaceous vegetation less than 1 m tall, with scattered trees)

1J1 Evergreen low savanna (Trees evergreen)	1 Evergreen orthophyll low savanna	e. Fern savanna (Palau Is.) Live-oak savanna (California)
	2 Evergreen broad-sclerophyll low savanna	e. Heavily grazed phases of evergreen broad-sclerophyll tall savannas
	3 Evergreen narrow-sclerophyll low savanna	e. California foothill savanna where *Pinus sabiniana* predominates More open phases of pinelands of South-eastern U.S. *Juniperus virginiana* savanna (Eastern U.S.)
	4 Evergreen narrow sclerophyll lichen savanna	e. Lichen muskeg with *Picea* dominant
	5 Evergreen narrow sclerophyll swamp savanna	e. *Sphagnum* muskeg with *Picea* dominant
1J2 Deciduous low savanna	1 Deciduous orthophyll savanna	e. *Quercus lobata* and *Q. douglasii* savanna (California) *Crataegus* pastures where trees are well developed (Eastern U.S.)
	2 [Deciduous broad sclerophyll low savanna]	
	3 Deciduous lichen savanna	e. Lichen muskeg with *Larix* dominant (most open phases)
	4 Deciduous swamp savanna	e. *Sphagnum* muskeg with *Larix* dominant (most open phases)

1K *Shrub savanna*

(Closed grass or other herbaceous vegetation with scattered shrubs)

1K1 Evergreen shrub savanna	1 Evergreen orthophyll shrub savanna	e. *Sida-Lepturus* savanna (Pacific coral islands)
	2 Evergreen broad sclerophyll shrub savanna	e. Lower phases of *Curatella-Byrsonima* savanna (South America)
	3 Evergreen narrow sclerophyll shrub savanna	
	(a) Resinous evergreen narrow sclerophyll shrub savanna	e. *Juniperus communis* savanna (North temperate zone)
	(b) Non-resinous evergreen narrow sclerophyll shrub savanna	e. *Casuarina* savanna, early stages (Guam)
	4 Evergreen microphyll scrub savanna	
	(a) Green evergreen microphyll shrub savanna	e. Sparse, grassy phases of subalpine scrub (Hawaiian volcanoes)
	(b) Gray evergreen microphyll shrub savanna	e. Grass with scattered sagebrush (*Artemisia tridentata*) (Western U.S.)
1K2 Deciduous shrub savanna	1 Deciduous orthophyll shrub savanna	e. Early stages of forest succession in old fields (Eastern U.S.)
	2 [Deciduous broad sclerophyll shrub savanna]	
	3 [Microphyllous deciduous shrub savanna]	

	4 Mesophyllous deciduous thorn shrub savanna	e. Pastures with *Crataegus* (Eastern U.S.)
	5 Microphyllous deciduous thorn shrub savanna	e. *Acacia* savanna (Caribbean)

1L *Tall grass*

(Closed herbaceous vegetation exceeding 1m in height, predominantly graminoid)

1L1 Evergreen tall grass (Shoots remaining green the year round)	1 Evergreen orthophyll tall grass	e. Stands of *Panicum purpurascens* and of *Panicum maximum* in Pacific Islands
	2 Tall evergreen graminoid marsh	e. *Scirpus* marsh *Typha* marsh Papyrus (Tropics)
	3 Evergreen orthophyll tall tussock grass	e. Snow-tussock (*Danthonia*), closed phases (New Zealand)
	4 Evergreen sclerophyll tall grass	e. *Saccharum spontaneum* stands (Western Pacific) *Miscanthus floridulus* stands, where these are not seasonally brown (Western Pacific) *Imperata* savanna, taller phases (Western Pacific)
1L2 Seasonal tall grass (Turning brown in dry season or winter, often burned)	1 Seasonal orthophyll tall grass	e. Tall grass prairie (Mississippi Valley) High veld (*Themeda*) (Transvaal) Patna, taller aspects (Ceylon)
	2 Seasonal sclerophyll grass	e. *Miscanthus floridulus* or sword grass, in areas with dry season (Western Pacific) *Themeda* stands (Western Pacific)

1M *Short grass*

(Closed herbaceous vegetation, less than 1 m tall, predominantly graminoid)

1M1 Evergreen short grass	1 Evergreen orthophyll short grass	e. Lalang, *Imperata cylindrica*, shorter phases (Western Pacific)
	2 Evergreen orthophyll graminoid marsh	e. *Paspalum* marsh; *Cyperus* marsh (Guam)
	3 Evergreen orthophyll short tussock grass	e. *Festuca* tussock (New Zealand)
	4 Evergreen short grass and sedge bog	(Growing from a mass of grass- and sedge-peat) e. *Oreobolus* bog
1M2 Seasonal short grass	1 Seasonal orthophyll meadows (short grass)	e. Patna shorter aspects (Ceylon) Short grass prairie (western Mississippi Valley) Grass and sedge tundras (Arctic) Most temperate zone pastures
	2 Seasonal orthophyll marsh	e. Marshes of Ganges Delta Rice field Salt marsh (temperate Atlantic coasts)
	3 Seasonal sclerophyll short grass meadow	e. Closed aspects of puna (Andes)
	4 Seasonal sclerophyll marsh	e. Saw grass everglades (Florida)

1N *Broad-leafed herb vegetation*

(Closed vegetation, predominantly of broad-leafed herbaceous plants)

1N1 Evergreen broad-leafed herb vegetation	1 Evergreen broad-leafed weedy vegetation	e. Pioneer weedy vegetation after clearing (Tropics)
	2 Evergreen broad-leafed marsh	e. Cultivated taro pits (Polynesia)
	3 Evergreen fern meadow	e. *Gleichenia* sward (Tropics) *Nephrolepis* meadow (Western Pacific)
	4 Evergreen giant herb thicket	e. Banana plantations and wild banana stands
1N2 Seasonal broad-leafed herb vegetation	1 Seasonal broad-leafed meadow	e. Early stages in old-field vegetation (Humid Temperate areas) *Trollius* meadow (Lappland)
	2 Seasonal fern meadow	e. *Pteridium aquilinum* brake

1O *Closed bryoid vegetation*

1O1 Closed bryophyte vegetation	1 *Sphagnum* bog	e. Raised bog Muskeg
	2 Moss meadow	e. *Rhacomitrium* meadow (Hawaii)
1O2 Closed lichen vegetation	1 Lichen bog	e. *Cladonia* muskeg
	2 Lichen meadow	e. *Cladonia* meadow (Subarctic)

1P *Submerged meadows*

(Vegetation of rooted aquatic herbs, adapted for permanent complete submersion except in some cases for floating leaves)

1P1 Evergreen submerged meadows	1 Evergreen watergrass	e. Turtle grass
	2 Macrophyllous evergreen submerged meadows	e. Canal vegetation (Thailand)
	3 Megaphyllous evergreen submerged meadows	e. *Victoria* meadows (South America)
1P2 Seasonal submerged meadows (plants disappearing, at least their shoots, in winter)	1 Seasonal watergrass	e. Eelgrass (*Zostera marina*) (Temperate coasts) Closed *Potamogeton* meadows (North-eastern U.S.)
	2 Broad-leafed seasonal submerged meadows	(Leaves of ordinary size, not narrowly linear or grasslike) e. Broad-leafed *Potamogeton* stands (eastern North America)
	3 Macrophyllous seasonal submerged meadows	(Water lilies and similar plants dominant) e. *Nymphaea-Nuphar* meadows (eastern North America)

1Q *Floating meadows*

(Closed vegetation of aquatic herbs, adapted to floating conditions, not rooted in bottom)

1Q1 Evergreen floating meadows	1 Broad-leafed evergreen floating meadows	e. Water hyacinth (*Eichhornia*) (Tropics and Subtropics) *Salvinia*, *Pistia*

	2 Evergreen floating grass	e. Gramalote (Amazonia)
	3 Thalliform evergreen floating vegetation	e. Duckweed (*Lemna*, etc.) (Tropics)
	4 Microphyllous evergreen floating meadows	e. *Azolla* (Tropics)
1Q2 Seasonal floating meadow	1 Thalliform seasonal floating meadow	e. Duckweed (*Lemna*, etc.) (Temperate Zones)
	2 Microphyllous seasonal floating meadow	e. *Azolla* (Temperate Zone)

2 Open Vegetation

(Plants or tufts of plants not touching but crowns not separated by more than their diameters; plants, not substratum, dominating landscape)

2A *Steppe Forest*

(Often called woodland or woodland-savanna) (Tree layer and lower layers open, lower layers may be open or sparse)

2A1 Evergreen steppe forest (Tree layers, at least, evergreen)	1 [Evergreen orthophyll steppe forest]	(Possibly does not exist with open lower layers)
	2 Evergreen broad sclerophyll steppe forest	
	(a) Megaphyllous evergreen sclerophyll steppe forest	e. *Borassus* forest (Northwest India, North Ceylon) *Phoenix* groves in oases (Arabia, North Africa)
	(b) Mesophyllous evergreen sclerophyll steppe forest	e. Mallee (*Eucalyptus*) bush (Australia)?

	3 Open evergreen sclerophyll swamp	e. Open phases of mangrove swamps (tropical coasts)
	4 Evergreen narrow sclerophyll steppe forest	
	(a) Resinous evergreen narrow sclerophyll steppe forest	e. Some open *Pinus* forests (Western U.S.)
	(b) Non-resinous evergreen sclerophyll steppe forest	e. Some open *Casuarina* forests (Saipan, Guam)
	5 Microphyllous evergreen steppe forest	e. Some *Prosopis* forests (Hawaii, Caribbean)
2A2 Deciduous steppe forest	1 Deciduous orthophyll steppe forest	e. Drier phases of deciduous dipterocarp forests (Thailand) At least some phases of 'foret-claire' (Indo-China) Open oak (*Quercus*) forest (sandhills of south-eastern U.S.)
	2 Microphyllous deciduous steppe forest	e. Denser aspects of Atlantic coastal forest (Colombia)
	3 Deciduous bamboo steppe forest	e. *Bambusa arundinacea* forest (near Bombay, India)
	4 Open deciduous thorn (steppe) forest	e. Open thorn forest (Caribbean lowlands)

2B *Steppe scrub*

(Like steppe forest, but with shrubs (over 0·5 m tall) instead of trees)

2B1 Evergreen steppe scrub	1 Evergreen orthophyll steppe scrub	e. *Sida fallax* scrub (Northern Marshall Is.)

	2 Evergreen broad sclerophyll steppe scrub	e. Scrub of serpentine areas (New Caledonia) Scrub of Mineral Belt (Near Nelson, New Zealand).
	3 Gnarled evergreen narrow sclerophyll steppe scrub (open krummholz)	e. Open *Pinus flexilis* krummholz (Rocky Mts.) *Picea-Abies* krummholz, open aspects (Northern Appalachians, Adirondack Mts. Eastern U.S.)
	4 Microphyllous evergreen steppe scrub	e. Open aspects of *Calluna-Erica* heath (Western Europe) Tola (*Lepidophyllum*) heath (Peru, Bolivia)
	(a) [Green evergreen microphyllous steppe scrub]	(Perhaps does not exist)
	(b) Gray evergreen microphyllous steppe scrub	e. Open sagebrush (*Artemisia tridentata*) (Great Basin, U.S.) *Krameria* or *Coleogyne* scrub (Western U.S.)
	5 Evergreen succulent steppe scrub	e. *Sarcobatus* flats (Great Basin, U.S.) *Tamarix* scrub (Arid areas, Mediterranean, Central Asia, South-western U.S.)
	6 Evergreen saltbush steppe scrub	(Shrubs with gray scurfy leaves) e. *Atriplex* flats (Great Basin, U.S.)
2B2 Deciduous steppe scrub	1 Deciduous orthophyll steppe scrub	e. Deciduous scrub-oak and shin-oak (*Quercus* spp.) (South-west U.S.) Denser phases of *Ulmus pumila* scrub (Central Asia)
	2 [Deciduous sclerophyll steppe scrub]	
	3 Microphyllous deciduous steppe scrub	e. Grassy phases of *Pentzia-Rhigozum* scrub (South Africa)
	4 Deciduous thorn steppe scrub (thornbush)	e. Open thornbush (Caribbean lowlands)

2C *Dwarf steppe scrub*

(Open predominantly woody vegetation less than 0·5 m tall)

2C1 Evergreen dwarf steppe scrub	1 Evergreen orthophyll dwarf steppe scrub	e. Early stages of evergreen steppe forest and steppe scrub
	2 Evergreen broad sclerophyll dwarf steppe scrub	e. Dwarf phases of garrigue (Mediterranean) Low woody aspects of puna (Southern Andes)
	3 Evergreen narrow sclerophyll dwarf steppe scrub	e. Low denser phases of tola (*Lepidophyllum*) heath (Peru, Bolivia)
	4 Microphyllous evergreen dwarf steppe scrub	e. Low open sand heath (Western Europe)
2C2 [Deciduous dwarf steppe scrub]		(Perhaps does not exist)

2D *Steppe savanna*

(Steppe with scattered trees)

2D1 Evergreen steppe savanna (Trees evergreen)	1 Evergreen orthophyll steppe savanna	e. *Guazuma* savanna (Northern South America)
	2 Evergreen sclerophyll steppe savanna	e. Sparse phases of *Curatella-Byrsonima* savanna (Colombia)
	3 [Evergreen microphyll steppe savanna]	(Perhaps does not exist)
2D2 Deciduous steppe savanna	1 Deciduous orthophyll steppe savanna	e. Sparser aspects of *Quercus lobata–Q. douglasii* savanna (California foothills)

	2 [Deciduous broad sclerophyll steppe savanna]	(Perhaps does not exist)
	3 Microphyllous deciduous steppe savanna	e. Deciduous savannas of northern Colombia
	4 Deciduous thorn steppe savanna	e. Low phases of *Acacia*-desert grass savanna (Africa)

2E *Shrub steppe savanna*

2E1 Evergreen shrub steppe savanna	1 Evergreen orthophyll shrub steppe savanna	e. *Polylepis* puna (Bolivia) *Sophora-Myoporum* savanna (High altitudes, Hawaii)
	2 Evergreen sclerophyll shrub steppe savanna	e. Desert grassland with *Yucca elata* (South-western U.S.) Desert grassland with scattered *Chamaerops* (North Africa)
	3 Evergreen microphyll shrub steppe savanna	e. Llareta (*Azorella*) puna (Bolivia, Chile)
	4 Evergreen succulent shrub steppe savanna	e. Desert grassland with *Opuntia* (South-western U.S.)
2E2 Deciduous shrub steppe savanna	1 Deciduous orthophyll shrub steppe savanna	e. *Ulmus pumila* scrub, open phases (Central Asia)
	2 [Deciduous sclerophyll shrub steppe savanna]	(Possibly does not exist)

	3 Microphyllous deciduous shrub steppe savanna	e. Mesquite (*Prosopis*) grassland (Texas–New Mexico)
	4 Deciduous thorn shrub steppe savanna	e. *Acacia macracantha* savanna (Andes)

2F *Dwarf shrub steppe savanna*

2F1 Evergreen dwarf shrub steppe savanna	1 [Evergreen orthophyll dwarf shrub steppe savanna]	(Possibly does not exist)
	2 Evergreen narrow sclerophyll dwarf shrub steppe savanna	e. Low phases of tola (*Lepidophyllum*) heath (Peru, Bolivia)
	3 Succulent dwarf shrub steppe savanna	e. Desert grassland with low *Opuntia* (South-western U.S.)
2F2 Seasonal dwarf shrub steppe savanna	1 Seasonal sclerophyll dwarf shrub steppe savanna	e. *Tetraglochin* puna (Andes)

2G *Steppe*

(Open herbaceous vegetation, tufts or plants discrete, yet close enough to dominate the landscape)

2G1 Evergreen steppe	1 Evergreen saltbush steppe	e. Herbaceous *Atriplex* steppe (Australia, Hawaii)
	2 Evergreen succulent steppe	e. *Suaeda* flats (Coastal or alkaline sand flats, (Africa, Thailand, New Mexico etc.) *Portulaca* steppe (Marshall Is.)
	3 Evergreen cushion plant steppe	e. *Pycnophyllum* puna (Bolivia) *Raoulia* steppe (New Zealand)

2G2 Seasonal steppe	1 Seasonal grass steppe	e. *Stipa* puna (Peru and Bolivia) Desert grassland (South-western U.S.) *Cenchrus-Lepturus* grassland (Canton Island and other dry coral islands)
	2 Annual herb steppe	e. Denser aspects of *Abronia*, *Eschscholtzia*, *Astragalus* and other ephermeral types on sandy areas (South-western U.S.)

2H *Bryoid steppe*

2H1 [Open bryophyte vegetation]		(Perhaps does not exist)
2H2 Open lichen vegetation	1 Lichen tundra	e. *Cladonia* heath (Lappland)

2I *Open submerged meadows*

2I1 [Evergreen open submerged meadows]		(Perhaps does not exist)
2I2 Seasonal open submerged meadows	1 Seasonal watergrass	e. *Zostera* (open phases)
	2 Broad-leafed seasonal submerged meadows	e. *Potamogeton* communities (Open phases) *Lobelia dortmanna-Littorella* communities (Scotland)
	3 Macrophyllous seasonal submerged meadows	e. *Nuphar lutea* community (open phases)

2J *Open floating meadows*

2J1 [Evergreen open floating meadows]		(Possibly does not exist)
2J2 Seasonal open floating meadows	1 Thalliform seasonal open floating meadows	e. *Lemna* communities (open phases)
	2 Microphyllous seasonal open floating meadows	e. *Azolla* (open phases)

3 Sparse Vegetation or Desert

(Plants so scattered that substratum dominates landscape)

3A *Desert forest*

(Scattered trees, subordinate shrub or herb layers very sparse, or absent)

3A1 Evergreen desert forest (may be evergreen because of persistent leaves or because of green stems)	1 Evergreen non-succulent desert forest	e. *Parkinsonia* desert (Arizona, Peru) *Olneya* desert (Arizona) (May be deciduous in coldest winters)
	2 Evergreen succulent desert forest	e. *Cereus giganteus* desert (Arizona)
3A2 Deciduous desert forest	1 Microphyllous deciduous desert forest	e. *Prosopis* desert (Arizona, Peru) *Idria* desert (Baja, California)

3B *Desert scrub*

(Scattered shrubs in an otherwise bare or only ephemerally vegetated landscape, not here differentiated into shrub and dwarf shrub classes.)

3B1 Evergreen desert scrub	1 Evergreen sclerophyll desert scrub	e. *Capparis* desert (Peru) *Welwitschia* desert (South-west Africa) *Phoenix arabica* desert (Arabia)
	2 Microphyllous evergreen desert scrub	e. *Larrea-Franseria* desert (South-western U.S.) *Coleogyne* desert (Arizona)
	3 Saltbush desert	e. *Atriplex* flats (South-western U.S.)
	4 Evergreen desert thorn-scrub	(Dominated by rigid green-stemmed, spinose, leafless or ephemerally leafy plants) e. *Koeberlinia* desert (New Mexico)
	5 Evergreen succulent desert scrub	e. *Opuntia fulgida* desert (Arizona) *Suaeda-Zygophyllum* desert (Arizona) *Tamarix* desert (Arabia)
3B2 Deciduous desert scrub	1 Microphyllous deciduous desert scrub	e. Sparse phases of *Pentzia-Rhigozum* scrub (South Africa)
	2 Deciduous desert thorn-scrub	e. *Fouquieria-Acacia gregii* desert (South-western U.S.) *Acacia constricta* desert (New Mexico)

3C *Desert herb vegetation*

(Scattered herbaceous plants only)

3C1 Evergreen desert herb vegetation	1 Evergreen succulent herb desert	e. Karroo succulent desert (South Africa)
	2 Evergreen psammophyte desert	(Plants with special adaptations enabling them to survive in shifting sand) e. *Cyperus conglomeratus* desert (Arabia)
	3 Lichen tundra sparse phases	e. Sparse *Cladonia* tundra (Arctic)
3C2 Seasonal desert herb vegetation	1 Seasonal desert grass	e. *Aristida* desert, where shrubs are lacking (South Africa)
	2 Ephemeral herb desert	(Vegetation principally of ephemeral annuals and geophytes, appearing only for short periods after infrequent rains) e. Loma vegetation (Peru)

3D *Sparse submerged meadows*

3D1 [Evergreen sparse submerged meadows]		(Possibly does not exist)
3D2 Seasonal sparse submerged meadows	1 Seasonal watergrass	e. *Zostera* (sparse phases)
	2 [Broad leaves seasonal submerged meadows]	(Possibly does not exist)
	3 [Macrophyllous seasonal submerged meadows]	(Possibly does not exist)

Glossary

Annual—Plants, or pertaining to plants, which complete their life span within a single year.

Aquatic—plants, or pertaining to plants, adapted for life submerged or partially submerged in water.

Arboreous desert—vegetation of scattered trees with bare ground or only sparse vegetation between them; desert forest.

Bamboo—a woody-stemmed grass belonging to the tribe Bambuseae.

Brake—vegetation of canes.

Broad-leafed—with leaves other than linear in outline; as opposed to grass-like or graminoid.

Broad-sclerophyll—vegetation of plants with stiff, hard, or coriaceous leaves that are other than acicular or narrowly linear in outline.

Bog—vegetation with lower layers dead, peaty, gradually changing upward to living plant materials.

Bryoid—a moss, liverwort or lichen.

Cane—as used here, a woody stemmed grass or plant resembling a grass.

Canopy—more or less continuous leafy layer formed by crowns of trees touching or almost touching each other.

Catena (vegetational)—a regular alternation of two or more types of vegetation.

Climber—a climbing, vine-like plant regardless of its mode of climbing.

Crown (of tree)—the leafy portion of a tree, as distinguished from the trunk and bare parts of principal limbs.

Cushion plant—a herbaceous or low woody plant so densely branched that it forms a dense resilient mat or cushion.

Deciduous—woody plants, or pertaining to woody plants, that seasonally lose all of their leaves and become temporarily bare-stemmed.

Desert—vegetation of plants so widely spaced, or sparse, that enough of the substratum shows through to give the dominant tone to the landscape; for practical purposes, where the plants are separated from each other by more than their diameter on the average.

Desert forest—desert vegetation with tallest layer of trees; arboreous desert.

Desert scrub—desert vegetation with tallest layer of shrubs; shrub desert.

Dwarf scrub—vegetation made up of dwarf shrubs, averaging less than 0·5 m tall.

Dwarf shrub—a shrub or woody plant less than 0·5 m tall.

Ecosystem—totality of an environment plus its included organisms.

Emerged or emersed aquatic plants—a plant adapted to life with its lower parts submerged in water, its upper parts in air.

Ephemeral—plants, or pertaining to plants, the shoots of which appear after a rain and complete their cycle of vegetative growth, flowering and fruiting within a few days or at most a few weeks; includes therophytes and geophytes.

Epiphyte—a plant utilising another living plant as a substratum, i.e. growing upon another plant.

Evergreen—plants, or pertaining to plants, which remain green the year round, either by retaining at least some of their leaves at all times, or by having green stems which carry on the principal photosynthetic functions.

Floating aquatic plant—plant adapted to a floating aquatic eixstence, not rooted in soil.

Floristic—pertaining to the species composition of vegetation.

Forest—open or closed vegetation with the principal layer consisting of trees, these averaging over 5 m tall.

Geophyte—perennial plant with an annual shoot; perennial underground parts.

Gnarled—pertaining to trees or shrubs with a twisted, bent or distorted appearance.

Graminoid—grass-like in appearance, with leaves mostly very narrow or linear in outline.

Grass—vegetation principally made up of grasses and other graminoid plants.

Grasses—plants belonging to the Gramineae or grass family.

Grove—a small forest or group of trees.

Hard-wood—broad-leafed trees of relatively slow growth and usually with relatively hard wood.

Herb—a plant with no significant woody tissue in the stem.

Immersed aquatic plant—a plant adapted for life submerged or almost submerged in water.

Krummholz—the twisted and distorted woody vegetation characteristic of mountain timberlines.

Layer—the aggregate of plants of a given and limited range of heights in a plant community, usually set off by a relative discontinuity from the layers above and below it, if any; also called a stratum.

Liana—a woody climbing plant.

Linear leaf—a leaf many times as long as wide and with essentially parallel sides at least in the middle portions.

Macrophyllous—used here in a special sense, only applied to aquatic plants, denoting approximately isodiametric leaves of the order of 10 cm across or larger, as opposed to leaves of other shapes and usually smaller; could be applied to a size category between mesophyllous and megaphyllous for any plants.

Mangrove—a sclerophyllous woody swamp vegetation type characteristic of tropical and subtropical saline shores.

Marsh—a herbaceous vegetation type composed principally of emersed aquatic plants.

Meadow—closed herbaceous vegetation, commonly in stands of rather limited extent, the term not usually applied to extensive grasslands.

Megaphyllous—with leaves 50 cm or more long and at least 5 cm wide.

Mesophyllous—with leaves of ordinary size; as used here includes the range between microphyllous and megaphyllous, can be subdivided if necessary.

Microphyllous—with very small leaves; as used here, for trees, leaves 25 mm in greatest diameter and less, for shrubs 10 mm and less, for dwarf shrubs 5 mm and less.

Mosaic—an irregular patchwork of two or more types of vegetation.

Multistratal—as applied to vegetation, having several distinguishable layers or strata.

Narrow sclerophyll—vegetation with needle-like or narrowly linear hard or stiff leaves or equivalent organs (photosynthetic branchlets, cladodes, phyllodes, etc.).

Open vegetation—vegetation with the plants or tufts mostly discrete rather than touching, but separated by less than their diameter on the average.

Orthophyll—vegetation with leaves mostly of ordinary texture, as opposed to sclerophyll.

Orthophyllous—texture of leaves ordinary, as opposed to sclerophyllous.

Papyrus—Marsh vegetation predominantly of *Cyperus papyrus*.

Peat—partially decomposed plant remains; as used here includes both bog or swamp peat (formed under waterlogged conditions) and heath peat, mor, or raw humus (formed under well-drained conditions).

Pneumatophore—knees or other outgrowths from the roots of swamp plants, presumably functioning as aerating organs.

Prop-root—roots originating on the stem of a plant or tree above ground level and entering the ground at a little distance from the stem, possibly serving as props or braces for the stem.

Psammophyte—a plant with special adaptations for survival on a substratum of shifting or moving sand.

Reed—a swamp grass with a somewhat woody stem; usually applied to the genera *Phragmites*, *Arundo* and *Gynerium*.

Saltbush—grayish or scurfy shrubs or herbs adapted for survival in saline or alkaline habitats; usually applied to the genus *Atriplex*, more rarely to other Chenopodiaceae.

Savanna—closed grass or other predominantly herbaceous vegetation with scattered or widely spaced woody plants.

Sclerophyll—vegetation with predominantly hard, stiff or coriaceous leaves or equivalent organs (photosynthetic branchlets, cladodes, phyllodes, etc.).

Sclerophyllous—texture of leaves hard, stiff or coriaceous, as opposed to orthophyllous.

Scrub—woody vegetation predominantly of shrubs, as used here ranging between 0·5 and 5 m in height.

Seasonal—applied to vegetation having a dormant season marked by conspicuous physiognomic change.

Secondary forest—forest vegetation following clearing or burning of the former cover.

Shoot—the above-ground portion of a plant, or the part of a plant above the roots or rhizomes.

Shrub—as used here, a woody plant under 5 m in height.

Shrub desert—desert vegetation predominantly of shrubs; desert scrub.

Shrub savanna—closed grass or other predominantly herbaceous vegetation with scattered or widely spaced shrubs.

Shrub steppe savanna—open grass or other herbaceous vegetation with scattered or widely spaced shrubs.

Soft-wood—trees of relatively fast growth and with rather soft wood.

Sparse vegetation—vegetation with the plants or tufts mostly more than twice their diameter apart.

Steppe—open grass or other herbaceous vegetation, the plants or tufts discrete but averaging less than their diameters apart.

Steppe forest—open forest with the lower layers also open, that is, the plants or tufts discrete but averaging less than their diameters apart.

Steppe savanna—savanna with the grass layer open rather than closed.

Steppe scrub—open scrub with the herb layers also open, that is, the average distance separating plants is greater than their average diameters.

Submerged aquatic plant—plant adapted to totally submerged aquatic existence or only leaves floating.

Substratum—the soil or other material in or on which plants are rooted or attached.

Succulent—having the stems or leaves conspicuously fleshy.

Swamp—woody vegetation predominantly of plants with their root systems adapted to withstand prolonged or permanent submergence or water-logged conditions.

Thalliform—pertaining to plants with the shoot not clearly differentiated into stem and leaves.

Thicket—tangled irregular vegetation predominantly of shrubs and small trees.

Thorn forest—forest predominantly of thorny or spiny trees.

Thorn scrub—scrub predominantly of thorny or spiny shrubs.

Tree—as used here, a non-climbing woody plant more than 5 m tall.

Unistratal—as applied to vegetation, of only one layer or stratum, or not distinguishably layered in structure.

Watergrass—submerged graminoid vegetation.

Weedy vegetation—vegetation of rapidly growing pioneer plants.

Proposed additional formations and subformations to the Fosberg classification

Code	Formation	Example
1A1 2	(d) Evergreen narrow sclerophyll swamp forest (Usually coniferous trees)	e. *Picea* swamp forest (Finland)
	(e) Evergreen microphyllous sclerophyll swamp forest (Small or scale-leaved trees)	e. *Chamaecyparis* swamp forest (USA)
1A2 6	Deciduous narrow sclerophyll forest (Usually coniferous trees)	e. *Larix* forest (North-eastern USA)
1B1 9	Gray mesophyllous evergreen scrub (Mesophyllous evergreen saltbush scrub)	e. *Halimione portulacoides* scrub (Britain)
10	Succulent evergreen thorn scrub	e. *Opuntia* scrub (Galapagos and California)
1C1 4	Evergreen succulent dwarf scrub	e. *Salicornia*, *Suaeda* flats (Temperate regions)
5	Evergreen narrow sclerophyll dwarf scrub swamp	e. *Juniperus horizontalis* swamp (North-east North America)
6	Evergreen narrow sclerophyll dwarf scrub	e. *Juniperus communis* spp. *nana* scrub (Britain)
1C2 2	Deciduous dwarf scrub bog	e. *Vaccinium–Betula–Myrica* with *Sphagnum* (Sweden)
1D1 6	Open evergreen succulent forest	e. *Euphorbia dawei* open forest (Uganda)
1E2 4	Deciduous orthophyll swamp scrub with trees	e. *Alnus–Cornus–Vaccinium* swamp, with *Pinus*, *Larix* and *Acer* (North-eastern USA)
1G1 3	Open resinous evergreen narrow sclerophyll scrub	
	(a) Open resinous straight evergreen narrow sclerophyll scrub	e. *Pinus halepensis–Juniperus* scrub (Tunisia) *Pinus* or *Picea* forest succession (N. Temperate)
	(b) Open resinous gnarled evergreen narrow sclerophyll scrub (krummholz)	e. Open *Pinus* or *Picea* krummholz (USA)
1H1 4	Open evergreen broad sclerophyll dwarf scrub	e. *Quercus coccifera–Erica–* Pistacia scrub (Tunisia)
1H2 3	Open deciduous dwarf scrub bog	e. *Vaccinium uliginosum–Betula–Myrica* with *Sphagnum* (Sweden)
4	(Open deciduous dwarf thorn scrub)	
1K1 5	Evergreen thorn scrub savanna	e. Scattered *Ulex* scrub (Britain)
1L2 3	Seasonal tall graminoid marsh	e. *Phragmites* marsh (Temperate)
1M1 5	Evergreen sclerophyll short grass	e. *Sclerodactylon* bunch grass (Aldabra)
1O2 3	Closed lichen crust	e. *Vesicaria–Rhigocarpus* crust (Sweden)
1P1 4	Evergreen submerged bryoid vegetation	e. *Fontinalis–Scapenia* (Sweden)
1Q1 5	Bryoid floating meadow	e. Bog depressions with floating *Sphagnum*
1Q2 3	Seasonal mesophyllous floating meadows	e. *Hydrocharis* meadows (Sweden)

1R Ephemeral herb meadows

Code	Formation	Example
(1R1)		
1R2 1	Microphyllous ephemeral grass meadow	e. *Panicum* meadow (Aldabra)
2	Microphyllous ephemeral succulent herb meadow	e. *Bacopa–Mollugo* meadow (Aldabra)

2B1	3	Resinous evergreen narrow sclerophyll steppe scrub	
		(a) Straight resinous evergreen narrow sclerophyll steppe scrub	e. Juniper scrub (Tunisia) Tola (*Lepidophyllum*) heath (Peru, Bolivia)
		(b) Gnarled resinous evergreen narrow sclerophyll steppe scrub (open krummholz)	e. Open *Pinus flexilis* krummholz (Rocky Mts.)
2C1	5	Evergreen succulent dwarf steppe scrub	e. *Suaeda–Salsola* and *Arthrocnemum* flats (Tunisia)
2C2	1	Deciduous orthophyll dwarf steppe scrub	e. High altitude *Salix* scrub (USA)
2F1	4	Evergreen broad sclerophyll dwarf steppe scrub savanna	e. *Rosmarinus–Stipa–Genista* (Tunisia)
2G1	4	(Evergreen graminoid steppe)	
2G2	3	Seasonal graminoid steppe marsh	e. Open *Carex*, *Equisetum*, and *Juncus* stands in fens (Sweden)
2H2	2	Open lichen crust (Open epilithic lichen vegetation)	e. *Verrucaria* crusts on calcareous rock (Britain)
3C1	4	Bryoid desert	e. Sparse epilithic moss cushions (see notes)
3C1	5	Broadleaf orthophyll desert	e. *Tribulus cistoides* stands in seabird rookeries (Galapagos)
3C2	3	Seasonal succulent herb desert	e. *Suaeda–Arthrocnemum* on intermittent lake beds (Tunisia)
	4	Seasonal broad-leaved herb desert	e. *Gutierrezia–Astragalus–Hymenoxis* on steep eroded slopes (Canada)
(3E)		See notes	
(3F)		See notes	

Notes

(a) 1A12(e) No record of this subformation has yet been received.
(b) 1B22(b) This subformation should include narrow sclerophyll deciduous swamp scrub.
(c) 1H24 and 2G14 No examples of these formations has yet been recorded.
(d) 1R1 By definition this does not exist.
(e) 3C14 Whether a community should be included in this formation or as a mosaic of 1O12 with bare ground is open to argument. Only if it is certain that each tuft of moss is derived from a single protonema should the example be included here.
(f) 3E Dwarf scrub communities with sparse individuals may be recorded under 3E. No formations and subformations will be defined until more records are received.
(g) 3F Sparse floating meadows. Examples of these communities may occur although no formation or subformation has yet been defined.

Appendix 3. Classification of soils

Surveyors are asked to indicate by the appropriate symbol in the left-hand column of Section 7 (2) the general category of soils found beneath each plant association.

No single internationally accepted system of soil classification is available. The adopted classification is primarily morphological and is essentially similar to that which is the basis of the system of Aubert and Duchaufour (Duchaufour, Ph., 1965, Précis de Pedologie, Masson, Paris). It is hoped it will be possible for the non-specialist to fit most soils into the broad categories given. Six categories are subdivided where necessary on either chemistry or causal factors, and categories S and F are further subdivided on profile horizon succession. To amplify other categories, the characteristic horizon succession is given for these also, and examples of soil groups considered to fall within the categories given. Thirteen categories are thus indicated by letter symbols with number subscripts. Horizon nomenclature is based on generally accepted usage (see, for example, KUBIENA W.L. 1953, *Soils of Europe*, Murby, London (English language version). Finer subdivisions have been avoided, because, although these give greater precision, they add complexity.

Horizon symbol	*Description of horizon*
A	Surface horizons of maximum organic matter content and/or maximum biological activity and/or horizons from which sesquioxides and clays have been transported to lower horizons in the soil profile.
B or (B)	Horizons lying between A horizons and the parent material C horizons, which show weathering and release of iron oxides without their transportation, (B); or which are horizons of deposition of transported sesquioxides and/or clay, B.
C	The parent material from which the soil profile has developed, whether rock *in situ* or unconsolidated deposits such as glacial drift, wind-blown sand or alluvium.
G	Horizon in waterlogged (poorly drained) soils in which reducing rather than oxidising conditions are dominant, therefore with typically grey colours resulting from ferrous oxides.

Appendix 3

Soil Categories

Soil category	*Sub-category*	*Profile type = horizon succession*	*Class symbol*	*Example soil groups in each category*
Saline soils	—	AC	S_1	Solonchak
		ABC	S_2	Solonetz
High sesquioxide (ferritic) soils	—	A(B)C	Fe	Terra rossa, rotlehm Laterite Ferritic brown earth
Organic soils	—	A	O	Peat, fen, bog
Well-drained non-saline non-ferritic soils with good profile development	Calcareous surface or sub-surface horizons	AC	F_1	Rendzina, chernozem, chestnut soil (kastanozem)
		A(B)C	F_2	Brown calcareous soil, Terra fusca
	Non-calcareous throughout profile	AC	F_3	Brown ranker, skeletal brown earth
		A(B)C	F_4	Brown earth (braunerde), sol brun acide, brown earth with gleying
		ABC	F_5	Podzol, peaty podzol, brown podzolic soil, sol brun podzolique, sol lessivé
Poorly-drained non-saline, non-ferritic soils with good profile development	Calcareous surface or sub-surface horizons	AGC	P_1	Calcareous gley, fen marl
	Non-calcareous throughout profile	AGC	P_2	Non-calcareous gley
Soils with weak profile development	Controlled by climate	AC	I_1	Serosem, burosem, desert soil
	Controlled by lack of time available for profile development	AC	I_2	Recent alluvium, raw warp soil, grey warp soil, regosol

A key to assist selection of the appropriate category is also given. Where a soil cannot readily be placed in a single category, two or more must be given; e.g. a shallow AC profile on rock in a mountain area might be doubtfully placed between F_1 or F_2 and I_2.

Key to Soil Categories

1	Soil containing a high concentration of alkaline salts.		2
	Soil without high concentration of alkaline salts.		3
2	Saline gley soil with water-soluble salts in upper horizons.	S_1	
	Saline soil with water-soluble salts in lower horizon and high exchangeable sodium in the surface.	S_2	
3	Soil with high concentration of iron oxides.	Fe	
	Soil with normal concentration of iron oxides.		4
4	Soil with dominantly organic surface horizon at least 50 cm deep. If total soil depth less than 50 cm, then surface organic horizon directly succeeded by unaltered rock.	O	
	Soil without dominantly organic surface horizon or with organic surface horizon succeeded by mineral soil horizon at less than 50 cm depth.		5
5	Well drained (i.e. no evidence of strong impedance or waterlogging (above 40 cm depth).		6
	Poorly drained (i.e. evidence in mottled colours or otherwise of strong impedance or waterlogging nearer surface than 40 cm).		11
6	Immature profile, that is with weakly developed and shallow soil formation, possibly with little biological activity.		7
	Well-developed horizon sequence with moderate to strong biological activity.		8
7	Immaturity resulting from climatic factors, e.g. very low rainfall and/or temperature.	I_1	
	Immaturity resulting from lack of time for soil formation to proceed, e.g. on recent alluvium, dune-sands and eroded surfaces.	I_2	
8	Calcareous in one or more soil horizons.		9
	Non-calcareous throughout profile.		10

9 Shallow or simple profiles of A horizons overlying parent material. F_1
A(B)C profiles. F_2

10 Shallow or simple profiles of A horizons overlying parent material. F_3
A(B)C profiles. F_4
ABC profiles, i.e. with a horizon accumulation of clay and/or iron oxides. F_5

11 Calcareous in one or more soil horizons P_1
Non-calcareous throughout profile. P_2

Notes

Organic Soils: The category of organic soils includes soils in which peaty material is thicker than 50 cm and those of shallower depth of peat directly overlying unaltered rock. Where mineral soil horizons occur above 50 cm depth the soil is to be classified in the appropriate mineral soil group, e.g. F_5, P_1, P_2.

Immature Soils: The category of immature soils (I) is transitional to other categories. 'Weakly developed shallow soil formations' at 6 in the key is a definition intended to refer to little-altered parent material, i.e. embro- or proto-soils. Shallow surface horizons with organic matter maxima overlying uniform unaltered parent material at less than 10 cm depth would be classed in category I. This is subdivided as to whether the immaturity is considered to result from climatic factors or from the youth of the surface on which soil formation is proceeding.

Calcareous and non-calcareous soils: The sub-categories of calcareous soils include those in which soft, easily weathered limestone fragments are present, and soils with a pH greater than 7·0 (in an unlimed state) in surface or subsurface horizons.

Appendix 4. Comparison of Fosberg and Westhoff–Den Held vegetational classifications

Fosberg coding	Westhoff–Den Held coding	Syntaxon (vegetation-type)
1A16	37Aa2	Fago-Quercetum pp.: *Quercus–Ilex* community
1A17	36Aa1	Leucobryo-Pinetum: coniferous forests with Quercion herb-layer; tree-layer closed
1A21	36Ab1	Betuletum pubescentis
	37Aa1	Querco roboris-Betuletum; deciduous phase
	2	Fago-Quercetum; excl. *Quercus–Ilex* community
	3	Convallario-Quercetum dunense
	38Aa1	Carici remotae-Fraxinetum
	2	Consortium of Carex remota and Populus nigra
	3	Pruno-Fraxinetum
	4	Macrophorbio-Alnetum
	5	Violo odoratae-Ulmetum
	6	Fraxino-Ulmetum
	7	Anthrisco-Fraxinetum
	8	Crataego-Betuletum
	Ab1	Stellario-Carpinetum
1A22	33Aa1	Salicetum triandro-viminalis; forest phase
	2	Salicetum arenario-purpureae; forest phase
	3	Salicetum albo-fragilis; forest phase
	35Aa1	Carici elongatae-Alnetum; older phase
	2	Carici laevigatae-Alnetum; older phase
1A26		Larix forests
1B17		Young coniferous plantation; coniferous layer closed
1B18	30Ba1	Genisto pilosae-Callunetum; taller scrub phase
1B21	32Aa1	Myricetum gale; scrub phase
	2	Frangulo-Salicetum auritae
	4	Salicetum pentandro-cinereae
	5	Salicetum pentandro-arenariae
	34Aa1	Carpino-Prunetum spinosae; closed phase
	2	Sambuco-Prunetum spinosae; closed phase
	Ab1	Hippophao-Ligustretum
	2	Hippophao-Sambucetum
	3	Ulmo-Clematidetum
	4	Orchio-Cornetum
	5	Solano-Rubetum ulmifolii
	Ac1	Polypodio-Salicetum; high scrub phase
	2	Thalictro-Salicetum; high scrub phase
	Ba	Sambuco-Salicion capreae
	b	Lonicero-Rubion sylvatici
	37; 38	coppice
1B22	32Aa3	Alno-Salicetum cinereae
	33Aa1	Salicetum triandro-viminalis; scrub phase
	2	Salicetum arenario-purpureae; scrub phase

Appendix 4

Fosberg coding	Westhoff–Den Held coding	Syntaxon (vegetation-type)
	3	Salicetum albo-fragilis; scrub phase
1C12	27Aa2	Caricetum trinervi-nigrae; scrub phase
	29Aa2	Ericetum tetralicis
	3	Empetro-Ericetum
	30Ba1	Genisto pilosae-Callunetum; low closed scrub phase
	2	Empetro-Genistetum tinctoriae; low closed scrub phase
	Bb1	Polypodio-Empetretum
1C13	29Ba1	Erico-Sphagnetum magellanici; closed evergreen scrub phase
	Bb1	Empetro-Sphagnetum rubelli; closed scrub phase
	Ba2	Sphagnetum palustri-papillosi; deciduous phase
1C22	29Ba1	Erico-Sphagnetum magellanici; closed deciduous scrub phase
	30Bb2	Pyrolo-Salicetum
	32Aa1	Myricetum gale; dwarf scrub phase
	34Ac1	Polypodio-Salicetum; dwarf scrub phase
	2	Thalictro-Salicetum; dwarf scrub phase
1D14		Coniferous forests with Quercion herb layer; open tree layer
1D21	26Aa4	Pellio-Conocephaletum; closed lower layer phase
1D22	26Aa2	Trichocoleo-Sphagnetum; closed lower layer phase
	3	Pellio epiphyllae-Chrysosplenietum oppositifolii; closed lower layer phase
	Ab1	Cratoneuro filicini-Cardaminetum; closed lower layer phase
1E21	37; 38	Coppice with scattered trees; Dutch 'Spaartelgenbos'
1F11	30Ba1	Genisto pilosae-Callunetum with scattered trees
	2	Empetro-Genistetum tinctoriae with scattered trees
1G13		Young coniferous plantation; coniferous layer open
1G21	34Aa1	Carpino-Prunetum spinosae; open phase
	2	Sambuco-Prunetum spinosae; open phase
1G22	17Bb1	Althaeo-Calystegietum sepium; open dune scrub phase
1G23	35Aa1	Carici elongatae-Alnetum; young phase with open scrub layer
	2	Carici laevigatae-Alnetum; young phase with open scrub layer
1H12	29Ba1	Erico-Sphagnetum magellanici; open evergreen shrub phase
	2	Sphagnetum palustri-papillosi; herb phase
	Bb1	Empetro-Sphagnetum rubelli; open scrub phase
1H21	16Ab7	Community of Ononis spinosa and Carex distans; open open shrub phase
	27Ba5	Junco baltici-Schoenetum nigricantis; shrub phase
	6	Ophioglosso-Calamagrostietum epigeji; shrub phase
1H23	29Ba1	Erico-Sphagnetum magellanici; closed deciduous scrub phase
1K13	34A Ass. 1	Squarroso-Juniperetum
	36Aa2	Dicrano-Juniperetum; closed phase
1L23	19Ba1	Scirpetum lacustris
	2	Typhetum angustifoliae
	3	Scirpo-Phragmitetum
	4	Typhetum latifoliae
	7	Sociation of Glyceria maxima

Fosberg coding	Westhoff–Den Held coding	Syntaxon (vegetation-type)
	8	Thelypterideto-Phragmitetum
	9	Scirpetum maritimi
	10	Scirpetum triquetri et maritimi
	24Ad1	Halo-Scirpetum maritimi
	27Aa4	Pallavicinio-Sphagnetum
	Ba1	Scorpidio-Caricetum diandrae
	Ass. 1	Scorpidio-Utricularietum
	29Ba2	Sphagnetum palustri-papillosi; young phase
1L24	19Ca1	Cladietum marisci
1M21	12Aa1–5 incl.	In cornfields, excl. maize: see 1N21
	12Ba2	Bromo-Hordeetum murini
	7	Sileno-Allietum vinealis
	Bc1	Medicagini-Toriletum nodosae
	13Aa1	Teesdalio nudicaulis-Arnoseridetum minimae
	Ab1	Papaveretum argemones
	2	Specularietum speculi-veneris
	Ba1	Linarietum spuriae
	2	Papaveri-Melandrietum noctiflori
	15Aa1	Euphorbio-Agropyretum juncei; closed phase
	2	Agropyretum boreo-atlanticum; closed phase
	Ab1	Euphorbio-Ammophiletum; closed phase
	2	Elymo-Ammophiletum; closed phase
	16Aa3	Juncetum tenuis
	16Ab1	Potentillo-Festucetum arundinaceae
	2	Rumici-Alopecuretum geniculati
	3	Junco-Menthetum longifoliae
	4	Potentillo-Menthetum rotundifoliae
	5	Community of Scirpus planifolius
	6	Community of Agrostis stolonifera subvar. salina and Trifolium fragiferum
	8	Poö-Lolietum
	9	Ranunculo-Agrostietum caninae
	10	Caricetum vulpinae
	11	Juncetum effusi
	17Ba3	Atriplici-Agropyretum pungentis
	19Ca3	Caricetum hudsonii
	5	Caricetum acuto-vesicariae
	6	Caricetum appropinquatae
	7	Sociation of Phalaris arundinacea
	20Aa1	Spergulo-Corynephoretum; closed herb phase
	Ba1	Ornithopodo-Corynephoretum; closed phase
	2	Airo-Caricetum arenariae; closed phase
	3	Festuco-Thymetum serpylli; closed phase
	4	Agrostietum tenuis; closed phase
	Bb2	Association of Euphorbia seguieriana and E. cyparissias
	3	Association of Dianthus deltoides and Herniaria glabra
	Bc1	Violo-Corynephoretum; closed phase
	2	Tortulo-Phleetum arenarii; closed phase
	3	Festuco-Galietum maritimi
	4	Taraxaco-Galietum maritimi
	5	Anthyllido-Trifolietum scabri

Appendix 4

Fosberg coding	Westhoff–Den Held coding	Syntaxon (vegetation-type)
	6	Anthyllido-Silenetum nutantis
	21Aa1	Koelerio-Gentianetum
	2	Medicagini-Avenetum pubescentis
	22Aa1	Violetum calaminariae
	24Ab1	Juncetum gerardii
	3	Junco-Caricetum extensae
	4	Scirpetum rufi
	25Aa1	Crepido-Juncetum acutiflori
	2	Scirpetum sylvatici
	3	Senecioni-Brometum racemosi
	4	Community of Hypericum tetrapterum and Lychnis flos-cuculi
	5	Community of Ophioglossum vulgatum and Orchis morio
	Ac1	Cirsio-Molinietum
	Ba1	Arrhenatheretum elatioris
	2	Fritillario-Alopecuretum pratensis
	3	Lolio-Cynosuretum
	27Ba4	Parnassio-Juncetum atricapilli
	5	Junco baltici-Schoenetum nigricantis; herb phase
	6	Ophioglosso-Calamagrostietum epigeji; herb phase
	30Aa1	Nardo-Gentianetum pneumonanthes
	2	Botrychio-Polygaletum
	3	Hyperico maculatae-Polygaletum
1M22	2Aa2	Zosteretum marinae stenophyllae; closed marsh phase
	3	Zosteretum noltii; closed phase
	6Aa1	Eleocharitetum multicaulis
	4	Eleocharitetum acicularis
	5	Ranunculo-Juncetum bulbosi
	6	Pilularietum globuliferae
	14Aa1	Spartinetum maritimae
	2	Spartinetum townsendii
	19Aa1	Spargano-Glycerietum fluitantis
	Ba5	Community of Acorus calamus and Iris pseudacorus
	Ca2	Caricetum paniculatae
	4	Caricetum ripariae
	24Aa1	Puccinellietum maritimae
	Ac1	Puccinellietum distantis
	2	Puccinellietum fasciculatae
	3	Puccinellietum retroflexae
	27Aa1	Caricetum curto-echinatae
	2	Caricetum trinervi-nigrae; herb phase
	3	Sphagno-Caricetum lasiocarpae
	Ba2	Parnassio-Caricetum pulicaris
	3	Juncetum alpino-articulati; closed phase
	28Aa3	Consociation of Carex rostrata and Sphagnum
	4	Consociation of Carex lasiocarpa and Sphagnum
	5	Sociation of Eriophorum angustifolium and Sphagnum apiculatum
	6	Sociation of Molinia coerulea and Sphagnum cuspidatum
1M24	17Bb4	Association of Juncus maritimus and Oenanthe lachenalii
1N12	24Aa3	Halimionetum portulacoidis
1N21	10Aa5	Isolepido-Stellarietum
	11Aa1	Polygono-Bidentetum; closed phase

Fosberg coding	Westhoff-Den Held coding	Syntaxon (vegetation-type)
	2	Ranunculo-Rumicetum maritimi; closed phase
	Ab1	Malachio-Bidentetum fluviatile; closed phase
	2	Polygono brittingeri-Chenopodietum rubri; closed phase
	3	Chenopodietum glauco-rubri; closed phase
	12Aa1	Chrysanthemo-Sperguletum; excl. corn-fields
	2	Mercuriali-Fumarietum; excl. corn-fields
	3	Veronico-Lamietum hybridi; excl. corn-fields
	4	Oxalido-Chenopodietum polyspermi subatlanticum; excl. cornfields
	5	Echinochloo-Setarietum; excl. corn-fields
	12Ba1	Chenopodio-Urticetum urentis
	5	Melandrio-Berteroetum

Appendix 5. Uniform classification of Scandinavian vegetation for use in mapping

Fosberg coding	Association	Ref. no.	Explanatory notes and descriptions
1A17a	Vaccinio-Pinetum (boreale)	Forest (5)	*Vaccinium*-rich mixed coniferous forest of *Picea* and *Pinus*
	Eu-Piceetum	Forest (10)	Vaccinium- and Dryopteris-rich forests of *Picea*
	Melico-Piceetum	Forest (11)	Low-herb *Picea* forests
1A21	Populo-Quercetum	Forest (12)	Poor oak forest with grasses, heaths and mosses
	Melico-Quercetum	Forest (13)	Richer oak forest with grasses and herbs
	Deschampsio-Fagetum	Forest (14)	Poor beech forest
	Alno(incanae)-Prunetum	Forest (15)	Forests of *Alnus incana* and *Prunus padus* along streams
	Alno(incanae)-Fraxinetum	Forest (16)	Rich *Fraxinus* forests under favourable conditions
	Equiseto-Fraxinetum	Forest (17)	*Equisetum*-rich forest occurring rarely along spring-lines
	Ulmo-Tilietum	Forest (18)	Forest with *Ulmus glabra* and *Tilia cordata*, rich herb-layer
	Ulmo-Quercetum	Forest (19)	Vigorous *Quercus* and *Carpinus* forest with *Fraxinus* and *Ulmus*. Much affected by agriculture
	Fraxino-Fagetum	Forest (20)	Rich beech forest on wet flushed soils. Dominated by *Fraxinus* and *Acer* in some instances
	Melico-Fagetum	Forest (21)	Pure beech forest, poor in species
1A22	Lycopo-Alnetum (glutinosae)	Forest (4)	Lush vigorous forest on mineral soils with high variable water-table on lake-shores, inlets by the sea, etc.
	Carici elongatae-Alnetum (glutinosae)	Forest (3)	Black alder swamps in marshes
1A12d	Chamaemoro-Piceetum	Forest (9)	Spruce swamp-forest
1B21a	Salicetum triandrae	Forest (1)	Willow thickets along streams and rivers, on mud and sandbanks
1B22a	Calamagrostis (purpurea)-Salicetum pentandrae	Forest (2)	Willow swamp-scrub, often in small stands surrounded by spruce swamp

Fosberg coding	Association	Ref. no.	Explanatory notes and descriptions
1C12a	Arctostaphyleto-Cetrarion nivalis	Fjeld (22)	Oligotrophic dwarf shrub heath
1C12c	Arctostaphyleto-Cetrarion nivalis	Fjeld (22)	As above but with *Empetrum* and with peat accumulation
1D12a	Vaccinio uliginosi-Pinetum	Forest (8)	Open pine bog-forest with heaths, mosses, etc.
	Chamaemoro-Piceetum	Forest (9)	Open spruce-swamp with heaths and mosses
1D14a	Cladonio-Pinetum (boreale)	Forest (6)	Open *Pinus* forest with closed heaths, lichens and mosses
	Barbilophozio-Pinetum	Forest (7)	Open pine forests with birch, heaths, lichens and mosses
	Eu-Piceetum	Forest (10)	Open spruce forests
1D21	Barbilophozio-Pinetum	Forest (7)	Birch phase of open pine–birch forests with heaths, lichens and mosses
	Eu-Piceetum	Forest (10)	Birch phase of open spruce–birch forests with *Dryopteris* and *Vaccinium myrtillus*
	Populo-Quercetum	Forest (12)	Open oak-aspen forest
1D2a	Chamaemoro-Piceetum	Forest (9)	Open spruce-swamp
	Carici elongatae-Alnetum (glutinosae)	Forest (3)	Open black alder, spruce and birch swamps
1F12	Barbilophozio-Pinetum	Forest (7)	Microphyll heath with scattered pine and birch
1H12/3	Oxycocco-Empetrion hermaphroditi	Fjeld (12)	Open dwarf scrub and dwarf shrubs on raised bogs
1H12	Arctostaphyleto-Cetrarion nivalis	Fjeld (22)	Open dwarf scrub of *Arctostaphylos* with *Juncus trifidus*
	Kobresieto-Dryadion	Fjeld (24)	Open dwarf scrub with sedges and *Dryas*
1H14	Arctostaphyleto-Cetrarion nivalis	Fjeld (22)	Open dwarf scrub with evergreen broad sclerophylls prominent
1H22	Phyllodoco-Vaccinion myrtilli	Fjeld (23)	Open phase of *Vaccinium myrtillus* heath
1J14	Cladonio-Pinetum (boreale)	Forest (6)	Open pine forest with lichens
1J15	Vaccinio uliginosi-Pinetum	Forest (8)	Open pine-swamp with mosses
1L21	Festucetum arundinaceae	Shingle (3b)	Tall fescue with herbs
1L23	Spartinetum townsendii	Saltmarsh (3a)	Pure stands of *Spartina* or *Spartina* with *Puccinellia* etc.
	Phragmito-Scirpetum maritimi	Saltmarsh (4a)	Tall reed vegetation of *Phragmites* with *Scirpus maritimus*
1M12	Puccinellietum maritimae	Saltmarsh (5a)	*Puccinellia* with *Aster*, *Limonium* and *Halimione*
1M11	Ammophilion	Sand (1)	Marram grass on sand-dunes

Fosberg coding	Association	Ref. no.	Explanatory notes and descriptions
1M21	Festucetum arundinaceae	Shingle (3b)	Shorter phases of tall fescue grasslands
	Phyllodoco-Vaccinion myrtilli	Fjeld (23)	Grass heath: *Deschampsia* phase of dwarf shrub heath
	Arctostaphyleto-Cetrarion nivalis	Fjeld (22)	Grass heath: *Juncus trifidus* phase of dwarf shrub heath
	Kobresieto-Dryadion	Fjeld (24)	Grass heath: sedge phase of calcicolous dwarf shrub heath
	Nardeto-Agrostion tenuis	Fjeld (26)	*Nardus–Agrostis* meadows and pastures
	Nardeto-Caricion bigelowii	Fjeld (17)	Snow-patch grassland
1M22	Magno-Caricion	Fjeld (6)	Sedge-swamps
1M25(?) (1M22)	Leuco-Scheuchzerion (Caricion rotundatae)	Fjeld (7)	'Poor fen' vegetation with sphagna and sedges prominent
	Stygio-Caricion limosae	Fjeld (8)	'Poor fen', less oligotrophic than the preceding, with *Juncus stygius* and sedges but few sphagna
	Salicion myrsinitis (Caricion atrofuscae-saxatilis)	Fjeld (9)	Eutrophic sedge-fen with moving water
	Caricion canescentis-nigrae	Fjeld (10)	Oligotrophic sedge-fen with sparse to dense willows
1M22(?)	Puccinellietum phryganodis	Saltmarsh (6a)	Pioneer stands of *Puccinellia* on seashores
	Caricetum subspathaceae	Saltmarsh (6b)	Sedges and *Agrostis* replacing *Puccinellia* at somewhat higher levels
	Eleocharetum uniglumis	Saltmarsh (7a)	Grasses and sedges in brackish marshes
	Sperguletum salinae (Puccinellietum distantis)	Saltmarsh (10a)	*Spergula* and *Puccinellia* on eroded marshes and in depressions
	Juncetum gerardi	Saltmarsh (8b)	Middle and upper levels of saltmarshes
	(Blysmo-) Caricetum pulchellae	Saltmarsh (8c)	Local on thinly covered rock or moraine
	Elytrigietum repentis maritimum	Shingle (3a)	Low grasses and herbs on old shingle-banks
1N21	Lactucion alpinae	Fjeld (15)	Tall-herb vegetation (mesotrophic to eutrophic)
	Ranunculeto-Oxyrion digynae	Fjeld (19)	Low-herb communities on unstable substrata
	Potentilleto-Polygonion vivipari	Fjeld (25)	Calciphilous herb-rich grassland
	Bidention tripartitae	Shingle (1)	Ruderals of drift on shingle by lakes or rivers
	Cakiletum maritimae	Shingle (2a)	Ruderals with *Cakile maritima* on sandy seashores
	Atriplicetum latifolii/litoralis (coll.)	Shingle (2a)	*Atriplex* and *Polygonum* spp. on seashores

Fosberg coding	Association	Ref. no.	Explanatory notes and descriptions
	Elytrigieto-Euphorbietum palustris	Shingle (3c)	Chiefly on stony drift with seaweeds
1N22	Allosoreto-Athyrion alpestris	Fjeld (16)	Oligotrophic–chionophilous fern communities on stable stony ground
(1N21 or 3C2 ?)	Sperguletum salinae (Puccinellietum distantis)	Saltmarsh (10a)	Degenerative phases with invasive *Puccinellia maritima* and *Agrostis maritima* etc., becoming sparse
1O11	Oxycocco-Empetrion hermaphroditi	Fjeld (12)	Dwarf shrub community of the later stages of the regeneration-cycle of raised bogs
1O12	Stygio-Caricion limosae	Fjeld (8)	Phase with closed moss layer
	Salicion myrsinitis (Caricion atrofuscae-saxatilis)	Fjeld (9)	Phase with closed moss layer
	Caricion canescentis-nigrae	Fjeld (10)	Phase with closed moss layer
	Sphagneto-Tomentypnion	Fjeld (11)	Eutrophic–calciphilous community with closed moss layer, on drier sites than Salicion myrtillis
1O13(?)	Cratoneuro-Saxifragion aizoidis	Fjeld (13)	Eutrophic–calciphilous spring vegetation
	Montio-Epilobion hornemanni	Fjeld (14)	Oligotrophic–calciphilous spring vegetation
	Cassiopeto-Salicion herbaceae	Fjeld (18)	Oligotrophic dwarf shrub communities with closed moss layers in late snow patches on unstable ground
	Polarion (coll.)	Fjeld (20)	Calciphilous dwarf willow communities on unstable ground, with closed moss layer
1O21	Arctostaphyleto-Cetrarion nivalis	Fjeld (22)	Phase with closed lichen layer
1P21	Zosteretum nanae	Saltmarsh (1a)	Tidal shore vegetation with *Zostera* and *Ruppia*
2C14	Arctostaphyleto-Cetrarion nivalis	Fjeld (22)	Open phase of dwarf shrub heath
2I21/2	Potamion eurosibiricum	Fjeld (5)	Seasonal open submerged aquatic vegetation
2I22	Litorellion	Fjeld (4)	Seasonal open submerged broad-leaved aquatic vegetation
3C12	Cakiletum maritimae	Shingle (2a)	Sparse community of mainly annual species of drift-covered sandy seashores
3C21	Luzulion arcticae	Field (21)	Sparse calcicolous high alpine vegetation of herbs only

Fosberg coding	Association	Ref. no.	Explanatory notes and descriptions
3C23	Salicornietum strictissimae	Saltmarsh (2a)	Sparse coastal vegetation of succulent halophytic herbs, often pure *S. strictissima*
	Salicornietum europaeae	Saltmarsh (2b)	As preceding but usually depressions with periodic heavy salt enrichment and with *S. europaea* replacing *S. strictissima*
3C24	Elytrigio-Euphorbietum Palustris	Shingle (3a)	Sparse vegetation of broad-leaved herbs on drift-covered stony seashores
3D21	Eleocharetum parvulae	Saltmarsh (1b)	Sparse submerged non-succulent aquatic plants
3D2?	Salicornietum strictissimae	Saltmarsh (2a)	Sparse submerged succulent herbaceous halophytes

Appendix 6. Detailed analyses of replies to individual questions

This appendix analyses the replies to individual questions so as to provide a background to the performance of the check-sheet as a tool for data-collection. It includes consideration of the design of questions and of certain aspects of information retrieval. For each question a final paragraph explains how the elicited information is stored, the technical terms employed, and their significance, being listed on p. 131.

Question 1: Source of information

This question comprised four parts, each giving rise to little comment concerning the style of completion. The points of particular interest were the background of the surveyors and the origin of the records they supplied.

From the name and address of the surveyor an attempt was made to categorize his background. No specific request was made for him to indicate this and so the figures cannot give more than an approximation. Seven backgrounds were recognized and the percentage frequency of each is given below:

Background of surveyor	A	B	C	D	E	F	G	H
Percentage frequency	35	47	1	1	4	6	4	2

A, Research organization; B, University; C, Wildlife consultant; D, Warden; E, Museum; F, Foresters; G, Individuals – unspecified; H, No entry made.

From the total collection of check-sheets it was obvious that those coming from a single co-ordinating source were the most consistently and in general the most reliably completed: returns from individuals varied enormously in quality and style. In countries such as the USA, where co-ordination was not effected through key personnel who controlled the consistency of approach, the total contribution was of variable quality. Where, as in Australia, co-ordination was much tighter and control was exercised on the format of certain free-style answers, the more rigid framework of expression gave little latitude for dealing with anomalies. The data were nevertheless a very good basis for comparative studies since no problems of the equivalence of vegetational units were involved, though some qualifications concerning the actual reliability of the data are made later in this section.

The table of surveyors' backgrounds shows that 35% came from research organizations and 47% from universities, the former including surveyors engaged in forestry research. Foresters or forest rangers completed 6% of the sample, museum staff 4%, while consultants, wildlife wardens and individuals with no identifiable background accounted for the remaining 8%. This presents a fairly encouraging picture, but it gives no clue as to the position an individual may hold within his organization. Check-sheets have been returned by directors of national parks, university professors, senior government biologists and also from postgraduate students and college lecturers whose main subject is unrelated to the life sciences. For a more satisfactory evaluation of the reliability of the data a specific request needs to be made for information on the background of surveyors.

The third section of Question 1 dealt with the source of the information provided, whether from the site or from records, a matter of some importance for assessing the reliability of the data.

Source	From records	From the site	From both	Unknown
Percentage frequency	40	18	38	4

It is clear that more check-sheets involved information from records than from the site alone, but about half of those reporting information from records also derived some from the site. What is not made clear is the extent to which each is involved. The component from records may comprise only the sections on location, status, climatological station and references. An indication of the origin of the biological information would have been of particular value here.

The Australian data on vegetation (see Question 7 below) were accompanied by an indication of reliability on a four-point scale:

A Good: based on aerial photographs and/or ground survey
B Fair: based on personal knowledge of the reserve or on reliable literature
C Poor: based only on knowledge of the district
D No information.

This index was processed as part of Question 1 so as not to increase further the complexity of Question 7. The variability of the vegetational data argues strongly for a compulsory indication of source.

The question was processed as five subsections: two strings for the surveyor's name and address, two binaries for the source of data, three integers for the date, and a fixed string A, B, C or D for the Australian reliability index, as a fifth subsection.

Question 2: Site names

The question of subdividing IBP areas so that a single check-sheet refers to only part of a site was discussed at some length in the *Guide* (pp. 11–14), and recommendations for the basis upon which subdivision should be carried out were put forward. It is of interest to examine how these recommendations were applied and to look at some of the characteristics of IBP subdivisions. The subdivision of IBP areas has been approached in several different ways, but in the majority of cases where it has been carried out the recommendations put forward in the *Guide* have been followed.

Only sixty-six IBP areas, less than 3% of the total number of check-sheet sites, were subdivided or sampled in some way. Of these thirty-eight were represented by conventional master/slave series. No more than six, however, had fully completed master check-sheets, so in this respect the *Guide* was not followed closely. For the remainder of the thirty-eight, dummy master check-sheets, comprising the first six questions, were compiled at Monks Wood (see Table A6.1). Deviations from the recommended procedure occurred in twenty-eight areas. Four of these had only one sub-division: two were cases where only part of a protected area had been surveyed, and a dummy master was compiled as a summary sheet for the whole of each area; in the other two cases a check-sheet for the whole of an IBP area was accompanied by a single slave for part of the area.

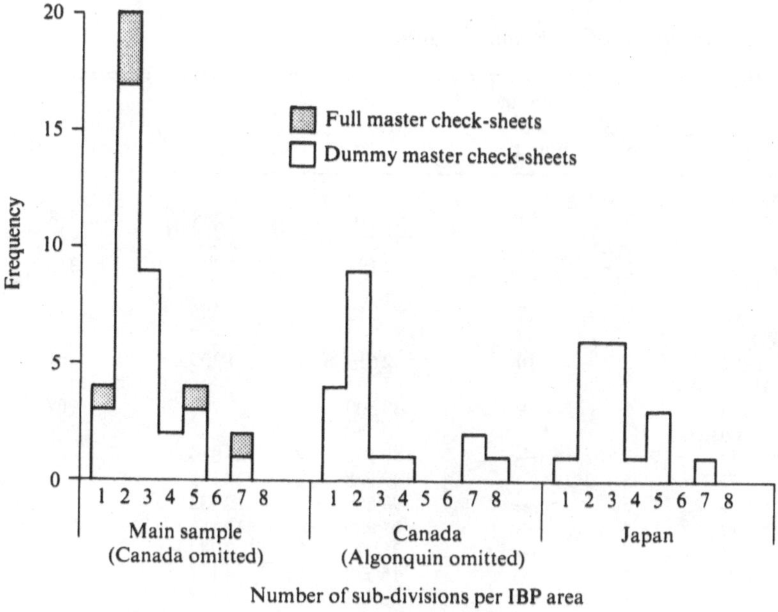

Fig. A6.1. Frequencies of IBP subdivisions in the main sample: Canada and Japan.

The French contribution included a further five subdivided areas but owing to the incompleteness of the entries they are omitted from this sample. The remaining nineteen areas are Canadian and the method of subdivision adopted was quite different. The basis of it is that, since none of the large areas could be surveyed satisfactorily, Intensive Study Areas (ISAs) were selected and a check-sheet completed for each. This resulted in a total of thirty-five ISAs from Algonquin Provincial Park (701 696 ha), and these are discussed below as an illustration of the procedure. Within thirty-eight conventional master/slave series the histogram of the frequency of number of slaves shows the expected skew from a mode of 2 (Fig. A6.1), the frequencies for Japan and Canada, excluding Algonquin Provincial Park, are also shown. No significant relationship between the size of whole sites and the number of subdivisions was evident, but it had been expected that using numbers of plant communities, or different types of land form, as an index of diversity would have shown consistent relationships at least within national contributions. The Japanese contribution, one of fairly high consistency in other respects, shows that this is not the case. Of 117 IBP areas from Japan and Japanese islands, eighteen were subdivided, the highest percentage for any country. Figure A6.1 shows two and three as equally the most frequent numbers of subdivisions.

The Japanese subdivisions range in area from 258 ha with ten plant communities up to 84 882 ha with eighteen; but ten are also found in subdivisions of various areas up to 18 741 ha and eighteen in others of various areas down to 2679 ha. The largest recorded number of communities is 104 for an undivided area of 231 929 ha in Japan. Correlation of area with diversity of land-form gave similarly inconclusive results.

The largest undivided area for which a check-sheet was received is Katmai National

Appendix 6

Table A6.1. *Summary of IBP sub-divisions*

Serial numbers	No. of slaves	No. of com-munities	Area (hectares)	Area/slave	Area/community
Full masters					
YQBE 02	7	55	197 876	28 268	3 598
OKGE 01	6	12	772 079	128 680	64 340
OLBE 06	2	9	372 183	186 092	41 353
MWOE 01	2	9	710	355	79
MWOE 02	2	10	2 610	1 305	261
MUHE 01	5	39	38 500	7 700	987
Dummy masters					
SKEJ 23	7	—	189 102	27 014	—
72	5	—	140 698	28 139	—
74	5	—	98 660	19 732	—
91	5	—	65 925	13 185	—
24	4	—	31 927	7 982	—
OMCE 07	4	—	16 524	4 131	—
SOEJ 02	3	—	3 866	1 288	—
SKEJ 15	3	—	83 351	27 783	—
70	3	—	55 231	18 410	—
71	3	—	77 137	25 712	—
92	3	—	122 309	40 769	—
96	3	—	16 903	5 634	—
OYGE 01	3	—	15 657	5 219	—
MKXE 08	3	—	26 350	8 783	—
YWEE 15	3	—	1 901 119	630 373	—
SKEJ 22	2	—	7 807	3 904	—
34	2	—	26 714	13 357	—
OSTE 04	2	—	30 751	15 375	—
05	2	—	343	172	—
OKGE 11	2	—	545	273	—
OTDE 04	2	—	161	80	—
OSIE 02	2	—	12 194	6 097	—
10	2	—	66 983	33 492	—
MLVE 01	2	—	43 600	21 800	—
13	2	—	6 094	3 042	—
MLAE 11	2	—	377	189	—
YPEE 03	2	—	40	20	—
WOFE 32	2	—	3 191	1 596	—
SKEJ 73	2	—	25 600	12 800	—
93	2	—	121 600	60 800	—
94	2	—	55 860	27 930	—
95	2	—	19 338	9 669	—

Monument (Alaska), covering 1 079 036 ha. The largest IBP subdivision is one of the three into which the 1 901 119 ha of the Kruger National Park (South Africa) is divided, and has an area of 997 181 ha. The smallest subdivision, of only 1.6 ha, is the smaller of the two areas into which the Ndola Forest Reserve (Zambia) is divided.

The examples given below illustrate the variety of criteria upon which the subdivision of IBP areas has been based.

Kruger National Park – South Africa

The park was divided into southern, central and northern parts using rivers which formed natural and complete boundaries. The result is three areas the largest of which is nearly three times the size of the smallest. Each has the same diversity of relief type and similar numbers of vegetation types, however.

Queen Elizabeth National Park – Uganda

Seven subdivisions were made of the total 220 000 ha, giving an average of 31 430 ha. In fact the largest subdivision was 42 217 and the smallest 7 770 ha. The lines of subdivision were based upon natural features: lakes, rivers and sharp relief.

Ndola Forest Reserve sample plots – Zambia

An experimental area of 40 ha subdivided into two on the basis of use. The larger subdivision concerns control plots completely protected from fire. The smaller area (1.6 ha) contains two plots; one burnt early, the other late in the dry season. The remainder of the IBP area (16.8 ha) was not included in the subdivisions.

Algonquin Provincial Park – Ontario, Canada

This park had check-sheets completed for thirty-five ISAs within its boundaries. The maximum size is 3 840 and the minimum 2 ha, supporting five and one recorded vegetation types respectively. The total area covered by the ISAs is 9 447 ha giving an average size of 269 ha. In this way a sample covering 1.4% of the total area has been described in much more detail than would be possible for the whole park.

The second and third parts of Question 2 relate to maps. From the sample only 53% provided a sketch map on the check-sheet as requested, but the proportion was made up to 79% by those with attached maps, some of which were hand-drawn sketches, but mostly were sections copied from published maps. Twenty-one per cent of the sample had no map. This question was reduced to two parts in the USA check-sheets: the name of the area, and a request for a map showing boundaries to be attached. The French version was modified to obtain the relevant map numbers in the 1:50 000 series for Metropolitan France, and requested a sketch map under Question 3 rather than 2. The question was processed as four subsections. The first is a compulsory string for the name of the IBP area with an optional extra string for an alternative name or language, the second subsection comprises an optional string for the name or other identification of the IBP subdivision. The third and fourth subsections are each of one binary item to indicate the presence of an appended map and a sketch map on the check-sheet, respectively.

Question 3: Location of IBP area

In several instances the latitude and longitude figures were reversed and occasionally quite wrong. The method for indicating North/South or East/West was mostly satisfactory because the country or state could be used as a check, but there were instances where the surveyor struck out the symbol that should have been left. A different design could help here and possibly avoid the problems which arose in countries through which the 0° meridian passes, where such an error may not be so easily detected. The French Check-Sheet requested the latitude and longitude to be given in grades rather than degrees. In fact both were used, although it was not always stated which.

The second part to the question gave rise to some confusion as to the precise manner of its completion. Entries were made in every case for the country and state but in several countries administrative units at a third level were not used. The confusion arose when more than one state or country was involved but could be straightened out in almost every case with the help of a gazetteer. Island groups gave rise to minor problems here.

In order to aid retrieval and to express geographical relationships between administrative units the geo-code was used in addition to each country and state, and, in some cases, county name. It was also incorporated into the serial number as described in an earlier section. Where two or more counties in any one state were involved the geo-code was generally omitted because of low sensitivity and the low probability of its use at this level for the present purpose.

The USA Check-Sheet in addition to the above information asked for the name and distance of the nearest city.

Question 4: Administration

This question was consistently well completed, within the limits of availability of the United Nations List of National Parks and Equivalent Reserves. Entries under 'Official Category' were all appropriately worded and usually the land tenure was given where no protection was in force at that time. The address of administration most often given was that of the national or head office concerned, and rarely the local seat. It would have been more useful to have the name of the organization or department with the address of its local office or representative in all cases. The international class was taken from the 1971 UN List. Later additions and an abridged list were published to which reference is made in the synthesis.

The table below shows the distribution of the sample check-sheets between the classes.

	1971 List		1973 List				
	A	B	A	C	D	Unknown	B+C
Percentage frequency	16	35	18	21	25	3	56

Class A, Accepted for inclusion in the UN List; Class B, considered for inclusion in the UN List but rejected; Class C, of some existing conservation status, but not included in A or B above; Class D, with no conservation status.

The class of five check-sheets – all slaves from one IBP area – was not given, nor was it possible to deduce it from the information supplied.

The criteria given in the text of the UN List for inclusion or rejection of a site do not allow for precise distinction between Classes B and C in the check-sheet. The Commission on National Parks requested countries to present information on sites which they wished to be considered for inclusion in the list. Those that were rejected were then listed in a separate chapter for each country. It is not clear upon what proportion of protected area individual countries submitted information as candidate entries for the list. For some countries sites of all degrees of protection appear to have been put forward while for others only those of high status were considered. Where possible the rejected sites are listed by name in the list but there are many entries such as '87 permanent faunal reserves' or '40 000 natural monuments'. For these reasons it was decided that there is no substantial difference between class B and C in the check-sheet. The situation is therefore simplified: sites are either protected or not and if they are, they may be included in the UN role of honour.

There is a need to indicate with some precision the category of protection that an area enjoys and the zones of use which are allowed in it. Detail to the level required to identify the categories in *Classification and Use of Protected Natural and Cultural Areas* (Dasmann, 1973*b*) would be very useful. This system goes a considerable way towards meeting the need expressed in the *Guide* for 'up to date internationally agreed definitions and standards'.

The USA check-sheet requests only the 'Title and address of administrator' in the actual question although the accompanying field instructions refer to two parts, the first being the National Category as in the Mark VII check-sheet. No reference is made to international class.

The question was processed in three subsections: a string for the national category, a string for the address of administration and a fixed string, A, B, C or D, for the international class.

Question 5: Characteristics of IBP area

This question was completed in almost all check-sheets. Occasionally the average altitude was given rather than a maximum and minimum. No specific request was made for the area covered by water bodies but the information was volunteered in a significant number of cases, sometimes under this question, but more frequently along with ice, bare rock and other unvegetated surfaces under question 7.1: Vegetation. Provision in the data processing was made to handle this information as an extra optional entry for area.

The question was processed as two subsections. The first consisted of a compulsory integer and a fixed string from the area units – 'acre', 'hectare', 'square mile' or 'square kilometre', and an optional integer and fixed string for water area; the second comprised two compulsory units each of an integer and a fixed string from the linear units 8 'feet' or 'metres', for the maximum and minimum altitudes.

Question 6: Climate

The question was omitted in only 3% of the sample; and climatological data, usually as monthly or yearly precipitation and temperature averages, were appended to 18%. Climatological stations are present on only 6% of all IBP areas in the sample, and on 11% class A (1973) sites.

In the French version of the check-sheet an additional request for an indication of the actual parameters recorded at the station was included.

The question was processed as five subsections: a string for the name, a binary item for presence or absence in the IBP area, an integer and a fixed string from the linear units – 'miles' or 'kilometres', a string for the compass direction and a binary item to indicate presence or absence of additional data.

Appendix 6

Question 7: Vegetation and soils

Vegetation

This part of the check-sheet is concerned with the description and evaluation of survey-sites by reference to characteristics of their vegetation and is therefore the most important of those dealing with aspects of their ecology. We shall consider here only the way in which the questions were answered in returned check-sheets: a lengthy discussion of the underlying principles and problems can be found in chapter 2.

During early trials leading to the present form of the check-sheet various methods of vegetational recording were used. At the time there were few available systems of objective recording and none seemed to provide a satisfactory basis for descriptive and comparative studies (pp. 14–16). Strong recommendations were put forward for the use of an *a priori* classification of plant formations, though it was realised that classification only to formation level would convey insufficient detail for an ecological inventory as a basis for decisions on nature conservation. It was, therefore, decided that the descriptive working unit should be the association but, there being no practical procedure for standardization, surveyors were invited to use any familiar recording system with the recommendation that it should be roughly at association level.

The term 'association' is used in a number of different ways (pp. 47–9) and the only specific reference to the kind of entry required was that 'surveyors should enter an informative, descriptive title, which may be a local name or could incorporate the names of dominant species'. In fact both types of designation were employed, but dominant species figured in over 75% of the entries.

Within the general definition of association given by I. F. Henderson (1963) – 'A plant community forming a division of a formation or larger unit of vegetation as of tundra, grassland, forest, and characterized by dominant species' – there is considerable scope for variation in scale. The size of an area, the degree to which it is already known, its diversity and the time available to the surveyor will determine whether or not associations of minor extent or significance are included in the inventory. The returned check-sheets showed considerable variation both within and between national contributions. Consistency was effectively achieved only in those countries with active co-ordination of the survey and an agreed policy on the general points about which no recommendation was possible in the *Guide* itself.

Table A6.2 gives examples of the principal types of community description in completed check-sheets.

Because the entry for community description was made free-style and almost no constraints were placed upon its nature, it was important to use a standard method of description at the formation level which would enable reference and cross-reference to be made. The Fosberg classification was eminently suitable for this purpose by fulfilling the roles both of a classification of plant formations and of a basis for coding the component associations in a simple manner. There were, nevertheless, a number of difficulties associated with its use, and these are considered in detail in Chapter 2.

The classification was interpreted correctly by the majority of surveyors for vegetation types corresponding fairly closely with the described units: 70% of the sample had full codes for each vegetation entry. The majority of these comprised a single code but certain special cases resulted in deviations from the normal entry. One of the first to be encountered was the question of intermediates and mosaics. In the *Guide* it was recommended that when examples on intermediate formations were met, an entry combining the two relevant formations was appropriate, but for mosaics separate entries of the different codes should be made. Surveyors tended to use the

Table A6.2. *Types of community descriptions*

Community description	Interpretation and comment
Mangrove swamp (1A1 2b)	Conveys less information than the Fosberg code
Strand forest (1A1 6a)	Gives the broad ecological setting only
Stunted forest – mixed species (1B1 4a)	Gives only structural qualification to the Fosberg code
Millettia sutherlandii (1A1 4)	Forest with single species dominance – minimum information
Podocarpus, riverine forest (1D1 5)	Ecological qualification to a floristically identified forest type – minimum information
Abies lasiocarpa, Juniperus communis, Vaccinium scoparium, Cladonia spp. (1A1 7a)	Forest with dominant species given for each main stratum – regarded as a practical number of specific names for an entry
Festuca altaica, Dryas octopetala, Salix reticulata Cladonia spp., extensive arid alpine tundra complex (3C1 3)	Highly informative, brief community description – regarded as an ideal type of entry where no reference to a published description is available
Pinus sylvestris, Calluna vulgaris, Vaccinium vitis-idaea, Pleurozium schreberi Hylocomium proliferum, heath forest. Modified, after Sterner (1929) (1A1 7a)	A complete entry, giving dominant species in each stratum, position in an ecological series and reference to a published description. The entry inevitably tends to be long
Querceto-Betuletum, Agrosto-Anthoxanthetum (1A2 1)	An obviously pseudosyntaxonomic name which conveys nothing beyond the four genera present
Caricetum elatae (1M2 1)	A syntaxonomic name with no reference to a publication. The community name can be accepted but is less useful than the dominant species
Salicetum albo-fragilis (1B2 1a) (Soó 1934) R.Tx (1948) 1955	The complete entry for a syntaxonomic name. Reference to the publication will give a great deal of background information but will only exceptionally allow a confident identification of dominant species

former procedure for both cases, and this was in fact encouraged on the grounds that *at the chosen scale of survey* recognition should be made of the different spatial relationships of communities wherever possible. The problem of scale is one of great consequence in all aspects of the survey, and particularly in relation to the inventory of vegetation-types.

At the formation level every attempt should be made to record the major structural types as they exist and at a constant scale. It was explained in the *Guide* that new formations and intermediates could be handled within the classification according to the recommendations made. This was an attempt to maintain consistency of scale by giving surveyors the opportunity to record anomalies either by using codes in combination or by bringing attention to the fact that a particular vegetation-type falls outside the present scope of the classification. Despite this, vegetation-types for which no adequate provision existed were found entered under formations which most nearly met the situation, or which had an example that appeared to match it. Thus *Phragmites* and *Typha* marshes in the North temperate zone were consistently coded as 1L1 2 ('Tall evergreen graminoid marsh'), whereas they clearly merited a new formation, 1L2 3 ('Tall seasonal graminoid marsh'). The inclusion of '*Typha* marsh' as an example under 1L1 2 dissuaded surveyors from proposing a new formation to

Table A6.3. *Deviations from single full code entries*

Fosberg code	Interpretation and comment
1A17a	The normal full entry for a vegetation sub-formation corresponding to a single 'node' in the classification. Resinous evergreen narrow sclerophyll forest
1A17	The code shortened to formation level. Evergreen narrow sclerophyll forest
1A1	Entry at level 1 giving the formation class. Evergreen forest
1A	Entry at level 1 giving the formation group. Forest
1A17a/1A21	Mostly an intermediate community between resinous evergreen narrow sclerophyll and winter-deciduous orthophyll forest types
1A17a/1D14a	An intermediate or mosaic community with a closed/open tree canopy. The formations both involve coniferous tree canopies
1A17a/1C17	A mosaic of coniferous forest and microphyllous evergreen dwarf scrub. This type of relationship would split into two separate communities as the scale of survey increases owing to their great dissimilarity

suit the observed situation. In the amended version of the classification the example is qualified by reference to the tropics, where it is evergreen.

Adopting the association as the fundamental unit of inventory involved a distortion of the original purpose of the Fosberg scheme. It was devised for application at a scale roughly equivalent to that suitable for vegetation maps at 1:1 000 000, allowing subdivisions of the formations, down to about association level, to be distinguished on the map. This presents no serious problems in regions where natural vegetation extends over large continuous areas, and here vegetation-types fitting badly into the system were seldom encountered. A different approach is needed where natural and semi-natural communities have been fragmented by man's activity in a landscape which may be almost wholly artificial. The most suitable procedure is then to list the communities recognizably distinct at association level and group them through their physiognomy into the named formations of the Fosberg system, this being used primarily to provide a physiognomically-based coding for groups of communities which individually may be of very limited extent. If a community cannot be fitted into any of the named formations it becomes necessary to propose a new one to accommodate it: it may be of high ecological importance even though it cannot be mapped at a scale of 1:1 000 000 or even 1:2 000 000. The formations of the Fosberg scheme are thus being used in two different ways: as a basis on the one hand for subdivision into associations and on the other for grouping associations into units of higher level. The problems that arise when the associations recognized in a given country or region are those of a floristically-based classification, not readily relatable to the Fosberg system, are considered in Chapter 2.

Table A6.3 gives examples of entries showing some of the deviations from the normal entry of a single full code for each community.

It will be seen that one of the deviations is a reduction of the coding to two or three characters representing a whole formation-class or formation-group, either because of lack of information or through use of the simplified alternative version of Question 7. This version (Level 1) was distributed in an attempt to increase the flow of completed sheets by reducing the amount of information requested. It was adopted in several countries, notably in Japan and for certain parts of the USA contribution. It asked for the truncated Fosberg coding together with the usual community descriptions but

Table A6.4. *Example of Fosberg coding in a transitional area*

Stratum	Formation class 1A	1D	1E	1F	1I	1J
Tree	C	O	S	S	S	S
Shrub	X	C	C	a	a	a
Dwarf shrub	X	C	X	C	a	a
Tall grass	X	C	X	X	C	a
Short grass	X	C	X	X	X	C

C, closed; O, open; S, scattered; X, closed/absent; a, absent.

without estimates of area, and the questions on soils and similar communities were omitted. At this level the Fosberg codings are of very little value, the information conveyed being merely that the community is, say, closed grassland or open scrub, possibly with an indication of its seasonality. A number of surveyors who used the Level 1 check-sheet completely misunderstood the accompanying key to formation-classes and their code-entries were unintelligible. As a means of expediting the return of check-sheets the idea was not very successful, and no indication can be given as to whether more would have been returned if this had been the only version distributed.

Examples of vegetation-types proposed as a basis for new formations or subformations were generally entered with a blank in the appropriate column, this being filled at the data-centre on assignment of a coding.

Compound codings were entered for two basic reasons: the vegetation was either intermediate between the two formations whose codings were cited, or was a mosaic of two or more formations. The most frequently entered was for a forest community dominated by a mixture of deciduous broad-leaved and coniferous trees, entered as 1A21/1A17a. This occurs over large areas in the transition zone between the boreal and eastern deciduous forest provinces of North America. Locally one or other component is solely dominant, and a single coding is then appropriate.

Two examples of mosaics may be mentioned. In open *Eucalpytus* forests the lower layers may range, irregularly over the area, from dense to open or sparse, and the appropriate coding is 1D13bn/2A12b. A more typical example of a mosaic is where there are scattered clumps of scrub in grassland, for which the appropriate compound coding might be 1B21a/1M21. Table A6.4 shows that the formation-classes 1A, 1D, 1E, 1F, 1I and 1J form a series in which the tree-layer passes from closed to scattered while successively lower life-forms assume dominance beneath and between the trees. It has been suggested that if the definitions of these formation-classes were extended to cover situations in which clumps rather than individuals are scattered, the need for compound coding would be much reduced. But this would involve some maximum size above which clumps are recognized as stands of a separate homogeneous vegetation-type, and it has not been thought practicable to draw such a line.

The question on vegetation was radically modified in both the French and the American check-sheets. In the former the Fosberg coding was replaced by estimates of cover for each layer of the vegetation, and the community description by the names of the dominant and main subdominant species. The layers were separated by height, using the same figures as those in the Fosberg classification on the assumption that this would facilitate interconversion at a data-collecting centre. In many instances, however, there was doubt as to the correct Fosberg coding. In Table A6.5 examples

Table A6.5. *Descriptions of vegetation and the appropriate Fosberg codings*

Community reference number	Percentage cover for each stratum					First dominant species	Second dominant species	Fosberg code assigned
	Tall woody	Medium woody	Low woody	Tall herbaceous	Short herbaceous			
1	100	—	10	20	80	*Carpinus betulus*	*Anemone nemorosa*	1A21
2	40	70	—	—	50	*Quercus toza*	*Koeleria gracilis*	Possibly 1D21
3	—	—	—	—	70	*Artemisia campestris*	*Silene otites*	1N21
4	—	—	—	—	30	*Cheiranthus fruticosus*	*Cymbalaria muralis*	2G25
5	—	—	60	10	40	*Ulex gallii*	*Erica cinerea*	1C17
6	—	—	—	—	100	*Festuca duriuscula*	*Brachypodium pinnatum*	1M21
7	80	20	20	70	30	*Alnus glutinosa*	*Carex pendula*	1A21 or 1A22
8	80	50	20	80	20	*Populus* spp.	*Arundo phragmites*	1A21 or 1A22
9	—	10	20	20	40	*Molinia coerulea*	*Ulex europaeus*	1K15 or 2E13
10	80	20	10	30	60	*Sphagnum* spp.	*Molinia coerulea*	None

are given of typical descriptions with the single or alternative codings which seemed most appropriate.

The USA Check-Sheet incorporated a greatly simplified question on vegetation. Surveyors were asked to enter the 'primary plant community' and to list 'other important plant communities', giving the approximate areas occupied by them. Community descriptions were requested in terms of the *Forest Cover Types of North America* (Society of American Foresters (SAF), 1964) or of the *Potential Natural Vegetation of the Conterminous United States* (Küchler, 1964), wherever possible. A list of dominant plants and important large animals was also to be given. The SAF categories for forest types were used slightly more frequently than Küchler's units for non-forest communities, but the majority of entries gave only the names of dominant species. No reference to the Fosberg classification was included.

The difficulty of estimating the area covered by a particular vegetation-type is illustrated in the summary table (p. 314) where it is seen that only two-thirds of all check-sheets have any entries for community area and only half gave areas for all communities. Many surveyors gave percentages of the total rather than absolute areas, and this option was incorporated into the processing.

The vegetation question was processed as the first subsection of two, the second dealing with soil. Each entry for a community was treated as a compound item to be repeated a specific number of times equalling the number of communities. The compound items each consisted of the following components: an integer for the reference number; two fixed strings, the second optional, for the Fosberg coding split at the formation-group level; a string for the community description; an integer for the area; a fixed string for the units 'acre', 'hectare', 'square mile', 'square kilometre' or for 'percent'; and a specified number of double fixed strings for Fosberg codings involved in any compound entries. The latter are identified by inserting 'comp' in place of the first part of the conventional single codings.

Soils

The section dealing with soils was presented in a very simple fashion on the check-sheet, although the *Guide* gives a clear account of the problems that were recognized in the design of the question. The result was that, subject to the inevitable lack of soils information compared with that available for the vegetation-types which they support, the data recorded were variable in nature between different contributors, but provided a more complete account than could be expected by asking for quantitative records, even of a few parameters. From experience with trials held in 1965 free-style descriptions were requested, in conjunction with a very broad classification into thirteen major soil groups. The scheme was used in all contributing countries with surprisingly little comment, and this is to a large extent a reflection of the simplicity of the question and the classification used in it, although the latter is discussed critically elsewhere in the context of this survey and in relation to other classifications.

Table A6.6 summarizing the completion of this question shows that in the 65% of the sample check-sheets with an entry for the question 49% had both soil codes and notes for at least some of the communities, and that 37% had full entries for all communities. The proportion of check-sheets with complete entries for both vegetation and soils is almost half that with complete vegetation entries alone, and roughly a third of the total with an entry of any description for vegetation.

One additional group was added to the soil classification to include unweathered rock and the most coarsely textured lithosol. It was given the code I3, with a single horizon of C. It is in effect the precursor of I2 and comes immediately after it in the

Table A6.6. *Summary of answers to Question 7 (USA)*

Vegetation	Full codes	Area	Soil	Code	Notes	Similar communities	Full entries	Percentage frequency
√	—	—	—	—	—	—	—	96
√	C	—	—	—	—	—	—	70
√	P	—	—	—	—	—	—	78
√	C	C	—	—	—	—	—	40
√	C	A P	—	—	—	—	—	30
√	—	C	—	—	—	—	—	52
√	—	P	—	—	—	—	—	14
√	—	—	√	—	—	—	—	65
√	—	—	√	P C	P C	—	—	49
√	—	—	√	C	C	—	—	37
√	C	C	√	C	C	—	—	23
√	—	—	—	—	—	√	—	65
√	—	—	√	—	—	√	—	60
√	—	—	—	—	—	√	C	45
√	C	C	—	—	—	√	C	23
√	C	C	√	C	C	√	C	19

C, complete set; P, partial set; A P, absent or partial set; P C, partial or complete set; √, presence of a check-sheet entry.

Table A6.7. *Soil notes and their associated coding*

Soil type	Other notes
O	Eutrophic peat with very high groundwater (Netherlands)
F4	Well-drained podsolic soils developed from residual sandstones and shales; silt loam texture (USA)
Fe	Dark brown sandy loam over red horizons overlying compact weathered siltstone, sometimes calcareous. Developed in sites on Karoo siltstone (Zambia)
Fe	Latosol with mild humus, mildly acid (Peru)
I2	Regosol (USA)
—	Krasnozem (Gn 3.11) with some shallow bouldery Krasnozem (Um 6.13). Parent material basalt – Profiles A, B & C (Australia)
P2	Gleyed orthic regosol, lacustrine clay overlain by sand, no organic matter, water table near surface (Canada)
—	Shallow sandy loam on hot slopes with occasional outcrops of gravel and rock (India)
S2	Fine silty loams, gypsaceous; brown to bleached brown; sometimes regosols on pseudosands (Jordan)

table of soil categories given in Appendix 2 of the *Guide* (p. 123). The examples consist largely of unweathered rock with a few records for block scree and boulder fields. The vegetation supported is principally of epilithic lichens and bryophytes, but scattered herbs and trees in rock fissures are also included where the weathered material is totally insignificant at the scale of survey.

The majority of soil code entries comprised a single coding, a few included two and, more rarely, three. Little difficulty with the classification was experienced although there was some inconsistency in the treatment of rooted and non-rooted aquatics. In both cases codings were sometimes assigned on the basis of the underlying soil or the sediment, and sometimes omitted, with notes restricted to the water conditions. Soil codings were more often entered alone than were soil notes, although all check-sheets from Japan omitted notes, and most French check-sheets omitted codings. The soil notes varied in their length and content but in general matched the recommendations of the *Guide* fairly well. Some entries were very brief or absent, but very few could be considered too long. Some typical entries for the question are given in Table A6.7, with their associated coding.

The *Guide* suggested that the notes should include some or all of the following: parent material, texture, structure, colour, base status, water regime, humus content, depth, and horizon sequence. Very few entries made reference to all the factors, and relatively few co-ordinated contributions made consistent reference to any group of them. This was to be expected owing to the very sketchy knowledge of soil in many countries. Some of these factors were used regularly however and it was considered worthwhile on the basis of the first 200 check-sheets returned to use a broad framework of categories in the data processing to aid retrieval and make the best use of the data given. The categories are listed below:

Soil group. An attempt was made to assign each entry to a slightly more detailed classification which appeared soon after the survey had started. The categories used were the twenty-three major units for the FAO/UNESCO Soil Map of the World (FAO, 1968).

Texture. Five categories were used corresponding to the examples given. The degree of correspondence is necessarily very loose, as it is for each of these soil factors owing to the nature of the question.
  Very fine – clays
  Fine – silts
  Medium – loams
  Coarse – sands
  Very coarse – gravel and larger particles

Colour. Seven classes chosen arbitrarily to fit the entries given regularly in the first check-sheets.

Black	Orange
Dark brown	Yellow
Light brown	White/grey
Red	

Only the extreme features in the descriptions and the most frequent entries could be classed into four categories.

Well drained	Waterlogged
Badly drained	Liable to flooding

Reaction. Reference was normally made to the three usual categories although pH values were given in several national contributions.
  Acid   Neutral   Basic

Depth. Figures for depth were only given occasionally but reference was frequently made to class a soil as deep or shallow. The moderate category was rarely used.
  Shallow   Moderate   Deep

The question was processed as a number of compound entries corresponding to the number of plant communities. Each entry consisted of the following: an integer for the community reference number, a specified number of fixed strings for the soil coding(s), five optional fixed strings for the categorized soil notes, and a string for the soil note in full.

Question 8: Similar communities

This question proved difficult to complete for many surveyors. It demands a very thorough knowledge of the region in which an IBP Area occurs and an insight into the overall dynamic status of the major communities. This is reflected in the figures for completion of the question in check-sheets from the sample. Roughly two-thirds of the check-sheets had some kind of entry for the question, but less than a quarter had a complete entry for all plant communities. Whereas 37% of the sample had full entries for both vegetation and soils, only 19% had full entries for vegetation, soils and similar communities. An entry was regarded as complete for this section if, for each community, one tick was given under each of the two main columns: 'protected' and 'protected and unprotected' sites. For a fully complete entry there should be two ticks, one for status: 'abundant' (A), 'infrequent' (IF) or 'none known' (NK) and one for trend: 'decreasing' (D) or 'increasing' (IC). The highly subjective nature of records given in this question can be illustrated by reference to the Netherlands check-sheets, which are among the most consistent of contributions received.

The records for five communities illustrate the extreme case of low consistency for this particular contribution. It must be recognized that similar irregularities may occur in every contribution even where considerable effort has been made to achieve a consistent approach to recording amongst surveyors. For countries with a generally

Table A6.8. *Examples of highly subjective records*

Community name	No. of occur-rences	Protected					Protected and unprotected				
		A	IF	NK	D	IC	A	IF	NK	D	IC
Alnion-glutinosae	7	7	0	0	0	1	3	4	0	3	1
Eleocharitetum multicaulis	5	3	2	0	1	0	2	3	0	4	0
Caricion acuto-nigrae	10	8	2	0	1	2	9	1	0	6	1
Crepido-Juncetum acutiflori	7	4	3	0	0	1	1	6	0	1	0
Medicagini-Avenetum pubescentis	8	2	6	0	0	0	1	7	0	6	0

A, abundant; IF, infrequent; NK, none known; D, decreasing; IC, increasing.

low record for completion of this question the information given must be regarded as of very dubious value.

The question of scale is again raised in this section. Whereas the majority of questions refer to land within the limits of the IBP Area, this one demands a much wider context. The surveyor is asked to give information on the status of a community within the national or regional units entered under the question on location of the IBP Area. Because no provision is made to indicate which, however, surveyors very rarely indicate the scale they are using. In less than 5% of cases where the question is completed is there any such indication. In European and other small countries the scale is clearly national, in larger countries the situation is not at all clear: some check-sheets are marked to the effect that they refer to a state or province, others are unmarked. The USA contribution illustrates the extreme case, since the check-sheets were distributed to a large number of individuals each of whom had access to the *Guide* with no more specific co-ordination as a national project. Within some states, however, there is evidence of co-ordination, if only because of the part played by one or two individuals in completing most of the check-sheets. The result is a very varied collection of check-sheets, some with Question 8 referring to the state, others referring to the country.

The question was processed as a specified number of compound items, each comprising an integer for the community reference number and four optional fixed strings representing the possible combinations of up to four ticks in each row.

Question 9: Landscape

The question was generally well completed in the 86% of the sample having entries. The brief descriptions were concise with very few examples of extremes either way. Some surveyors referred to vegetation types almost to the exclusion of land form despite the explicit instruction to the contrary in the *Guide*.

The sample records for relief type by percentage figures showed the expected skewed distribution in the frequency table from one type as the mode, to six as the maximum.

Number of relief types	1	2	3	4	5	6	
Percentage frequency	51	20	8	3	1	1	Maximum: 84%

The scale of application was therefore much as intended, i.e., at the level of major landscape types. The situation tended to be oversimplified in some areas of known high diversity, however, suggesting that the areas might have been usefully subdivided. In a small number of cases, 2% of the sample, the surveyor entered ticks for the presence of relief type but omitted to give percentages. Certain countries, including Hungary and Czechoslovakia, and one or two individuals, expressed their need to record flat ground with no element of dissection at all. Provision was made to accommodate such entries in the processing, but it was found that with the option in use it was often misapplied despite the direction that it should refer exclusively to large tracts of flat land such as the central European plains where no appreciable drainage pattern is discernible. The category was in fact applied, in some cases, to small flat peat bogs in otherwise hilly regions.

It was reported that the examples in the *Guide* were found to be helpful in illustrating the altitudinal and erosion classes.

Special features of the landscape were recorded in 48% of the sample check-sheets. The intention had been to develop a thesaurus as terms accumulated but the number and variety of features was great and insufficient information was available at the time that processing guidelines had to be laid down. The same situation applied to the section on special features for coastline and freshwater (Questions 10 and 11) with which there was some overlap, e.g., waterfall and coastal cliffs.

Processing was done in three subsections: a string for the brief description; a matrix of binary items for the relief types and integers for the percentages in each row and column; and thirdly a string for special landscape features.

Question 10: Coastline

This question was generally not well completed. Where a coastline was recorded as present, 14% of the sample, in the majority of cases one part of the question at least was omitted or incorrectly answered. Indications of presence were often given in place of percentages for the type of substrate and/or physiography, and the processing rules were altered to take account of this. Percentage figures were given for both features in 57% of the positive records in the sample, for one feature in 17%, and for neither in 26%. The tidal range was frequently omitted. Special features of the coastline were recorded in 8% of the sample check-sheets, i.e., 61% of those with a positive record. The range of features was not great, mostly involving offshore stacks and islands, accretions, and estuarine features.

The French check-sheet omitted the categories of coral and ice as possible coastal substrates since the questionnaire was aimed at Metropolitan France.

Processing was done in six subsections: a fixed string 'many', 'few' or 'none' for protected bays; eight integers and eight binary items for the percentage and/or presence of each substrate type; three integers and binaries for physiography; a string for special features: an integer for tidal range, with a fixed string for linear units – 'metres' or 'feet'; and a fixed string for length of coastline – 'under 1', '1 to 10', or 'over 10'.

Question 11: Freshwater

The question on freshwater was generally well completed with the exception of the first part which was inadequately explained in the *Guide* for most surveyors. The result was that the philosophy of the question had to be re-interpreted before the data could be used. This caused the loss of some information where the question had been correctly answered, in the interest of consistency throughout. The example in the *Guide* of the coding to be given in the case of a river reducing through a stream to a series of pools was sufficient to explain the intention of the first line in the subsection, but not the others. Many surveyors were consequently unsure, and the two lines for standing and running water had to be considered as summaries for their separate subsections. If all entries for running water were intermittent the summary showed intermittent, if any were permanent the summary showed permanent. In this way the ambiguities were resolved.

From the sample 66% of the check-sheets had a positive entry for freshwater on the site: 24% had special features listed; the range of these was high but rapids, waterfalls and glacial features predominated.

The French check-sheet reduced the first section to two rows, one for standing the other for running water. No specific instructions were given for the method of completion, but fewer examples of incorrect entries were found than in the Mark VII check-sheet.

The question was processed in four subsections; the first three as matrices of binary items, the fourth as a single string for special features.

Question 12: Salt and brackish water

Little comment need be made on this question since the number of entries was low and involved no particular problems. The space for extra features was not often needed but 'salt pans' were occasionally entered here. From the sample only 7.5% of check-sheets had an entry. The French version restricted the entry to the named categories: salt lakes, estuaries and salt pools (less than one hectare).

The question was treated as a single section with four binary items, and the provision to add the name and a binary item for each of a specified number of additions.

Question 13: Adjacent water bodies

This question had a misprint. The label for the first box should have been a sub-heading – 'Fresh'. The box itself was intended for 'Pond' records and so some information was lost here.

The sample showed that 43% of check-sheets had positive entries for the question, mostly under fresh water. It is not clear how far the double entry of water bodies under Questions 11 and 13 contributed to this figure, because the only instances where this could be identified with certainty were those with one or two water bodies, and a map supplied with the site boundary marked. Entries under salt and brackish water bodies occasionally had 'straits' or 'sea' as extra categories, indicating the inappropriate connotation of the word 'ocean'. 'Open sea' might have been better, thereby including all areas in direct communication with the oceans, even through relatively narrow channels.

The question was processed as two subsections; the first as three binary items – freshwater, the second as five – salt and brackish water, with the provision to add the name and a binary item for each of a specified number of additions.

Table A6.9. *Proportion of positive, negative and omitted entries for Questions 10–15*

Question	Percentages of sample check-sheets: Recorded presence	Recorded absence	Question omitted	Probability of significant differences
10 Coastline	14	35	35	
11 Freshwater	67	8	9	
12 Salt and brackish water	8	33	43	$P = 0.05$
13 Adjacent water bodies	44	17	23	
14 Outstanding floral and faunal features	71	11	2	$P = 0.01$
15 Exceptional interest	75	7	2	

Question 14: Outstanding floral and faunal features

This question is another of a highly subjective nature, depending not only upon past records available for the site and the knowledge and interest of the surveyor, but also upon the interpretation of several concepts of an extremely complex nature. The first point which must be made is that this is the only question in the check-sheet which asks for the absence of a feature to be recorded. Of six questions concerning features which may be absent the proportions of positive, negative and omitted entries are given in Table A6.9.

The trend for each question shows that in general there is no significant difference between numbers of questions with a recorded absence and those that are left blank. The difference for salt and brackish water is just significant at $P = 0.05$, with more questions omitted. The same trend is not significant for adjacent water bodies although the figures are in the same proportions. No reason can be given for the difference. For Question 14 the position is clear, however, and all questions should incorporate such a provision in order to differentiate at least at this level between the two classes of negative information.

The practice was not carried through to distinguish between faunal and floral records and those entries which were voluntarily made to indicate lack of knowledge or time for survey are the only basis upon which it may be said that in general records for floristic interest are more widely available than for faunistic interest. This is not evident from the sample check-sheets, as reference to the table will show. The difference between entries under 14.2 and 14.4 is not significant.

The standard of completion varied a great deal as is to be expected in questions of this nature. There were two main causes for this variability. Firstly, there was an obvious bias towards groups with which a surveyor is familiar. This is difficult to judge: it can be detected in cases where a site is well known and groups of known interest are omitted, but for those sites undergoing primary survey there is no club. This is another pointer in favour of having the surveyor's background.

In all sites there is a tendency for records to be most complete for the higher groups of both plants and animals. If standards are to be kept high, as recommended in the *Guide*, the number of ticks under flora and fauna would be expected to be rather low except in a few outstanding areas. Check-sheets with a large number of ticks spread

Table A6.10. *Proportion of positive entries for Question 14 in the sample check-sheets* ($P = 0.05$)

Percentage frequency of check-sheets from the sample with:	
a positive entry for the question	70
a positive entry in 14.2 – fauna	52
a positive entry in 14.4 – flora	60
a positive entry in 14.3 – animal species	40
a positive entry in 14.5 – plant species	40
a positive entry in 14.2 and 14.3	39
a positive entry in 14.4 and 14.5	40
a positive entry in 14.2 and 14.4	42

over all groups and under several types of interest therefore immediately arouse suspicion. This constitutes the second source of variability and is easy to detect in its extreme form, but less obvious examples are inevitably missed. The number of such entries which have been detected and identified as unreliable, together with those which stem from a misunderstanding of the intention, is less than 1% of the total number of check-sheets. Variations in the style of completion included the use of terms such as 'few' and 'abundant', and actual figures. These were used principally for 'species diversity', and the number of rare species in any group. In the former case there was often no indication of whether the number was outstandingly high, giving a net loss of information which was explicitly warned against in the field instructions.

The recommendations given in the *Guide* were rather perfunctory for such a complex question and use was made of examples to illustrate the intention. A more detailed consideration can be made certain of the more difficult aspects of the question, based on the data received. The discussion is intended to refer to both plant and animal records where columns in the questions overlap. The points are dealt with in order of their occurrence in the questions, with reference to the requested lists of species names where appropriate.

Species diversity

This should have been 'species richness', but there was no evidence of misunderstanding. The intention is quite clear from the *Guide* in which both 'rich' and 'diverse' are used. In those check-sheets in which number of species in each group were given the range for animals was from 1 to 300, excluding invertebrates.

Abundance of particular species

The heading given in the *Guide* under Fauna was 'Abundance of individuals', which fits in more closely with the definition given. This applies equally to plants.

Super-abundance of individuals

No mention is made of this column in the *Guide*, much less a definition for its use. The Australian CT Committee made reference to this omission and suggested that the term be interpreted subjectively as any animal population which is overgrazing an area at the time of survey. It is difficult to see the need for such a category at this level of detail when the second one can be used generically.

Appendix 6

Rare species, threatened/relict species, species of biogeographical interest

Rarity in a species may prevail for a number of reasons and these are considered in detail in another section. Several conceptual problems directly concern the entries in the check-sheet however and these are considered here in relation to scale and general biogeographical interest. Rarity, for whatever reason, involves some consideration of numbers and distribution. No simple and general threshold, however, can be used for either plants or animals below which a species becomes 'rare'. The interpretation in the survey was therefore entirely subjective. However, the phrasing of the questions could have been made more straightforward to ease the task of surveyors who found difficulty in using these three columns in conjunction with the request for species names in the third and fifth parts of the question. No explicit reference was made by any contributor to particular problems associated with this question although all those spoken to agreed that it would be an improvement to adopt the modification referred to below.

The first problem identified from completed check-sheets was that surveyors were inconsistent in their categorization of a species as 'rare' or 'threatened/relict', both in the same and widely separated localities. The reason for this was suspected to be a wrong interpretation of the latter category possibly through not reading the field instructions or *Guide*. Species such as the osprey and bald eagle in the USA were recorded as 'threatened/relict' almost as frequently as 'rare'. Nationally common species occurring in outliers were often recorded as 'threatened/relict'. This situation was further confounded by the category of 'biogeographical interest' which may be interpreted as generic to the first two.

The following headings would have helped to clarify the situation:

Nationally rare species
Globally rare species
Species of other biogeographical interest

The first category reduces the problem of what is meant by a region. Ideally the units should be of biological significance or integrity, or failing this, roughly equal in area. Rarity at the national level is more straightforward to apply and is not without relevance in a programme such as IBP which depends largely on national contributions. The second category purposely avoids all reference to the entirely separate concepts of threat and historical origin which were previously used in the heading and yet were quite inappropriate to the philosophy of the question as expressed in the *Guide*. 'Globally rare' exactly matches the *Guides*' explanation. The third category is intended to include relict species, for whatever reason, if of sufficient importance, and endemic species as particular types of biogeographical interest.

The second problem concerned the presentation of the part questions dealing with named species of interest. The field instructions requested the use of the scientific name in all cases, with the vernacular name as an optional addition. No request was made for the status of the species, however, or for its taxonomic group, consequently only a proportion of surveyors added these items. Taxonomic groups could be assigned to all but a very few animal species, which were assumed to be either wrongly quoted or from the lower order, for which up-to-date lists of genera are not readily available. The status of species was given in a greater proportion of check-sheets than was the taxonomic group – roughly 70% – however, the intention of the surveyor was not always clear from the use of the appropriate columns in the other parts of the question, and many entries were processed without status as rare, threatened, relict or endemic.

The suggested modification here is to request an indication of the status for each species by appending an appropriate symbol to the record. Since the categories are not mutually exclusive more than one symbol may be used. In keeping with the modified column headings the categories should be altered to: nationally rare, globally rare, relict and endemic. The presence of species in the last two categories would be indicated by ticks under the column for biogeographical interest, whereas in the Mk VII check-sheet relict species may have been recorded under either of two columns, depending upon the surveyor's interpretation.

Exceptional associations

Of the extremely few records given under this category no indication was given as to the nature of the feature. The examples given in the *Guide* would all qualify for explicit mention under Question 15: Exceptional interest. The value of this category must be brought into question in view of its very limited use.

Breeding or nesting populations, migrating populations, overwintering populations

These categories of faunal interest were correctly used and deserve no special comment. The species involved and their numbers were very occasionally given.

Outstanding specimens

Records of outstanding plant specimens were almost without exception restricted to trees, many to individuals found in remaining areas of virgin forest. Heaviest use of the category was from the USA, predictably in those check-sheets completed by the staff from forestry research stations.

The use of extra rows to the two matrices in this question was rather limited. Extra groups which did feature included: Arachnida, Lepidoptera, Coleoptera, Odonata, Mollusca, Arthropoda and, in the plant kingdom, Fungi. Where extra columns were used to specify particular subgroups within the listed taxa, the generic row was used as a summary, including all records. The first row in the flora matrix was used in the same way to summarize all the angiosperm records. Extra columns were rarely used but occasionally endemic species were recorded here rather than under 'species of biogeographical interest'.

The term 'grass' was interpreted as referring to members of the Gramineae and records for sedges and other graminoid plants were processed as herbs. No recommendation was made in the *Guide* and although the term as defined by Fosberg includes 'vegetation principally made up of grasses and other graminoid plants' most surveyors kept to the systematic definition when dealing with floristics.

The question was processed as five subsections; the first being a single binary item for a positive or negative entry for the question as a whole. The second and fourth subsections comprised a matrix of binary items corresponding to the rows and columns of the check-sheet with provision to add headings and binary entries for a specified number of extra columns and/or rows. The third and fifth subsections were designed to include a specified number of species names, both scientific (obligatory) and vernacular (optional) as strings, with the status and taxonomic group of each both as fixed strings. The status entry was made optional to cater for those species that were unqualified as discussed above.

Appendix 6

Question 15: Exceptional interest

Seventy-four per cent of check-sheets in the sample had an entry for this question. For most protected areas surveyors were generally well aware of the answer to the question: 'What, if anything, is particularly interesting about this area?' The possibility of hitherto undiscovered qualities must be ignored as in all facets of the survey. For sites undergoing primary survey the answer was less readily obtained, particularly concerning the degree of interest and importance in an international or national context. Comments tended to refer to the regional or local setting. The actual categories of interesting features, however, were more clearly separable than was anticipated in the *Guide*, and while it must be accepted that the possible combinations of interest are numerous the recurrence of particular types of entry made worthwhile the development of a rough framework of principal interest to help in retrieval. This comprised eight headings most of which were further subdivided to indicate more specific interests. The scheme was adopted very early in the survey on the basis of the first 200 returned check-sheets, but although in retrospect it could have been designed to suit the situation more precisely it has proved to be useful in narrowing retrieval searches to sites with the general type of interest in question.

The framework is as follows:

(1) Geology – Rocks, fossils
(2) Geomorphology – Formations, scenery
(3) Botany – Virgin stands, diversity, rare species/communities, phytogeographic interest
(4) Zoology – Diversity, rare species, breeding populations, zoogeographic interest
(5) Pedology
(6) History – Archaeology, legend
(7) Education – Teaching, research
(8) Recreation – Tourism, sport

The sub-divisions are not themselves mutually exclusive nor are they intended to be comprehensive. In view of all the possible combinations however it was necessary to limit the coding to one sub-division only within each of the eight categories.

In the French version the question was specifically limited to 'particular scientific interest'; although the Mark VII check-sheet had never been intended to collect information of non-scientific interest such as scenic and sporting attraction, entries of such interests were often made under this question. The French question was sub-divided into six parts as follows:

(1) General
(2) Current scientific use
(3) Scientific gain if protection is given:
  (*a*) Foreseeable development under management
  (*b*) Potential for scientific research
  (*c*) Interest organization
(4) Foreseeable development if protection is not given
(5) Tourist interest
(6) Urgency for protection

The Mark VII question was processed as a single section comprising a string for the whole entry and up to eight fixed strings taken from the main scheme. For pedology the entry is in fact a binary item.

Question 16: Significant human impact

This question gave few problems of literal interpretation but there were several points arising from its underlying philosophy and the way in which it was presented. Eighty per cent of the sample check-sheets had an entry for the question. A number of contributors felt that the first part of the question, dealing with the general extent of human impact, was too insensitive. The Scandinavian countries recommended the use of extra columns in the main part of the question to indicate the extent of each particular impact. This system was used in the Finnish contribution and it gave a much clearer account of impact but added considerably to the complexity of the question. The Finnish contribution comprised twenty-six check-sheets from twenty-four IBP Areas. Comments here relate to the number of check-sheets only. The extent of each impact type was given in thirteen check-sheets, the remaining being completed in the conventional manner for the Mark VII version. Two different surveyors were involved both of whom contributed to the completion of one of the check-sheets in which the refinement was used, and therefore had some measure of agreement on the approach that was used.

From the row totals in Table 6.11 it is clear that most of the impact types extended over only part of the area. Of those that cover the whole area the specific records for extent show that they do so in almost every case, in fact with only one exception in thirty-one records. The four impact types involved are: research, which although frequently entered in check-sheets must rarely be of significance in its effect on a site, grazing, removal of predators and hunting. Reindeer grazing is a feature of vast tracts of Scandinavian countryside and is practised as a right by Lapps in many statutorily protected areas. Removal of predators has had an effect over the whole of Europe, although the more remote parts of Scandinavia might be expected to have suffered less than the more south-westerly countries. The *Guide* states that 'a very high proportion of IBP Areas will have suffered this impact in the past' and yet entries in other European countries were very low. This illustrates the tendency to emphasize the more recent events and to ignore those of the past. Hunting here includes trapping and fishing and contributes to the removal of predators, All other impact types were recorded as extending most often over only a part of an IBP Area.

The figures for impact types in each check-sheet (column total) showed a trend which may illustrate a major difference in approach to the modified and unmodified question. In all check-sheets completed by the surveyor who used the refinement at least one impact type was recorded as covering the whole area. The other surveyor recorded a general impact over the whole area in eight of his thirteen check-sheets. The remaining five were recorded as having a general impact over only part of the area despite having records for hunting, removal of predators and research in every case. Each of these impacts was individually recorded as affecting the whole IBP Area by the other surveyor. This strongly suggests that since the single entry for general impact is so insensitive, in comparison with the refinement, surveyors may be tempted to ignore the light impacts in order to convey the more accurate impression that effects are 'significant' over only a part of most sites. It should be noted that grazing is not recorded for any of the five sites. More reliable interpretation of this point would be possible if a single surveyor had applied both systems.

From the main sample, 80% of the check-sheets had an entry for the question including only 3% with a record of no significant human impact over the whole area, 27% with an impact over part of the area, and the remaining 50% over the whole area. The section on particular impacts gave rise to some concern over the significance and

Table A6.11. *Summary of answers to Question 16 (Finland)*

MLVE	1S1	1S2	002	003	004	005	006	007	008	009	010	011	012	13S1	13S2	014	015	016	017	018	019	020	021	022	023	024	T	X	P	E
Cultivation	N	N	N	P	N	N	X	N	N	P	N	N	N	N	N	X	N	N	N	N	N	N	N	P	X	N	6	3	3	0
Drainage	N	N	N	N	N	P	X	N	N	P	N	N	N	N	N	X	N	N	N	N	N	N	N	P	X	N	6	3	3	0
Other soil disturbance	N	N	P	P	N	P	N	N	N	N	N	N	N	N	N	N	P	X	X	X	X	N	X	P	X	N	11	6	5	0
Grazing	E	E	E	E	E	N	X	N	X	N	E	E	E	N	N	X	N	N	N	N	N	N	N	P	X	N	13	4	1	8
Selective flora disturbance	P	P	P	P	P	N	X	X	X	E	N	P	N	X	X	X	N	X	X	N	N	X	N	P	X	N	18	10	7	1
Logging	P	P	P	P	P	N	X	X	X	P	N	P	N	N	X	X	P	X	N	N	N	N	X	E	X	N	17	8	8	1
Plantation	N	N	N	N	N	N	N	N	N	N	N	N	N	N	N	N	N	N	N	N	N	N	N	N	N	N	0	0	0	0
Hunting	E	N	E	N	N	N	X	X	X	N	E	N	N	X	X	X	N	X	X	X	X	X	X	E	N	E	17	12	0	5
Removal of predators	E	E	E	N	E	N	X	X	X	N	E	E	E	X	X	X	N	X	X	X	X	X	X	E	N	N	20	12	0	8
Pesticides	N	N	N	N	N	N	N	N	N	N	N	N	N	N	N	N	N	N	N	N	N	N	N	N	N	N	0	0	0	0
Introduced plants	P	N	P	P	P	P	N	X	X	P	P	N	N	N	N	X	P	N	X	N	N	N	X	P	X	N	15	6	9	0
Introduced animals	N	N	N	P	N	N	N	N	N	N	N	N	N	N	N	N	N	N	N	N	N	N	N	N	N	N	1	0	1	0
Fire	P	P	P	P	P	P	X	X	X	P	N	N	N	N	X	X	N	X	X	X	N	N	N	P	X	N	17	9	8	0
Permament habitation	P	N	N	P	P	N	N	X	N	P	N	N	P	N	N	X	N	N	N	N	N	N	X	P	N	N	9	3	6	0
Recreation and tourism	P	P	P	P	P	P	X	X	X	P	P	P	N	N	N	N	P	N	N	X	N	N	N	P	N	N	15	4	11	0
Research	E	E	E	E	E	E	X	X	X	N	E	N	N	X	X	X	E	X	X	X	X	X	X	E	X	N	22	13	0	9
Roads and paths	—	—	—	—	—	—	—	X	X	—	—	P	—	—	—	X	—	X	X	X	X	N	N	N	X	N	9	8	1	0
Mowing	—	—	—	—	—	—	—	—	—	—	—	—	—	—	—	—	—	—	—	—	—	X	—	—	—	—	1	1	0	0
Navigation aids	—	—	—	—	—	—	—	—	—	—	—	—	—	—	—	—	—	—	—	—	—	—	—	—	—	P	1	0	1	0
Total	10	7	10	11	9	6	10	10	10	8	6	6	3	4	6	12	5	8	8	7	5	5	7	13	10	1				
Part	6	4	6	9	6	5	—	—	—	7	2	4	1	—	—	—	4	—	—	—	—	—	—	9	—	0				
Entire	4	3	4	2	3	1	—	—	—	1	4	2	2	—	—	—	1	—	—	—	—	—	—	4	—	1				
General	—	—	—	—	—	—	E	E	E	—	—	—	—	E	P	E	—	P	E	P	P	E	P	—	E	—				

T, total; X, none; P, part; E, entire; N, no recorded impact.

degree of impact. The use of symbols to denote different degrees was suggested independently by a number of contributors including the Scandinavian countries and the system was put into use in the French check-sheets. The categories used were: weak, moderate and strong. The argument to support this categorization is that it allows a more sensitive interpretation to be made of the 'significance' of an impact. Under such a scheme it is feasible to omit the term 'significant' from the question heading and to record all impacts which, according to the *Guide*, might be considered insignificant, or of at least a weak character. Explicit guidance on this point would be required. The French version used the same categories in association with the trend of an impact to qualify the rate of change, if any.

The combination of all these refinements results in a formidable question. Of the 80% of sample check-sheets with an entry for the question it should be stated that just over a third included present records for some impacts which had no record of trend. To complicate the question would inevitably raise this proportion. All questions of refinement should be considered in relation to the total performance of the check-sheet and the fact that of 13 000 distributed, 3000 were returned. While it would be desirable to include extra refinements of the questionnaire for use in countries with a firm commitment and capability to carry out the survey, general modifications should clearly be made with the aim of simplifying the task of completing the questionnaire.

A point which deserves special mention is that a large number of surveyors recorded an impact as 'past' but gave an entry for trend, usually 'decreasing'. This possibility of misinterpretation was not foreseen in the *Guide*, although it is quite logical. The effects of some impacts are slow to disappear despite cessation of the impact itself: there may always be some detectable evidence of the impact. The surveyor is in fact recording the disappearing evidence rather than the impact itself in such cases. In a few instances surveyors placed a tick in the 'no information' column under trend to indicate those impact types about which they were unable to supply any information, either positive or negative. Reference to negative information is made elsewhere, including examples from other questions.

Rather few entries appeared as extra rows in the list of impact types in view of the enormous range of possibilities. Of those that did appear many were particular examples of impact types such as 'Selective flora disturbance' or 'Other soil disturbance'. In such cases the entries were duplicated in the generic row so that an overall summary was maintained for each impact type. Ten per cent of the sample check-sheets contained extra rows, the majority of which were considered too restricted or insignificant aspects of their generic impact type to stand alone, and were consequently transferred to the generic row. The title however was included as a comment against the row and processed as described below.

The request for additional details was met in one of three positive ways: extra notes were appended with specific comments under each impact type, a general statement of human impact in the site was given, or the check-sheet entries themselves were briefly annotated. From the sample 16% of check-sheets gave extra notes of some kind. It was decided early on in the survey to incorporate brief notes in the data base since they were provided in almost one fifth of all entries. Comments pertinent to specific impact types and the general style of their recording are given below with particular reference to the Finnish contribution.

Cultivation and drainage

Many protected areas showed past influences of these impacts. They gave rise to little comment and were often omitted in any extra notes appended to a check-sheet.

Other soil disturbance

Because this is such a widely embracing category it was often annotated in the check-sheet or some specific aspect of it was made the heading of an additional row. Road, track and pipeline construction, and army manoeuvres were among the most frequently recorded activities. Mining and surface extraction were less frequent.

Grazing

Entries here were mostly legitimate although some surveyors entered records for what was obviously an example of overgrazing by wild animals, possibly caused by a reduction in habitat through man's influence elsewhere. The *Guide* states quite clearly that only domestic animals are meant to be included.

Selective flora disturbance

Aspects of this type of impact were recorded in many sites. Among the most frequently occurring were sedge, reed, rush and grass cutting and mowing, berry picking, and selective removal of timber for commercial use or in subsistence cropping for construction and fuel.

Logging and plantation

No special comment attaches to these impacts beyond the frequency of their occurrence. Logging is recorded as an impact in many sites and considered to be ecologically important.

Hunting and removal of predators

These have been discussed briefly in an earlier section.

Application of pesticides

It was felt by Swedish contributors that special mention should be given to herbicides and forest fertilizers since, at least in northern Europe, they are widely used. This category was infrequently used in all regions and no reason was found to make a special case for herbicides. None of the Finnish sites had such a record. The case of fertilizers may be considered as one facet of a general category for pollution which is an obvious omission from the list. It was felt that the effects of pollution were far too complex to consider at this level and the decision was taken to treat it as a separate project. During early stages of check-sheet design, pollution was not considered as an entry and it is an indication of how quickly the importance of the problem has grown that it would now be amongst the first in a line of candidate entries.

Introductions – plants and animals

Relatively few records have been given under these headings. It was with some surprise that the high number of records for introduced plants in the Finnish sites was discovered. Closer inspection showed that the records were largely for aliens and weed species appearing along roads and tracts, and that this appeared to be another example of emphasis on recent events. In less remote areas, where they are often accepted as part of the semi-natural flora, the same species might have been disregarded. In such areas entries in the check-sheets tend to be of deliberate introductions. This kind of phenomenon must constantly be borne in mind when interpreting the check-sheet data.

Fire

This section attracts no special comment concerning its style of completion. Fire was recorded as an impact in many sites but few suggestions were made as to the most probably responsible agency.

Permanent habitation

Records encompassed tourist facilities, residence of native population, research staff accommodation and huts for seasonal occupancy by nomads.

Recreation and tourism

Among the most frequently recorded of impact types, recreation and tourism were rarely qualified into particular pursuits, presumably because of their variety.

Research

Records of research were easily the most frequent. It is inconceivable however that in more than a few cases the impact can be 'significant' within the context of the check-sheet question and in relation to the much more drastic effects of low but uncontrolled intensities of activities like selective flora disturbance, removal of predators and general soil disturbance, each of which might be involved to some extent in a research programme. It is more likely that surveyors are considering the research itself to be significant rather than the effect it is having on the site.

The question was processed as two subsections; the first as a single fixed string, the second as sixteen compound items with provision for a further specified number, and a final string to include a general statement on impact. Each compound item consists of two binary items for past and present impact, a fixed string for trend and a string for any brief annotation of the impact type. Extra rows have the same structure, with the title in addition.

Question 17: Conservation status

The situation in any protected area is rarely so simple that it can be summarized accurately by this question. Management strategy and utilization in large areas are often zoned in some way. It was for this reason that such activities were put forward as a basis for possible sub-division of an area. In fact this was done in very few cases (see Question 2, Ndola Sample Plots example) and consequently some problems arose concerning the question. Eighty-three per cent of the sample check-sheets had an entry for the question. Extra rows to the question were entered in 6% of the sample and almost without exception concerned game animals, especially deer. No recurrence of any particular group of plants or features of abiotic interest was evident.

The section on protection gave rise to little comment. It was completed consistently, and where the whole question was only answered in part, 28% of the sample, this one was usually present. In sites with a generally high protection status, those in which hunting or fishing is allowed showed up by an entry under 'partial protection' for fauna, or the specific animal group, and under 'total protection' for the flora and non-living elements. This does not show up in sites with a lower overall status since all ticks may appear under 'partial protection'. The problems of interpretation of the question were foreseen in the *Guide*, which gave clear recommendations on the approach to be used. Despite the request for qualification of any entry when this might explain the situation in full, only very few surveyors did this outside the Canadian contribution, in which notes on this section were frequently given.

The section on utilization resulted in a small amount of misinterpretation of the words 'controlled' and 'uncontrolled'. This was ascribed to the situation that would be expected if some surveyors completed the question without prior reference to the *Guide* for definitions. The Australian IBP/CT Committee pointed out that the term 'utilization' might be construed by planners and politicians to mean that National Parks and other protected areas could be legitimately used for timber extraction and other activities *de facto*. They suggested the phrase 'use for non-conservation purposes', but this excludes the whole concept of management involving the perpetuation of traditional practices, some of which may well include selective timber extraction. With zoned utilization of protected sites this section becomes insensitive, as described above.

Conservation management proved to be the most difficult section, again through insensitivity, but also through some uncertainty on the part of surveyors as to whether they could enter more than one tick per row. No explicit guidance on this latter point is given in the *Guide* although the example itself contains a double entry. No more than 3% of the sample check-sheets – 3.5% of those with an entry – contained ticks under both categories of management. This greatly oversimplifies the case and, from the nature of entries given and the problems that underlie them, the only reliable inference that can be taken is whether or not there is active management in the area for the elements in each row. The section was quite often left blank although entries were made in the other three.

The section on permitted research was well completed. Double ticks were often given to indicate that both kinds of research were permitted, but in processing 'experimental' was considered to include 'observational' study.

The question was processed as a single section, each row with a series of fixed strings, as appropriate for each of the four headings. Provision to add a specified number of extra rows was made.

Questions 18 and 19: References and other relevant material

These questions received a very mixed response. No overall trend in the amount and quality of appended information was apparent. Some countries went no further than to answer questions in the check-sheet for each area, although supplementary information was easily to hand. Others appended notes and references to almost every check-sheet, and the Canadian contribution must be given all possible credit here for the thorough nature of its treatment. For some sites of long standing, bibliographies were provided as a key to the enormous amount of work that has been done on them. The table below summarizes the answers to these questions.

Percentage of sample check-sheets						
			With aerial photographs available			With extra information (Question 19)
With an entry for Question 18	With a reference listed	With maps listed	Whole site	Part of site	None	
73	31	34	51	2	20	29

Question 18 was omitted from the USA check-sheet; Question 19 was worded to include notes on features of ecological, geological and taxonomic interest with reference to 'endemic or endangered plants and animals, etc.' In the French version an extra part was added to Question 18, asking for the names of people to whom reference should be made for further information.

The material appended to the check-sheet was examined during editing and if necessary appropriate entries were made in each of the sections concerned. Question 18 was processed as three subsections; the first two of single binary items, the third as a fixed string. Question 19 was edited to a fairly brief statement summarizing the most important points in any appended material. It was processed as a single string.

References

Acocks, J. P. H. (1953). *South Africa Botanical Memoir 28.*

Avery, B. W. (1968). General soil classification: hierarchical and co-ordinate systems. In *Transactions of the Ninth International Congress of Soil Science*, Adelaide, vol. 4, pp. 169–74.

Ball, D. F. (1967). Classification of soils. In *Guide to the Check Sheet for IBP Areas* (ed. G. F. Peterken), IBP Handbook no. 4. Blackwell, Oxford.

Braun (Braun-Blanquet), J. (1915). *Les Cévennes Méridionales* (*Massif de l'Aigoual*). Études sur la Végétation Méditerranéenne 1. Geneva.

Braun-Blanquet, J. (1928, 1951, 1964). *Pflanzensoziologie*. Springer-Verlag, Berlin.

Brewer, R. (1968). Clay illuviation as a factor in particle-size differentiation in soil profiles. In *Transactions of the Ninth International Congress of Soil Science*, Adelaide, vol. 1, pp. 489–99.

Brockmann-Jerosch, H. & Rübel, E. (1912). *Die Einteilung der Pflanzengesellschaften nach ökologisch-physiognomischen Gesichtspunkten.* Leipzig.

Burbidge, N. T. (1960). The phytogeography of the Australian region. *Aust. J. Bot.*, **8**, 75–212.

Butler, B. E. (1964). Assessing the soil factor in agricultural production. *J. Aust. Inst. agric. Sci.*, **30**, 232–40.

Cajander, A. K. (1916). *Metsänhoidon perusteet*, vol. 1, Porvoo.

Cajander, A. K. (1917). *Metsänhoidon perusteet*, vol. 2, Porvoo.

Ceska, A. & Roemer, H. (1971). A computer program for identifying species-relevé groups in vegetation studies. *Vegetatio*, **23**, 255–77.

Champion, H. G. (1936). A preliminary survey of the forest types of India and Burma. *Ind. For. Rec.* (NS) **1** (1), 1–286.

Clements, F. E. (1928). *Plant Succession and Indicators.* Wilson, New York.

Cline, M. G. (1949). Basic principles of soil classification. *Soil Sci.*, **96**, 81–91.

Comanor, P. L. (1971). Analysis of woody scrub vegetation of Ruhuna National Park, Ceylon. *Trop.Ecol.*, **12**, 209–22.

CPCS (1967). *Classification des Sols.* Commission de Pédologie et de Cartographie des Sols. INRA, Paris (Mimeo).

Dale, M. B. & Webb, L. J. (1974). Numerical methods for the establishment of associations. *Vegetatio*, **30**, 77–87.

Dansereau, P. (1951). Description and recording of vegetation upon a structural basis. *Ecology*, **32**, 172–229.

Dansereau, P. (1957). *Biogeography, an Ecological Perspective.* Ronald Press, New York.

Dansereau, P. (1958). A universal system for recording vegetation. *Contrib. Inst. Bot. Univ. Montreal*, **72**, 1–58.

Dansereau, P., Buell, P. F. & Dagon, R. (1966). A universal system for recording vegetation. 2. A methodological critique and an experiment. *Sarracenia*, **10**, 1–64.

Dasmann, R. F. (1972). Towards a system for classifying natural regions of the world and their representation by national parks and reserves. *Biological Conservation*, **4**, 247–55.

Dasmann, R. F. (1973*a*). *A System for Defining and Classifying Natural Regions for Purposes of Conservation.* IUCN Occasional Paper no. 7. IUCN, Morges, Switzerland.

Dasmann, R. F. (1973*b*). Classification and Use of Protected Natural and Cultural Areas.

References

Downes, R. G. (1969). Conservation in relation to the land and its use. In *The Last of Lands* (ed. L. J. Webb, D. Whitelock, and J. Le Gay Brereton), pp. 11–17. Jacaranda Press, Brisbane.

Duchaufour, P. (1965; third edition, 1970). *Précis de Pédologie*. Masson, Paris.

Du Rietz, G. E. (1921). *Zur methodologischen Grundlage der modernen Pflanzen soziologie*. Holzhausen, Vienna.

Duvigneaud, P. (1946). La variabilité des associations végétales. *Bull. Soc. roy. Bot. Belg.*, **78** (second series, **28**), 107–34.

Eberhardt, E., Kopp, D. & Passarge, H. (1967). Standorte und Vegetation des Kirchleerauer Waldes im Schweizerischen Mittelland. In *Vegations- und bodenkundliche Methoden der forstlichen Standortskartierung* (ed. H. Ellenberg). *Veröff. Geobot. Inst. Rübel, Zürich*, **39**, 13–134.

Eisenberg, J. F. & Lockhart, M. (1972). *An Ecological Reconnaissance of Wilpattu National Park, Ceylon*. Smithsonian Contributions to Zoology, no. 101. Washington, DC.

Ellenberg, H. (1959). Können wir eine gemeinsame Plattform für die verschiedenen Schulen in der Waldtypenklassification finden? *Silva Fennica*, Helsinki, **105**, 26–32.

Ellenberg, H. (1963). Vegetation Mitteleuropas mit den Alpen. In *Einführung in die Phytologie* (ed. H. Walter), vol. 4, part 2. Eugen Ulmer, Stuttgart.

Ellenberg, H. (ed.) (1967). *Vegetations- und bodenkundliche Methoden der forstlichen Standortskartierung* (with English summary). *Veröff. Geobot. Inst. Rübel, Zürich*, 39.

Ellenberg, H. (1973). Die Ökosysteme der Erde: Versuch einer Klassifikation der Ökosysteme nach funktionalen Gesichtspunkten. In *Ökosystemforschung* (ed. H. Ellenberg), pp. 235–65. Springer-Verlag, Berlin.

Ellenberg, H. & Klötzli, F. (1967). Vegetation und Bewirtschaftung des Vogelreservates Neeracher Riet. *Ber. Geobot. Inst. Rübel Zürich*, **37**, 88–103.

Ellenberg, H. & Mueller-Dombois, D. (1967). Tentative physiognomic–ecological classification of plant formations of the earth. *Ber. Geobot. Inst. Rübel Zürich*, **37**, 21–55.

Ellenberg, H. & Mueller-Dombois, D. (1967). A key to Raunkaier life forms with revised subdivisions. *Ber. Geobot. Inst. Rübel Zürich*, **37**, 56–73.

Elton, C. S. (1973). The structure of invertebrate populations inside neotropical rain forest. *J. Anim. Ecol.*, **42**, 55–104.

Emberger, L. (1960). *Le climat méditerranéen au point de vue biologique*. Institut Botanique, Montpellier.

Emberger, L. & Long, G. (eds.) (1967). *Carte Phyto-Écologique de la Tunisie Septentrionale: Notice Détailée*. Inst. Sci. Econ. Appl. (Centre d' Afrique du Nord).

FAO (1974). *Soil Map of the World, 1:5,000,000*. Vol. 1. *Legend*. UNESCO, Paris.

Fosberg, F. R. (1961). A classification of vegetation for general purposes. *Trop. Ecol.*, **2**, 1–28.

Fosberg, F. R. (1967). A classification of vegetation for general purposes. In *Guide to the Check Sheet for IBP Areas* (ed. G. F. Peterken). IBP Handbook no. 4. Blackwell, Oxford.

Galloway, R. W., Gunn, R. H. & Pedley, L. (1967). Land systems of the Nogoa-Belyando area, Queensland. *CSIRO Aust. Land Res. Series*, **18**, 21–67.

Gams, H. (1918). Prinzipienfragen der Vegetationsforschung. Ein Beitrag zur Begriffsklärung und Methodik der Biocoenologie. *Vierteljahrsschr. Naturforsch. Ges. Zürich*, **63**, 293–493.

Gaussen, H. (1957). Interpretation of data by means of vegetation maps. In *Proceedings of the Ninth Pacific Science Congress*, Bangkok.

Gaussen, H., Legris, P., Viart, M. & Labroue, L. (1964). *International Map of Vegetation, Ceylon*. Special sheet, Survey Department of Ceylon (Sri Lanka).

Gibbons, F. R. (1961). Some misconceptions about what soil surveys can do. *J. Soil Sci.*, **12**, 96–100.

Gibbons, F. R. (1968). Limitations to the usefulness of soil classification. In *Transactions of the Ninth International Congress of Soil Science*, vol. 4, pp. 159–67.

Gibbons, F. R. & Downes, R. G. (1964). *A Study of the Land in South-Western Victoria*. Soil Conservation Authority, Melbourne.

Good, R. (1947, 1953, 1964). *The Geography of Flowering Plants*. Longmans, Green & Co., London.

Grabau, W. E. (1968). An integrated system for exploiting quantitative terrain data for engineering purposes. In *Land Evaluation* (ed. G. A. Stewart), pp. 211–20. Macmillan of Australia.

Gunn, R. H. (1967). A soil catena on denuded laterite profiles in Queensland. *Aust. J. Soil Res.*, **5**, 117–32.

Holdgate, M. (1970). The national strategy for nature reserves. In *SPNR County Trusts Conference Proceedings*, 1970.

Holdridge, L. R., Grenke, W. C., Hatheway, W. H., Liang, T. & Tosi, J. A. Jr. (1971). *Forest Environments in a Tropical Life Zone: a Pilot Study*. Pergamon Press, Oxford.

Howard, P. J. A., Lindley, D. K. & Hornung, M. (1974). *Recording soil profile descriptions for computer retrieval*. Institute of Terrestrial Ecology, Merlewood Research & Development paper No. 55 (Mimeo).

Humboldt, A. von (1806). *Ideen zu einer Physiognomik der Gewächse*. Tübingen.

Hynes, R. A. (1973). Ecology of *Nothofagus* forest in Central New Guinea. MSc. Thesis, University of Papua New Guinea.

Hynes, R. A. (1974). Altitudinal zonation in New Guinea *Nothofagus* forest. In *Altitudinal Zonation in Malesia* (ed. J. R. Flenley). Third Aberdeen–Hull Symposium on Malesian Ecology.

IUCN (1971). *UN List of National Parks and Equivalent Reserves*. Hayez, Brussels.

IUCN (1974). *Biotic Provinces of the World. IUCN Occasional Paper no. 9*. IUCN, Morges, Switzerland.

IUCN, *Red Data Books:* vol. 1, *Mammalia*, H. A. Goodwin and C. W. Holloway (1972, 1974; revisions and additions by J. Thornback, in press); vol. 2, *Aves*, W. B. King (1978; new edition in press); vol. 3, *Amphibia and Reptilia*, R. E. Honneger (1975); vol. 4, *Pisces*, R. R. Miller (in press); The IUCN Plant Red Data Book, G. Lucas and H. Synge (in press); all published by IUCN, Morges.

Kato, M. (ed.) (1967). *Research Methods for Describing the Fauna*. Meirinsha, Sendai.

Kershaw, A. P. (1970). A pollen diagram from Lake Euramoo, north-east Queensland. *New Phytol.*, **69**, 785–805.

Kershaw, A. P. (1971). A pollen diagram from Quincan Crater, north-east Queensland. *New Phytol.*, **70**, 669–81.

Kershaw, A. P. (1973). Late Quaternary vegetation of the Atherton Tableland, north-east Queensland. PhD Thesis, Australian National University, Canberra.

Kikkawa, J. & Webb, L. J. (1967). Niche occupation by birds and the structural classification of forest habitats in the wet tropics, North Queensland. In *Proceedings of the Fourteenth IUFRO Congress*, section 26. Munich.

Kikkawa, J., Webb, L. J. & Tracey, J. G. (1974). A multidisciplinary project to establish criteria for selection and management of national parks in North

Queensland. In *Institute of Foresters of Australia Seventh Triennial Conference*, Caloundra, Queensland.
Köppen, W. (1936). Das Geographische System der Klimate. In *Handbuch der Klimatologie* (ed. W. Köppen and R. Geiger), vol. 1, part C. Borntraeger, Berlin.
Kovda, V. A., Lobova, Y. V. & Rozanov, V. V. (1967). *Classification of the World's Soils*. Soviet Soil Science, pp. 851–3.
Krajina, V. J. (1969). Ecology of forest trees in British Columbia. In *The Ecology of Western North America* (ed. V. J. Krajina & R. C. Brooke), vol. 2, part 1, 1–147. Department of Botany, University of British Columbia, Vancouver.
Krajina, V. J. (1974). *Biogeoclimatic Zones of British Columbia*. Map at 1:1.9 million. British Columbia Ecological Reserves Committee, Department of Lands, Forests and Water Resources, Victoria, BC.
Kubiena, W. L. (1953). *Soils of Europe* (English language version). Murby, London.
Küchler, A. W. (1964). Potential Natural Vegetation of the Conterminous United States. Manual to accompany the map. *American Geographical Society, Special Publication* no. 36. (Revised edition of map, 1965).
Küchler, A. W. (1967). *Vegetation Mapping*. Ronald Press, New York.
Küchler, A. W. (1974). A new vegetation map of Kansas. *Ecology*, **55**, 586–604.
Leeuwen, C. G. van (1966). A relation theoretical approach to pattern and process in vegetation. *Wentia*, **15**, 25–46.
Le Houerou, H.-N. (1969). La végétation de la Tunisie steppique. *Ann. Inst. Nat. Rech. Agron.*, Tunisie, **42**, 1–620 (with maps).
Lippmaa, T. (1939). The unistratal concept of plant communities (the unions). *Amer. Midl. Naturalist*, **21**, 111–45.
Margalef, R. (1968). *Perspectives in Ecological Theory*. University of Chicago.
Mirov, N. T. (1967). *The Genus Pinus*. Ronald Press, New York.
Mueller-Dombois, D. (1965). Eco-geographic criteria for mapping forest habitats in south-eastern Manitoba. *Forest Chron.* (Vancouver, BC), **41**, 188–206.
Mueller-Dombois, D. (1969). *Vegetation map of Ruhuna National Park: scale 1:31610*. Survey Department of Sri Lanka. Sheets 1–5.
Mueller-Dombois, D. (1972). Crown distortion and elephant distribution in the woody vegetation of Ruhuna National Park, Ceylon. *Ecology*, **53**, 208–26.
Mueller-Dombois, D. & Ellenberg, H. (1974). *Aims and Methods of Vegetation Ecology*. Wiley, New York.
Mulcahy, M. J. & Humphries, A. W. (1967. Soil classification, soil survey and land use. *Soils and Fertilizers*, **30**, 1–8.
Nicholson, E. M. (1968). *Handbook to the Conservation Section of the International Biological Programme*. IBP Handbook no. 5. Blackwell, Oxford.
Northcote, K. H. (1971). *A Factual Key for the Recognition of Australian Soils*, third edition. Rellim, Glenside, South Australia.
Northcote, K. H. & Skene, J. K. M. (1972). *Australian Soils with Saline and Sodic Properties*. CSIRO Australian Soil Publication no. 27.
Numata, M., Yoshioka, K. & Kato, M. (eds.) (1975). *Studies in Conservation of Natural Terrestrial Ecosystems in Japan*, Part 1: *Vegetation and its Conservation*. JIBP Synthesis Volume 8. University of Tokyo Press.
Oertel, A. C. (1968). Some observations incompatible with clay illuviation. In *Transactions of the Ninth International Congress of Soil Science*, vol. 4, pp. 481–8.
Olney, P. J. S. (1965). *List of European and North African Wetlands of International Importance*. IUCN Publications, NS 5, pp. 83–5. IUCN, Morges, Switzerland.
Pedley, L. & Isbell, R. F. (1972). Plant communities of Cape York Peninsula. *Proc. Roy. Soc. Qld.*, **82**, 51–74.

Peterken, G. F. (1967). *Guide to the Check Sheet for IBP Areas.* IBP Handbook no. 4. Blackwell, Oxford.

Peterken, G. F. (1968). International selection of areas for reserves. *Biol. Conservation,* **1**, 55–61.

Petersen, A. (1927). *Die Taxation der Wiesenländereien auf Grund des Pflanzenbestandes.* Berlin.

Poore, M. E. D. (1964). Integration in the plant community. *J. Ecol.*, **52** (Suppl.), 213–26.

Poore, M. E. D. & Robertson, V. C. (1964). *An Approach to the Rapid Description and Mapping of Biological Habitats.* IBP Central Office, London.

Radford, G. L. & Peterken, G. F. (1971). Report on field trials in Tunisia. In *IBP/CT Progress Report 1971*, by E. M. Nicholson and G. L. Douglas, London.

Raunkiaer, C. (1937). *Plant Life Forms.* Clarendon Press, Oxford.

Richards, P. W. (1952). *The Tropical Rain Forest.* Cambridge University Press.

Richards, P. W. (1971). Some problems of nature conservation in the tropics. *Bull. Jardin Bot. Nat. de Belgique*, **41**, 173–87.

Rübel, E. (1936). Plant Communities of the World. In *Essays in Geobotany in Honor of William Albert Setchell.* University of California Press.

Schimper, A. P. F. (1898). *Pflanzen-Geographie auf physiologischer Grundlage.* Gustav Fischer, Jena.

Schmithüsen, J. (1968). Vegetation maps at 1:25 million of Europe, North Asia, SW. Asia, Australia, N. Africa, S. Africa, N. America, Central America, South America (north part), South America (south part). in *Grosses Duden-Lexicon*, vol. 18, 321–46, Bibliog. Inst., Mannheim.

Shimwell, D. W. (1971). *The Description and Classification of Vegetation.* Sidgwick & Jackson, London.

Society of American Foresters (1964). *Forest Cover Types of North America.*

Soil Survey Staff (1960). *Soil Classification. A Comprehensive System. Seventh Approximation.* US Dept. of Agriculture, Washington DC.

Soil Survey Staff (1975). *Soil taxonomy: a basic system of soil classification for making and interpreting soil surveys.* US Dept. Agric. Handb. No. 436 (Govt Printer, Washington DC).

Spatz, G. (1972). Eine Möglichkeit zum Ersatz der elektronischen Datenverarbeitung bei der pflanzensoziologischen Tabellenarbeit. In *Basic Problems and Methods in Phytosociology* (ed. E. van der Marel and R. Tüxen). Ber. Intern. Vereinigung f. Vegetationskunde 1970, pp. 251–61.

Specht, R. L. (1970). Vegetation. In *The Australian Environment* (ed. G. W. Leeper) (Fourth edition), CSIRO–Melbourne University Press, Melbourne.

Specht, R. L., Roe, R. M. & Boughton, V. H. (eds.) (1974). Conservation of major plant communities in Australia and Papua New Guinea. *Aust. J. Bot.*, Supplement no. 7.

Steenis, C. G. G. J. van (1953). On the hierarchy of environmental factors in plant ecology. In *Proceedings of the 7th International Botanical Congress*, pp. 637–44.

Tansley, A. G. (1939). *The British Islands and their Vegetation.* Cambridge University Press.

Thornthwaite, C. W. (1948). An approach towards a rational classification of climate. *Geogr. Rev.*, **38**, 55–94.

Tiurin, I. V. (1965). The system of soil classification in the USSR. Main stages in the development of the soil classification in the USSR. In *The third International Symposium on Soil Classification, Pedology*, pp. 7–24.

Udvardy, M. D. F. (1975). *A Classification of the Biogeographical Provinces of the World.* IUCN Occasional Paper no. 18. IUCN, Morges, Switzerland.

References

UNESCO (1973). *International Classification and Mapping of Vegetation*. UNESCO Ecology and Conservation Series no. 6.

US Forest Service (1972). *Atlas of United States Trees*, vol. 1, *Conifers and Important Hardwoods*. US Government Printing Office, Washington, DC.

Walter, H. (1957). Wie kann man den Klimatypus anschaulich darstellen? *Die Umschau in Wissenschaft und Technik*, **24**, 751–3.

Walter, H. (1971). *Ecology of Tropical and Subtropical Vegetation* (transl. D. Mueller-Dombois, ed. J. H. Burnett). Oliver & Boyd, Edinburgh.

Weaver, J. E. & Clements, F. E. (1929, 1938). *Plant Ecology*. McGraw-Hill, New York.

Webb, L. J. (1959). A physiognomic classification of Australian rain forests. *J. Ecol.*, **47**, 551–70.

Webb, L. J. (1966). The identification and conservation of habitat-types in the wet tropical lowlands of North Queensland. *Proc. Roy. Soc. Qld.*, **78**, 59–86.

Webb, L. J. (1968). Environmental relationships of the structural types of Australian rain forest vegetation. *Ecology*, **49**, 296–311.

Webb, L. J. (1978). A structural comparison of New Zealand and south-east Australian rainforests and their tropical affinities. *Aust. J. Ecol.* **3**, 7–21.

Webb, L. J. & Tracey, J. G. (1975). *Key to the vegetation of the humid tropical region of North Queensland, plus 15 maps*. Rainforest ecology unit, Division of Plant Industry, CSIRO Long Pocket Laboratories, Indooroopilly 4068, Australia.

Webb, L. J., Tracey, J. G., Kikkawa, J. & Williams, W. T. (1973). Techniques for selecting and allocating land for nature conservation in Australia. In *Nature Conservation in the Pacific* (ed. A. B. Costin and R. H. Groves). ANU Press, Canberra.

Webb, L. J., Tracey, J. G. & Williams, W. T. (1972). Regeneration and pattern in the subtropical rain forest. *J. Ecol.*, **60**, 675–95.

Webb, L. J., Tracey, J. G. & Williams, W. T. (1976). The value of structural features in tropical forest typology. *Aust. J. Ecol.* **1**, 3–28.

Webb, L. J., Tracey, J. G., Williams, W. T. & Lance, G. N. (1970). Studies in the numerical analysis of complex rain forest communities. 5. A comparison of the properties of floristic and physiognomic–structural data. *J. Ecol.*, **58**, 203–32.

Westhoff, V. & Den Held, A. J. (1969). *Planten Gemeenschappen in Nederland*. Thieme, Zutphen.

Whittaker, R. H. (1962). Classification of natural communities. *Bot. Rev.*, **28**, 1–239.

Worthington, E. B. (ed.) (1975). *The evolution of IBP*. IBP Synthesis Series 1. Cambridge University Press.

Index

Note: references in *italic* are to figures

Index

Index

Index

Index

www.ingramcontent.com/pod-product-compliance
Lightning Source LLC
LaVergne TN
LVHW090305210926
848480LV00001B/144